Living with the Unexpected

Springer
Berlin
Heidelberg
New York
Barcelona
Hong Kong
London
Milan
Paris
Singapore
Tokyo

Anja K. Possekel

Living with the Unexpected

Linking Disaster Recovery to Sustainable Development in Montserrat

With 66 Figures
and 29 Tables

Springer

Dr. Anja K. Possekel
University of Hamburg
Institute for Geography
Bundesstraße 55
D-20146 Hamburg
Germany
e-mail: possekel@geowiss.uni-hamburg.de

Translator
Ines Adams
Sandstraße 20
D-49186 Bad Iburg
Germany

This monograph was a Ph. D. dissertation approved by the Geographical Faculty of the University of Hamburg in October 1998

ISBN 3-540-65709-6 Springer-Verlag Berlin Heidelberg New York Tokyo

Library of Congress Cataloging-in-Publication Data applied for
Die Deutsche Bibliothek – CIP-Einheitsaufnahme
Possekel, Anja K.: Living with the unexpected: linking disaster recovery to sustainable development in Montserrat; with 29 tables / Anja K. Possekel [Transl. Ines Adams]. – Berlin; Heidelberg; New York; Barcelona; Hong Kong; London; Milan; Paris; Singapore; Tokyo: Springer, 1999
Zugl.: Hamburg, Univ., Diss. 1998
ISBN 3-540-65709-6

Cover photography: David Lea (Eruption of August 6th, 1997. Pyroclastic flows going down Mosquito and Gages Ghauts. Hollender Estate in the foreground)
Cover-Design: Erich Kirchner, Heidelberg
Typesetting: Camera-ready by the author

SPIN 10696489 30/3136-5 4 3 2 1 0 – Printed on acid-free paper

To Götz, Reiner, Ulla and Montserrat!

Preface

In 1980 my parents took me to St Lucia for the spring holidays.These vacations had – in the sense of the butterfly effect – considerable influence on my further curriculum. In the course of time I carried out many other journeys to the Caribbean. My colleagues even speak of an advancing "Caribbeanisation". This publication is the product of a lot of people who have supported and accompanied me during all this time.

Most of my visits in the Caribbean were made possible by the DFG (Deutsche Forschungsgemeinschaft). It was of great advantage to my insight into Caribbean matters that I was able to participate in the DFG-financed research project "Complex resource mangement on small Caribbean islands" for five years. The final phase of my dissertation about Montserrat was also financially supported, this time by the local funds of the University of Hamburg.

Within the scope of the DFG project, I stayed on Montserrat for some months. The island fascinated me from my first encounter. This is especially true for the people I met there. They not only inspired me again and again with their ideas, knowledge and abilities. They also considerably contributed to this publication. Thanks also goes to the participants of the scenario workshop and the staff of the Physical Planning Unit, who invested much time and energy into Montserrat's future. The participants of the scenario project in Hamburg also put enormous efforts into the work. My colleague Wolf-Dietrich Sahr did not even shy away from the long journeys between Tübingen and Hamburg in order to participate – and he did not miss one single session.

The students involved in the DFG-project have over the past few years all helped me in their own special way, especially Ines Adams, Jan Bruns, Alex Nitzsche and Birte Wichmann. The same can without doubt be said about many people at the Department of Economic Geography, University of Hamburg.

Looking at the maps and figures, you will easily detect that a creative person with a fascinating sensibility for typography and aesthetics was at play. Claus Carstens was always available to give me the right advice at the right time. Many of the photographs were taken by Douglas Darby and David Lea from Montserrat. The others were digitalised and processed by Bernd Hartmann. Jens Rassmuss is responsible for the beautiful depictions of the scenarios constructed in Hamburg. Klaus Wohlberg did an excellent job while checking the final printout.

Prof. Dr. Jürgen Oßenbrügge and especially Prof. Dr. Robert Geipel assisted me at a critical bifurcation point and showed me a challenging and interesting direction which I might not have taken without their help.

I would like to take this opportunity to thank all these people for their support.

Immense thanks to Ines Adams for her never-ending energy while translating almost every single word of the original German text. With her excellent linguistic skill, pool of ideas and critical review of the text she also contributed considerably to this publication.

My special gratitude to Prof. Gerhard Sandner who has for a decade continuously inspired me and been a vivid example for the realisation of visions.

Last but not least: this work would never have found its way to the publisher without the continuous support and creativity of Beate, Götz, Reiner and Ulla. They have never given up encouraging me and have shown inexhaustible patience with my highwire act between chaos and self-organisation. I would like to express my sincere and deep thanks to the four of you.

Hamburg, February 1999 Anja K. Possekel

Contents

There is no reason why we should wait passively for the next natural disaster to strike. We should mobilize scientific knowledge and technological know-how. We should give a positive response to the strong political interest and commitment of the international community (Boutros–Ghali).

1 Introduction

Deadly typhoon, Ash Wednesday in paradise, China's merciless floods, mass escape from Bonnie, Bangladesh is drowning... When reading the daily headlines, one gets the impression that natural disasters have in the whole world risen dramatically. But nobody really knows whether this is the case or whether the international press and better information channels have just sharpened our awareness; maybe it is a bit of both. However, some problems point at our growing vulnerability to natural disasters. Our environment is degenerating, the global climate is changing, desertification threatens us, the number of species is diminishing, and urbanisation and population growth are common phenomena of our century, above all in developing countries. That much is certain: the damage caused by natural disasters has on a worldwide scale, despite higher spending for preventative measures, increased considerably. Natural disasters have between 1960 and 1970 caused material damage of around US$ 50 milliards. This amount has in the decade following risen to about US$ 70 milliards, and the costs between 1980 and 1990 were reported to have reached US$ 120 milliards. This disturbing trend has accumulated in material damage of US$ 360 milliards in the first half of the current decade (Eikenberg 1998). Damage in industrialised nations is unproportionally higher since their material assets are usually far higher. It has to be kept in mind, however, that the extent of relative economic damage and fatalities is by far the largest in the developing countries, and the majority of natural disasters occur in these countries.

The General Meeting of the United Nations has in the light of these developments declared the 1990s as *International Decade for Natural Disaster Reduction* (IDNDR). The main objective of the programme is the reduction of fatalities, damage to property and the socio-economic consequences caused by extreme natural events. To this end the following tasks have been defined by the General Meeting.

1. Each country's abilities to effectively and promptly mitigate the consequences of natural disasters are to be enhanced. Developing countries are to receive specific assistance with a view to estimating the potential extent of damage and the provision of early warning systems and disaster-resistant structures, when and wherever necessary.
2. Appropriate guidelines and strategies for the application of the existing scientific and technical knowledge are to be published, taking into consideration the cultural and economic variety amongst the nations.
3. It is further envisaged to encourage scientific and technical endeavours to close gaps in specific knowledge in order that damage to lives and property may be reduced.
4. Established and new technical information for estimates, predictions and the mitigation of disasters have to be spread on a broader scale.
5. Measures for the assessment, predictions, prevention and mitigation of natural disasters are to be worked out in the scope of technical assistance and technology transfer, teaching projects and education programmes. The programmes have to meet the local requirements and specific potential disasters. They continuously have to be evaluated with a view to their effectiveness (Resolution 44/236 of the General Meeting, United Nations, 22.12.1989).

The United Nations also urged all countries to prepare sustainable development plans until the year 2000. These should integrate the risks involved and give priority to the implementation of a comprehensive hazard management. The year 1999 will be the last of the IDNDR. In the meantime 140 national IDNDR committees have been established, pursuing the concept of disaster reduction. Motivation and potentiality for the implementation of the different measures do, however, vary greatly. The endeavours in countries such as Japan, Peru, New Zealand or the United States are quite far-reaching, whereas many developing countries still depend on foreign assistance. The international and also national endeavours therefore have to continue, especially since or even despite the IDNDR coming to a close (White et al. 1998).

This is especially true for small islands. Most of the islands are not only located in the tropics or its periphery. They are often also located in seismically active areas, as for example in the Caribbean, the Pacific or the Indian Ocean. They are exposed to a vast range of natural threats, amongst them volcanic eruptions, earth quakes, hurricanes, tsunamis, coastal erosion, land slides, rising sea-levels, droughts and floods. All of these are either creeping processes or sudden events that are related to living on an island. These threats can turn into complex natural disasters when they strike the often extremely vulnerable society of a small island. The vulnerability of the society is usually conditioned by factors such as land utilisation, migration, social structure, population growth, a small market, dependency on the exterior or the general political organisation (Possekel 1997; Mossler 1996; Cross 1992; Vermeiren 1991). Often the resources and spatial constraints are the reason why natural disasters cause much more damage on small islands than in larger area states. Anyone can easily imagine a small island being completely devastated by a hurricane. The situation is aggravated by the fact that small islands have often only just recovered from one disaster when the next one strikes. Mossler resumes the situation as follows: "Proportional impact is of primary concern to small islands as events with statistically few deaths and/or financial damage may have devastating impacts on isolated societies with small populations and limited financial resources" (Mossler 1996, p. 90).

Despite this knowledge, the urgent necessity to invest more time, energy and resources into the hazard management of small islands is often not recognised. There are two significant reasons, one being acute problems which often prevail in the social, political or economic sector. This is often perceived as far more serious than a potential threat. The other reason being the uncertainty as to when the next disaster will actually occur. The population and decision-makers naturally find it difficult to accept that a proportion of the limited resources of their small island should be invested into hazard management, especially when they have not experienced a natural disaster for quite some time. Nevertheless, the threats and their potential damage to long-term development endeavours, increased insurance premiums and the extremely high reconstruction costs following a natural disaster usually force the affected people to change their attitude.

These circumstances have evoked that increasing attention is paid to small islands since the middle of the IDNDR. In 1994 the *United Nations Conference on Natural Disaster Reduction* (UNCDR) passed the "Yokohama Strategy for a Safer World". It explicitly emphasises that priority be given to promoting the capacities of the *Small Island Developing States* (SIDS) for their national hazard management (Eikenberg 1998). Many other international and national organisations and programmes also emphasised the necessity to reduce the vulnerability of small islands to natural disasters, amongst them the *United Nations Commission for Sustainable Development* (UNCSD), the *United Nations Disaster Relief Organization (UNDRO)*, the *United Nations Environment Programme* (UNEP), the *South Pacific Disaster Reduction Programme* (SPDRP) and the *Pan American Health Organization* (PAHO). The SIDS themselves also called for more international support. In August 1998 altogether 42 SIDS appeared before the United Nations and presented their idea to develop a vulnerability index for small islands. They intended to illustrate the economic struggle of small islands when they have to recover from recurrent natural disasters.

The development achievements of the SIDS in the Caribbean are also constantly crushed by natural disasters. The geophysical and socio-economic characteristics make the Caribbean one of the most vulnerable regions of our planet (Cross 1992). In the past two centuries alone about 77,000 fatalities have been reported in connection with earthquakes, hurricanes and volcanic eruptions. The best-known eruption is that of the volcano Mt. Péle on Martinique in 1902. It devastated the town St. Pierre and buried its 29,000 inhabitants. Almost at the same time 1,600 people died on St. Vincent when the Soufrière Volcano erupted. And an earthquake in 1907 had devastating consequences for Jamaica. Kingston was destroyed, more than 1,000 people died and 90,000 were left without shelter. The history of the past 20 years is no less worrying. Hurricane David devastated the Eastern Caribbean island Dominica in September 1979. 80 % of its 90,000 inhabitants were left without shelter. St. Vincent's Soufrière volcano resumed activity in the same year. It erupted and the affected area had to be evacuated. Hurricane Gilbert devastated great parts of Jamaica in the autumn of 1988. The total damage was estimated at about US$ 1.5 milliards. Only a year later Hugo, one of the century's most violent hurricanes, struck the Eastern Caribbean and caused a total damage of US$ 3 milliards (Cross 1992). The hurricane seasons of 1995 and 1996 also had devastating consequences. There were 34 violent storms and nine hurricanes. Numerous islands were shattered, many people were injured or killed and thousands remained without shelter for months. The damage alone incurred in 1995 amounted to US$ 7.7 milliards (McGregor 1997). And Hurricane Georges was responsible for more than 300 fatalities in the Caribbean region in 1998. Damage of many hundreds of millions dollars was incurred, the Dominican Republic, Haiti, Puerto Rico, Antigua, St. Kitts and Nevis being particularly affected.

The negative experience with natural disasters in the Caribbean area led to intensified endeavours for a regional cooperation in hazard management. The *Pan-Caribbean Disaster Preparedness and Prevention Project* (PCDPPP) was established in the mid 1980s by joint cooperation between the UNDRO, the PAHO, the *Caribbean Economic Community* (CARICOM) and the Red Cross. A group of experts was set up to serve as a consultant agency for the Caribbean countries. The experience with the hurricanes Gilbert and Hugo has led to the formation of another organisation, the *Caribbean Disaster Emergency Response Agency* (CDERA). It provides disaster aid to its 16 member states. However, despite these endeavours there are still considerable shortcomings since purely technical measures such as building regulations, risk zoning and evacuation plans or relief actions in an acute emergency do not meet the entire range of requirements. Such aid neglects decisive factors: the understanding of the local conditions of vulnerability, the participation of the population and the integration of a comprehensive hazard management into a sustainable development strategy. The following questions thus play an important role.

1. How do the people perceive their environment, in this case resources and hazards, and how do they evaluate them?
2. Which decisions do the people make when they deal with resources and hazards?
3. Which are the structures that dominate a society?
4. How do human beings think and act in disaster situations?
5. Which methods could increase the participation of the local population in planning and management?
6. How can hazard management ideally be integrated into sustainable development planning?

These questions mark the frame for the subject matters this publication deals with. The main objective of the considerations is a practical linkage of hazard management and sustainable development planning. Focal points are (1) the analysis of the vulnerability and complexity of an insular system and (2) the examination of appropriate methods that facilitate the affected population to cope with a natural disaster. The potentials of the scenario method for a sustainable and participative reconstruction planning following a natural disaster will be examined in detail.

Regional case study is the Eastern Caribbean island Montserrat. The island is characterised by its small size, its extremely sensible ecosystem, its limited resources and recurrent natural disasters, such as floods, hurricanes and earthquakes. All these factors hinder a sustainable development of the island. The past decade has seen a number of promising attempts to its economic development, amongst others ecotourism and offshore banking. The initial success has again and again been set back by natural disasters. Hurricane Hugo, for example, caused serious damage on the island, and its economic development was severely affected. Large financial transfers of the British mother country and the endeavours of the local government and the population contributed to a relatively successful reconstruction. In 1994 the gross national product started rising again, mainly as a result of the booming tourism and construction industry. But recurrent eruptions of the Soufrière Hills Volcano between July 1995 and December 1997 shattered all hopes. The Montserratians have since lived with uncertainty.

Even though the publication concentrates on Montserrat, it describes phenomena, processes and interactions which are just as valid for other small islands and even larger area states. The research ties on to a project which has been carried out by the Institute of Geography at the University of Hamburg from 1993 till 1998. The project *Complex Resource Management on Small Caribbean Islands* is sponsored by the *Deutsche Forschungsgemeinschaft* (DFG) and examines the problems and potentials of resource management for the sustainable development of small islands. Hazards were intentionally excluded from the research objectives of the project. Whilst the project analyses the perception, evaluation and management of resources, this work follows the same basic objective, but focuses on hazards. The considerations thus supplement the research project in its attempt to examine the correlation between resources and hazards, normality and uncertainty, present and future, planning and management and between decision-makers and population in a multi-perspective approach. Its main intention is to encourage a better understanding of a complex system and to present options for decisions and activities directed at the shaping of Montserrat's future.

The work is structured in six chapters. Chapter 2 outlines the theoretical frame and concentrates on the focal subject of the publication – *coping with complexity and uncertainty.* Some (natural) scientific concepts, such as the theories of chaos, complexity and self-organisation, are presented, and it is investigated how these concepts might be transferred to social sciences. In a second step the development of two practically oriented approaches – *resource and hazard management* – are examined, again focusing on complexity and uncertainty. At the end of the chapter an attempt will be made to show the practical implications of these basic theoretical reflections for geography.

The third chapter deals with the methodical consequences of the theoretical reflections. It concentrates on strategic planning as an organisation of processes that examine and adapt to complex processes. The *scenario method* has been selected as approach to strategic planning since it is readily available and there exists a fund of practical experience with the method. First of all, the different possibilities to construct scenarios are outlined. In a further step the problems and potentials of the scenario method in the context of complexity and uncertainty are analysed. This analysis bases on the methodological reflections and practical experience of geographers, futurologists and economists. The methodical approach is outlined in detail at the end of the chapter.

Chapter 4 contains the *empirical implication of the basic theoretical considerations*. The complex system Montserrat is approached in a two-fold way. Both approaches are based on several field studies on Montserrat between 1994 and 1997. The first approximation (Chapter 4.1) is undertaken in an apparently safe situation, that is during the reconstruction phase five years after hurricane Hugo. The second approximation (Chapter 4.2) is carried out in an uncertain and disorderly situation, namely during the volcanic crisis. Both chapters attempt to explain structure, function and development of the complex system Montserrat. To this end the basic correlations of the system, the interests and perspectives of the decision-makers and the local population and the interactions between these levels are examined. Mass-statistical representation is not of consideration in this context. To the contrary, the author intends to follow the rules of qualitative social research and thus places priority on the perceptions of the inhabitants. The experience of their everyday lives is considered the key to the island. It is therefore the purpose to let the people speak, to listen closely to what they say and to examine and translate their statements, as Sandner (1994) has requested.

Chapter 5 first of all outlines some research derived from reports that summarise the practical experience scientists have made in a concrete *reconstruction phase* following a natural disaster. The deficiencies of purely centralistically organised and short-term oriented reconstruction programmes are explained. The necessity for visionary and participative reconstruction planning is consequently outlined. Also, the potential contributions of the scenario method are in this context investigated into. Chapter 5 is in this sense a continuation of the third chapter since it illustrates potentialities for the implementation of scenarios in hazard management, particularly in the phase of reconstruction.

Chapter 6 first of all contains a *critical examination of the scenario method.* It is based on a scenario project carried out in Hamburg with scientists, students and representatives of the public and private sector. The project was undertaken as a pre-test for the later practical implementation of the scenario method. Based on the experience gathered, the scenario method was then practised on Montserrat during the volcanic crisis. The participants of the two-day workshop were guided through the method and developed scenarios and strategies for a sustainable development of the island. The reports about the scenario project in Hamburg and the workshop on Montserrat will not only contain results as regards contents, but also consider the communication processes between the respective participants. This facilitates a comparative evaluation of the outsider and the insider perspective.

The concluding evaluation in Chapter 7 links the empirical findings and the experience outlined in Chapters 4 and 6 with the theoretical considerations of Chapter 2. *Five central principles* will be suggested which are decisive for a constructive coping with complexity and uncertainty. Special emphasis will obviously be placed on coping with the volcanic crisis on Montserrat, but the results are, nevertheless, applicable to a wide range of complex and uncertain situations. One thing will, above all, become clear: whatever the situation, it is absolutely vital to adjust to living with the unexpected, and the five principles will be the tool to consciously shape this life.

We live forward but learn backward and are constantly on the brink of the unknown (Kierkegaard).

2 Theoretical Reflections

Living at the end of a century encourages us to reflect about the past and to direct our thoughts towards the future. The past is often either critically reviewed or romantically idealised whereas the future offers both chances and risks which call for new concepts. Mankind is today faced with the challenge of ecological, social and economic problems on a global scale. The reasons are many – the Brundtland Report "Our Common Future" (World Commission on Environment and Development 1987) is one of the publications concerned with the topic. Dovers and Handmer (1992) summarise the global problematic as follows:

1. Environmental changes and destruction have reached an extent which endangers the entire biosphere. For the first time in the history of mankind the total destruction of life-supporting systems, in one word the earth, has become a possibility.
1. Societies are only beginning to recognise the global effects of their daily activities. We just have to remember that we have only recently begun to assume that global climatic changes are to a great deal caused by human activities.
2. Uncertainty is growing despite all technological and scientific progress, not least because our ability to change our environment is growing.
3. The world population is rising continuously and with it the proportion of inhabited areas. A considerable number of people are living in endangered regions. Changes in the environment are these days followed by steadily growing ecological, social and economic problems.

These rather urgent problems of the present age call for innovative solutions, and it is with good reason that the concept of *sustainable development* is on everyone's lips. It has by now advanced to *the* instruction for the 21st century. The concept is generally seen as the way to an optimised utilisation of resources, without overexploiting them and thus not endangering their future availability. The historic roots of these thoughts go back to the beginning of the 19th century when sustainability was used as a management concept in German forestry. The present discussion is characterised by two major components, one being the management of resources, the other our moral responsibility towards the next generations. This becomes evident when reading standard definitions of the concept, a good example being that of the Brundlandt Report: "Sustainable Development is development which meets the needs of the present without compromising the ability of future generations to meet their own needs" (World Commission on Environment and Development 1987, p. 41).

Human needs, security, harmony and the future are key elements of sustainable development. However, exactly this is the problem of the concept. It aims at maintaining a balance. The objective of sustainable development is a secure and stable situation in which the utilisation and protection of resources is balanced with the supply of today's population and future generations. Sustainable development is generally based on the idea that humans decide consciously and as such strive for a sensible use of resources. However, the environmental problems mentioned prove quite the reverse. Reality is characterised by great uncertainty, sudden changes and growing conflicts. This is especially valid for long-term concepts such as that of sustainable development. Complexity and uncertainty are therefore focal points in the discussion and the practical application of the concept. These two factors will in the next century be of a significance never experienced before (Mitchell 1997; Merry 1995; Gell–Mann 1994; Oßenbrügge 1993; Dovers and Handmer 1992; Rosenhead 1989; Bayrische Rückversicherung 1987; Beck 1987).

2.1 Complexity and Uncertainty

A major concern of scientists is the examination and analysis of complexity and uncertainty. Generally, there are two approaches to the topics. The first starts from the assumption that cause and effect of processes are easily comprehensible. Linear processes serve as an explanatory pattern for unidimensional and functional dependences. This is easily explained with billiards. When ball 'A' hits ball 'B', this then moves proportionally to the effective force applied by billiard ball 'A'. This type of Newtonian-based science intends to predict processes. To this end models are calculated and created, in order to then exert specific influence on the environment.

On the other hand, there are modern systematic scientific approaches. They start exactly from the opposite assumption, i.e. that of non-linearity, complexity and sudden changes. These approaches reach back to the 1940s and 1950s when criticism was uttered about the reductionism dominating at the time. Focus of this criticism was the strict analysis of single and isolated objects, i.e. reducing complex phenomena to simple terms (Schneider and Kay 1994; Morfill and Scheingraber 1991). It was a time of radical change with important contributions coming from the science of cybernetics, after the mathematician Norbert Wiener, and from the general systems theory, developed by the biologist Ludwig von Bertalanffy. Objective of cybernetics was and is the analysis of self-regulating mechanisms in living organisms and machinery. Focus of the examination are the functions of a system, i.e. its control mechanisms, its exchange and communication with other systems.

The general systems theory in contrast concentrates on the interactions and interdependences between the elements of a system. It tries to depict the structure of the system with the help of models (Rosnay 1979). Since the structure and function of a system cannot be viewed separately, both approaches have in practice formed more and more of an entity to become what is known as the systems approach.

What exactly is a system? In general, a system is defined as a set of elements and relations between these elements, forming the structure of the system (Albert et al. 1992; Funke 1992; Götz 1992). Isolated, closed and open systems are distinguished. In isolated systems neither matter nor energy is exchanged with the environment. They are non-existent in reality since they are theoretically postulated constructs which have their specific importance in physics. A closed system, on the other hand, does exchange energy with its environment. A good example for such a system is a cell. It has to be remarked, though, that the borderline to what is considered an open system is still being discussed. An open system then exchanges matters, information and energy with its environment. It can be defined as follows: "An open system is a grouping of component parts that individually establish relationships with each other and that interact with their environment as individuals and as a collective" (Kiel 1994, p. 4). An open system therefore consists of three components: (1) a set of elements, (2) a set of relations between the elements and (3) a set of relations between the system or its individual elements and its environment. A classic example of such an open system is the weather. A number of elements interrelate in a complex relation, e.g. humidity, air pressure, wind, irradiation, temperature, relief, vegetation, rivers etc.

The concept of an open system is nothing new. Innovative is the dynamic perspective with which a non-linear and complex world is described. The science of chaos has contributed considerably to the development of this dynamic perspective. Chaos is in this context not to be understood in its everyday sense, i.e. as a state of utter confusion. Chaos is rather the unpredictable transition into new patterns: "The science of chaos often deals with discontinuous forms of change. Discontinuous changes are those changes where something transforms itself completely. These changes occur when a system reaches a critical bifurcation point that leads either to disintegration or to radical change" (Merry 1995, p. 11). The theory is of such relevance because it recognises the dynamics of change, be it the biological evolution, radical changes in society or – to keep with our example – tomorrow's weather. According to the science of chaos, tiny changes within a system can have considerable effects. Systems are not linear. They do not react directly proportional to changes in their parameters, disruptions or fluctuations. And chaos is not an accidental effect, but follows certain laws. This is why the technical term *deterministic chaos* is often applied (Kay 1993; Lewin 1993; Morfill and Scheingraber 1991; Gleick 1987).

The meteorologist Edward Lorenz was the first to note deterministic chaos in the 1960s when he used computer simulation to calculate irregularities in the dynamics of convection. After he had completed a rather lengthy simulation he decided to repeat it. He entered the data once again, this time correct to three decimal places instead of six. After a short coffee break he returned to his desk, only to note that the second calculation differed considerably from the first. At the end of the day he got two completely different results. A tiny discrepancy in the initial conditions triggered off a completely different course of the simulated dynamics. At the beginning of the simulation there was an extremely sensitive balance, decisively changed by a tiny initial difference. What Lorenz described as the *butterfly effect* is in technical terminology called the sensitivity to initial conditions. (Peak and Frame 1995; Gell–Mann 1994; Lewin 1993; Gleick 1987). Lorenz' *term* depicts the meaning of chaos very well. Therefore, many authors refer to the following example: a butterfly flaps its wings in the Amazonian rain forest and triggers off events which cause a storm in Chicago. Another time the butterfly flaps its wings, but that time no meteorological effects are caused at all (Lewin 1992, p. 23).

At this point the question arises whether the butterfly effect implies that we cannot predict anything and that the deterministic understanding of our world, i.e. order, security and control, is an antiquated concept. The answer has to be 'no, this is not quite correct'. Lorenz' simulation shows that the output of chaotic behaviour is determined by its past – in this case by the decimal places which were, after all, changed. Thus chaos is deterministic, i.e. a prediction would on the basis of exact knowledge about the initial conditions and the effective forces be possible. As Peak and Frame (1995) have expressed it, chaos is not an act of caprice or coincidence. What bewilders us is the way the inaccuracy of the measurement increases. When dealing with chaotic systems and an insufficiently determined starting point, the initial uncertainty grows quickly and hinders long-term predictions (cf. Peak and Frame 1995, p. 154).

Morfill and Scheingraber (1991) with the following example succinctly explain that a chaotic system does not act accidentally. It follows certain rules which do not allow for coincidence. It is possible to predict the behaviour of a system for a certain time, using its rules. This period of time is the longer, the more we know about the initial conditions of the system. However, the inaccuracy grows exponentially, i.e. to be able to predict something over twice as long a period of time, the initial conditions of the system have to be known 100 times more exactly. For a period of time three or four times as long the measurements have to be 1,000, respectively 10,000 times as exact. A really long-term prediction of the behaviour of a chaotic system is in effect impossible because of the exponential growth of the inaccuracy (cf. Morfill and Scheingraber 1991, p. 54). As the authors point out, our ability to predict the behaviour of such systems is limited. At this point the weather once again serves as a good example. Based on the knowledge we have about the initial conditions and the forces at play, the weather forecast for the next day is up to a certain point usually quite accurate. Nevertheless, precise forecasts for a whole week are despite modern computer technology not possible (Kay and Schneider 1994).

The butterfly effect has by now had far reaching scientific consequences. A considerable number of disciplines are nowadays focusing on the behaviour of dynamic, non-linear systems. The knowledge about deterministic chaos forces us to thoroughly review our previous theories and concepts, such as 10-year plans for regional development or long-term strategic plans for companies

(Peak and Frame 1995). Apart from this, the analysis of non-linear systems promises to help us understand the world better and to allow us to develop alternative strategies for living with and in this world. This is the challenge the science of complexity takes up.

2.1.1 What Exactly is Complexity?

The everyday meaning of complexity generally refers to complicated matters. This is the case when the steps to find a solution to a problem appear incomprehensible. The preparation of an extensive meal might be just as complicated as programming a new software or constructing a statically demanding high rise building. Surely, all this is complicated. And we are generally confronted with this type of problem every day of our lives. Nevertheless, complicated and complex is just not the same.

It is apparent that the problem of complexity is already linked with the term itself. Even complexity researchers have not yet agreed on a standardised definition. Horgan (1995) published a list of altogether 31 definitions of complexity in his publication 'From Complexity to Perplexity'. The definitions cited often even contradict one another. Horgan concludes: "... such problems highlight the awkward fact that complexity exists, in some murky sense, in the eye of the beholder" (Horgan 1995, p. 76). It is not part of the scope of this work to analyse the various scientific definitions of complexity. The starting points of natural and social sciences will in the following be presented in order to extract those aspects which are of interest in the context and scope of this work.

Complexity in its origin is a mathematical concept: "Basically, complexity refers to two major aspects of a system: a) the mathematical structure of the irreducible component subsystems of the process and b) the manner in which the components are connected to form the system. Complexity is an attribute of the system itself ..." (Casti 1979, p. 41). Complexity is therefore the characteristic of a system. It emerges when a number of components act simultaneously within a system. These interacting components, in the theory of complexity often called agents, have a certain degree of local information and local influence. They are, nevertheless, not in a position to determine the whole state of the system. Complexity is therefore inherent to the structure of the system itself. Thus the spatial organisation of the individual elements within the system has to be the focus of any evaluation (Waldrop 1992; Slocombe 1990; Rosnay 1979). In the sense of Bertalanffy's general systems theory the structure of a system can be mathematically defined, depicted in models and calculated. "Complexity is the amount of computational resource (usually time and memory) that it takes to solve a class of problem" (Edmonds 1997, p. 7). In this case we are dealing with the resources necessary to solve a problem – here especially the time factor and the capacity of a computer. This type of complexity is also referred to as "rough complexity" and does not represent more or less than a mathematical description of a system's structure, trying to be as precise as possible.

The Santa Fe Institute in New Mexico (USA) is one of the most significant research institutes concerned with mathematically oriented research about complexity. Its staff not only examines "rough complexity", but above all the ability of complex systems to put things in order. Anyhow, what exactly does order mean in this context? Lewin (1993) has clarified this in his popular book *Complexity. Life at the edge of chaos*[1]. He refers to Langton's (1989) *Bottom up* hypothesis (Figure 2.1). The basic idea of this hypothesis is that the components of a system interact and create an emergent global structure. This structure in turn affects the behaviour of individual components by a process of feedback. The emergent global structure is a specific feature of a system, going beyond the abilities of its individual components (the so-called emergent characteristics). A new order is put up as a result of interactions and feed back within the system. This leads us to Gell–Mann's definition which roughly characterises effective complexity as the length of a succinct description of the system's regularities (cf. Gell–Mann 1994, p. 93). The Santa Fe Institute focuses exactly on these regularities when evaluating complex systems. It analyses systems which, in Langton's sense, create new order. The processing of energy and information in systems is of special interest in this context. The scientists believe that they can obtain results as to the laws governing these processes, giving them an insight into the basic patterns of change in a system over a period of time. This type of research is mathematically oriented and its objective is the predictability and control of processes which determine transformations.

The purely mathematically oriented research is nowadays facing criticism from different directions, amongst others from the following sciences: ecology, futurology, sociology and practically oriented planning and management (Shackley et al. 1996; Kiel 1994; Schneider and Kay 1994; Luhmann 1988; Rosen 1978). These critics say that the mathematical understanding of complexity bases purely on computer simulations and is therefore too simple a depiction of real life. The conception is as such judged as a step backwards to traditional reductionistic strategies because once again it intends to understand all activities of a system as mechanical interactions between the individual components, with the ultimate aim of control. Schneider and Kay explain why such an approach does not do justice to complexity: "... for this version of scientific method to work, an artificial situation of consistent reproducibility must be created. This requires simplification of the situation to the point where it is controllable and predictable. The very nature of this act removes the complexity, that leads to the emergence of the new phenomena which make complex systems interesting ..." (Schneider and Kay 1994, p. 640).

Schneider and Kay (1994), just like Funtowicz and Ravetz (1993; 1994) and a number of other scientists approach complexity in a different way. Their centre of attention is the interaction between two systems: the society and the environment. Objective of their research is a better understanding of the processes. The methodical challenge is therefore not to be sought in the mathematical calculation of simulated worlds but in the permanent crossing

[1] In the following the German edition *Die Komplexitätstheorie - Wissenschaft nach der Chaosforschung* will be referred to.

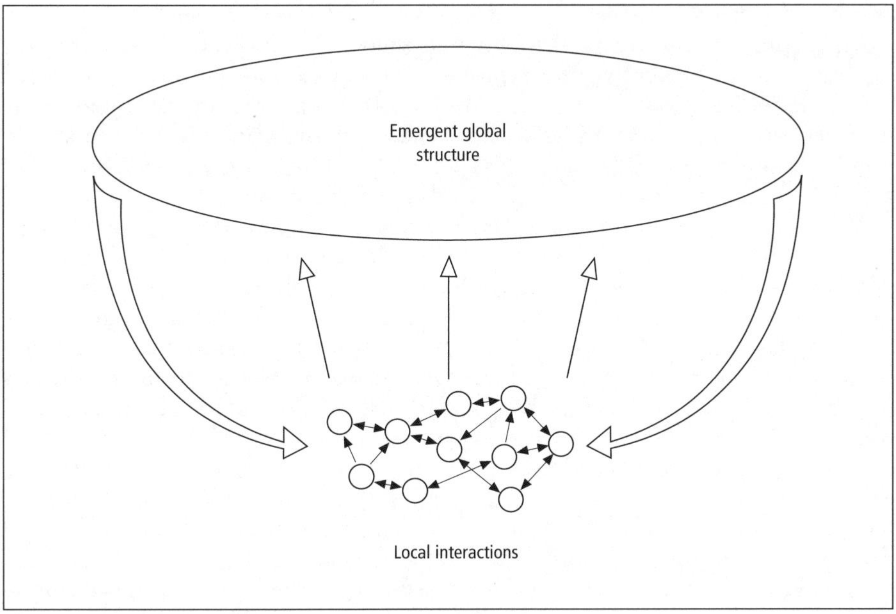

Figure 2.1. Chris Langton's conception of emergence in complex systems. (After Lewin 1993, p. 25)

of borders between the natural and the social, the scientist and his research object, the expert and the layman (Shackley et al. 1996). It is a matter of integration, changing perspectives and a continuous questioning. This strategy is supposed to enable scientists to research into the functions of a system, especially the processes of exchange, transfer and evolution, i.e. the temporary organisation of a system. They therefore define complexity as follows: "A system is complex if we can describe it in a variety of different ways, each of which corresponds to a distinct subsystem. Complexity then ceases to be an intrinsic property of a system, but it is rather a function of the number of ways in which we can interact with the system and the number of separate descriptions required to describe these interactions. Therefore, a system is simple to the extent that a single description suffices to account for our interaction with the system; it is complex to the extent that this fails to be true" (Mikulecky 1995, p. 4).

The sociologist Luhmann (1988) also examined the interaction between humans and their environment. However, environment here refers to the societal context and not to the natural environment. When a person interacts with his environment this then appears to him a very complex matter since the environment of any system is always much more complex than the system itself. Man can therefore only selectively interact with his environment. He has to simplify and reduce the complexity of his environment. Since he perceives just a part of his environment, he can accordingly interact with it in a restricted manner only. Luhmann therefore considered complexity as nothing else but the need to select. According to his definition a matter is complex when it consists of so many elements that these can only establish relations on a selective basis. Operative and reflected complexity both require a process of reduction which determines a pattern of selecting the relations and temporarily rules out other possibilities of how the elements might be interlinked as pure possibilities (cf. Luhmann 1988, p. 267). Luhmann in this context illustrates the limited capacity of mankind to grasp complexity.

However, when Luhmann (1988) refered to complexity, he addressed something completely different than Schneider and Kay (1994) or Funtowicz and Ravetz (1993; 1994). Luhmann focused on a selective choice of observations and experiences an individual has with its environment. Luhmann's method also intends to construct models, but not in the sense of simulating physical or biological systems. His intention was rather to apply a so-called formal system to order thoughts, observations and experiences. Schneider and Kay as well as Funtowicz and Ravetz in contrast focus (1) on the wide range of possibilities in which the individual can interact with natural systems, (2) on the dynamics of the interaction human-environment. This basic difference has been pointed out since it illustrates a problem often encountered when dealing with complexity. Some researchers talk about complexity when they address human interactions with natural systems, others refer to the complexity of a formal system. Mickulecky (1995) has also addressed this problematic: "... we are talking about complexity with respect to our interactions with the natural system. On the other hand, most authors seem to find it easier to speak about the complexity of the formal system. This is an entirely different issue (...) The multiplicity of views of complexity is a direct failure to recognise the reality of the modelling relation, whether it being used implicitly or explicitly, and thereby focusing on the formal system as if it were the natural system" (Mikulecky 1995, p. 10).

However, the complexity of a natural system cannot be reduced or simplified. It might be helpful to use a formal system in order to enhance our consciousness and sensitivity towards complexity. Examining and analysing a choice of agents and interactions may transmit at least a feeling for this complexity to the observer. This method is unproblematic as long as the model constructed is not misunderstood as the complexity of a natural system. It is of utmost importance that we are at all times aware that the complexity of a natural system will, despite all endeavours, not be ascertainable. Any attempt already constitutes a reduction of the phenomenon we are trying to depict. We also have to remember that there is always a number of models, all dependent on the individual perspectives and interests of their observer. Our mental models should therefore, in the sense of Shackley et al. (1996), be constantly reflected, questioned and adapted.

This necessity for a continuous reflection and adaptation of our mental models is also emphasised by the theory of complexity. It deals with dynamic processes and sudden transformations in the evolution of a system. Complexity in this context means the unfathomability and unpredictability of a system's behaviour, reactions and evolution, being a function of their structure, dynamics and interaction (Casti 1994; Funtowicz and Ravetz 1994; Waldrop 1992). Humans, in the sense of Luhmann (1988), strive to make complexity understandable with the help of mental and statical models. The theory of complexity, in contrast, focuses on the emergence, the becoming, the evolution of new structures and patterns in natural systems. Both approaches are significant for this research. But what exactly are these dynamic processes that create new structures and patterns? The precise meaning of this becomes clear when we address the self-organisatory mechanisms of complex adaptive systems. Understanding them will facilitate the approximation to our initial question "what is complexity?".

Self-Organisation and the Edge of Chaos

Complex adaptive systems are a specific group of complex systems. They refer to a variety of different processes, such as the transformation of human societies, the strategies of investors in financial markets, the immune system of a mammal or the learning processes of a person. The question arises what *adaptive* means in this context. Adaptive actually describes something all the above processes have in common: complex adaptive systems absorb information from their environment and through each individual interaction with it. This information is then examined for regularities which again can be worked into a model or scheme, at last managing the activities in the real world. This way a complex adaptive system can use suitable strategies to adapt to the requirements of its environment (Schretzenmayr 1996). Adaptive, in summary, is the ability to learn by experience, to specifically process information and to adapt.

Ilya Prigogine (1980) received the Nobel price for Chemistry for his work about the imbalance in open systems. In contrast to the understanding of classical sciences, he proved that chemical and physical systems do not operate from a state of equilibrium but from one of imbalance. The technical term applied to this state is *far-from-equilibrium* (Merry 1995; Peak and Frame 1995; Scott 1991). Complex adaptive systems are subject to a permanent process of change: "A far-from-equilibrium system does not return to some fixed stable state. It is forever in a continuous flux of change: never being the same, always becoming. Prigogine spoke about his theory as *from being to becoming*. A far-from-equilibrium system is like the flame of a candle or a whirlpool in a river. Its wholeness, structure, and form can only be maintained by the endless flow through it. Its existence depends on its flux" (Merry 1995, p. 31). Adaptive systems are systems that evolve and adapt to new conditions and influences. The evolution of an adaptive system is determined by its history and its environment. The different agents of a system process information in order to adapt to new circumstances. Adaptation is therefore the process of gradual change in a system in order to coexist with changes outside its environment. Characteristic of far-from-equilibrium systems is that they can suddenly exit from apparently stationary phases. These reactions are sometimes even caused by minor changes in their critical parameters.

Flux in this context refers to the flow of energy and takes us to thermodynamics, the science which has for a long time explained energy transformation patterns in closed and isolated systems.[2] This initial version of thermodynamics was then broadened by Prigone's results about the behaviour of open systems. When energy is supplied to an open system, an imbalance incurs. The system reacts to the new condition with the spontaneous development of a new structure. It dissipates the energy supplied to create a new order. This order can then be maintained through a continuous supply of energy to the system. Any change in the flux of energy as a consequence affects the order of the system. Prigogine based his theory of self-organisation on this typical behaviour of open systems (Nicolis and Prigogine 1989).

The example most commonly quoted when addressing the subject of self-organisation of an open system is the Bénard phenomena. A saucepan is filled with a liquid and heated on a stove. The liquid at the bottom of the pan heats up more quickly than at the top, the thermodynamic equilibrium thus being affected. The system now tries to even out the difference in temperature. As long as the temperatures vary only slightly, this can be realised by a simple flux of heat. The liquid itself stays in quite a stable state and thermal radiation extends from below to the surface. A state of *near-the-equilibrium* has been reached. However, the difference in temperature also implies a variation of its density. An unstable condition appears – the hotter and lighter liquid is covered by the cooler liquid with the higher density. Supplying further heat makes the liquid itself start moving. The molecules of the cooler and more dense part sink downward, following the laws of gravity, whereas the molecules of the hotter and lighter part move upwards. The liquid has now transformed to the state of *far-from-equilibrium.* The more energy is supplied, the more chaos grows. The liquid moves more and more turbulently until it reaches a completely chaotic condition. This is the point where a further phase is entered. The heat cannot distribute quickly enough so that a compensatory effect comes into play, the so-called convection cells which are developed. The energy supplied is used for the spontaneous forming of hexagonal molecules. Chaos has transformed to a new order. However, this order does not remain since it will again be subject to new phases of transition (Merry 1995; Mikulecky 1995; Schneider and Kay 1994; Morfill and Scheingraber 1991).

The theory of self-organisation no longer describes the behaviour of a system as a purely mechanical interaction between the individual components, but as the spontaneous forming of new structures under changing conditions. The following characteristics are typical of *far-from-equilibrium systems*:

1. They are inherently unstable, i.e. each change in the supply of energy is followed by a spontaneous self-organisation into a new structure;

[2] There are three laws to thermodynamics: The first two are of significance in this context and will thus briefly be explained. Most simply stated, the first law says that energy cannot be created or destroyed and that despite the transformations that energy is constantly undergoing in nature, the total energy within a closed or isolated system remains unchanged. The second law requires that if there are any physical or chemical processes underway in a system, then the overall quality of the energy in that system will degrade. The second law can also be stated in terms of the quantitative measure of irreversibility, entropy, which for any process is greater than zero (cf. Schneider and Kay 1994).

2. They are dissipative, i.e. they depend on a continuous exchange with their environment;
3. They are autopoetic, i.e. they have a certain degree of self-determination, or, to put it a different way, they have an ability of self-organisation since they are capable of spontaneously creating a new order (Nicolis and Prigogine 1989; Prigogine 1980).

There is one further characteristic which scientists are today concerned with: complex adaptive systems may possibly have their best capacity in an interim state between order and chaos. This brings us back to Langton's research (1989) in the field of artificial life. He simulated complex adaptive systems with so-called cellular automata, and discovered phases of transformation between chaos and order, i.e. processes of self-organisation. In 1990 he put up the hypothesis of the *edge of chaos*. According to this theory a system is extremely efficient and productive in exactly the interim state between order and chaos. At this point the system has its best capacity for the processing of energy and information and is thus capable of optimising its fitness. The edge of chaos is according to Langton a turbulent state between chaos and order, in which the systems due to their extreme adaptability function with maximised efficiency.

Strictly speaking is the edge of chaos again a purely mathematical function of simulated complex systems. However, transition phases in practice come up quite regularly so that scientists to a growing extent believe in the existence of this phenomenon. In the 1960s, for example, the biologist Stuart Kauffmann analysed the Boolean mathematical models of gene networks and discovered a phase of transition between chaos and order. He also realised that systems in this phase show a maximum capacity for evolutionary adaptation, practically complementary to the Darwinian theory of natural selection. Kauffman (1995) furthermore put up the hypothesis that it is immanent to a complex adaptive system to strive for the transition between chaos and order, since this is the state of their best fitness. The physician Per Bak (1991), too, plays an important role in this debate. His work focuses on the so-called self-organised criticality. Bak's hypothesis is that dynamic systems naturally head for a critical condition, as for example an ecosystem or the interaction of tectonic plates at the time of an earthquake. Bak gives the layman a good example of self-organised criticality, namely a heap of sand. Let a thin jet of sand trickle onto a flat surface. A heap is going to slowly grow which will soon extend to the edge of the surface. The small heap is growing bigger and bigger until suddenly one further grain of sand triggers of an avalanche. The heap of sand at this point does not accept any additional sand. The pile of sand is self-organised in the sense that it reaches the steady state all by itself. At the same time it is in a state of criticality, as defined by Bak, since the sand grains on the surface are not stable. One further grain can at any time cause a landslide (Lewin 1992; Waldrop 1992).

In summary we can say: complex adaptive systems are *far-from-equilibrium*. They are subject to a continuous transformation and exchange energy, matter and information with their environment. Internal fluctuations evolve at critical points (bifurcation points), triggering off processes of self-organisation which lead to new structural orders. Each new order is more complex than the previous one since the number of variables, the interactions and the vast range of possible patterns of behaviour continually increases. Complex adaptive systems possibly have their highest level of capacity in the interim state between order and disorder, a state which they themselves strive for.

The Relevance to Social Sciences

The *edge of chaos* is a very abstract model and the processes of self-organisation are still difficult to grasp. However, many scientists believe in their general relevance and keep looking for practical examples. Prigogine has investigated whether his theory of self-organisation is appropriate to biological or social systems. He has come to the conclusion that irreversible changes, transition phases and fluctuations also happen in ecological or social systems (Nicolis and Prigogine 1989; Prigogine and Stengers 1984). Processes of self-organisation are nowadays used for a variety of applications – explaining anything from a termites' nest, a traffic jam or urban development to the evolution and break-down of cultural and political systems (Schretzenmayr 1996; Merry 1995; Gell–Mann 1994; Lewin 1992; Scott 1991). Common examples are the splitting up of the former Soviet Union or the unexpected reunification of East and West Germany. And this is the lesson to learn: "Common sense, not to mention recent political experiences, suggest that healthy economies and healthy societies alike have to keep order and chaos in balance (...) they have to regulate themselves with a dense web of feedbacks and regulation, at the same time that they leave plenty of room for creativity, change and response to new conditions" (Waldrop 1992, p. 294). It is not only the political changes of the past which make us suspect that processes of self-organisation do not only exist but are absolutely vital for the effective functioning of a system which is subject to permanent changing conditions.

Funtowicz and Ravetz (1994) have developed a more precise approach for the application of mathematically derived theories to social systems. They first of all differentiate between "ordinarily complex systems" and "emergent complex systems". The former are characterised by simple structures and regulatory mechanisms which can be described mechanistically and functionally. The above mentioned models, i.e. Kauffman's gene network (1995) and Langton's cellular automata, are part of the group of ordinarily complex systems. A simple mathematical formula is on the other hand not sufficient to sum up emergent complex systems: "... in them, some at least of the elements of the system possess individuality, along with some degree of intentionality, consciousness, foresight, purpose, symbolic representations and morality" (Funtowicz and Ravetz 1994, p. 584). Emergent complexity characterises social systems in which information, symbols, values and interests are just as important as the interactions between individual and society.

Based on these reflections, Funtowicz and Ravetz (1994) call for a different scientific approach. They call their approach *post-normal science.* It is to be understood as a combination of social and scientific ideals, identities and interests which are often ambivalent, in reciprocal conflict to one another or subject to a constant transformation. The main reason for change being the erosion of apparent certainty, and just this factor makes the participation of as many people as possible in a decision process inevitable: "Emergent complexity provides a theoretical justification for post-normal science, in which the peer group for quality assurance is expanded beyond the certified experts to include all those with a stake in the issue. This concept helps us to appreciate that there is no single perception providing a comprehensive or adequate vision of the whole issue, nor any particular criteria of quality that can hegemonically exclude all others" (Funtowicz and Ravetz 1994, p. 578). This is the way the edge of the chaos also gains significance for social systems. Funtowicz and Ravetz transfer it as follows: the tendency to chaos is called fragmentation, i.e. the disorder of different opinions and interests which do not yet form a recognisable pattern and often end in conflict. The opposite tendency, that to order, is called diversity, i.e. the co-existence of different opinions and interests which through their dialogue have already formed visible groups. These different groups liven up the social structures and debates in a creative way.

In how far can such an approach help to deal with a complex world? The current debate about sustainable development gives us a good example. Nobody will dispute that fragmentation and diversity are basic to the theoretical research and practical application of sustainable development. After all, very different types of interest groups try to obtain new knowledge which is to be applied both locally and globally. One thing has by now become apparent: new knowledge about sustainable development can only be gained when different social groups and different countries in effect work and communicate on a cooperative basis (Evers and Nowotny 1988).

Since modern science strives for objective truth, control, predictability and certainty, it can serve to only a limited extent. The problematic rather calls for very different, subjective interests which go through the processes of participation, discussion and consensus and guarantee a continuous reflection about complexity, about new patterns and structures. This is the context where post-modern science might contribute as it roots in the knowledge about uncertainty. It looks for tools to help us accept uncertainty and to be creative when dealing with it. And post-modern science intends to offer us points of reference in a complex world. It already has become an important tool. Scientific work these days focuses more and more on matters such as uncertainty, adaptability to constantly changing conditions and consensus about qualitative values and objectives of a society: "... issues of risk and the environment present the most urgent problems for science, uncertainty and quality are moving in from the periphery, one might say the shadows of scientific methodology, to become the central, integrating concept" (Funtowicz and Ravetz 1993, p. 739). Nevertheless, science remains under pressure. The theoretical debates about uncertainty are far from concluded and the search for new or even revived methods, being applied to the new context, continues.

2.1.2 What is Uncertainty?

The term *uncertainty* is just as multilayered as *complexity*. We encounter a wide range of possible definitions which all depend on individual perceptions. The amount of definitions available does not actually facilitate the planning or decision processes when dealing with uncertainties (Knight 1991). It is therefore inevitable to have a closer look at the definitions available and the perceptions they are based on. First of all, there are intentions to strictly distinguish between certainty and uncertainty. Wildavsky (1988) uses this approach. Certainty reflects a situation in which the consequences of any action can be predicted. Uncertainty, in contrast, refers to a situation in which possible events as such are known beforehand, but it is not possible to estimate the likelihood or the time when things might happen. This distinction between certainty and uncertainty is mainly used in the theory of planning as it is based on the hypothesis of a certain knowledge. However, this understanding is facing growing criticism since experience has shown that especially decisions in planning are permanently subject to uncertainty.

A further possibility to define uncertainty refers to its sources. This theory differentiates three types of uncertainty. First, uncertainty conditioned by the environment (*environmental uncertainty*) and, second, uncertainty having its source in the evaluation or assessment (*value uncertainty*) and, third, uncertainty caused in the context of the coordinating processes (*related choice*). The first type expresses a lack of information about the environment, in this case understood as the environment of the system. The system dealt with is subject to external influence which cannot be controlled by man. Such an influence can be a natural disaster, technical defects or human error, all of them causing uncertainty. Value uncertainty is related to appropriate value judgements regarding expected outcomes of choices, i.e. it refers to the problems that arise when assessing preferences which might prevail in the future. This is especially valid for the development of social norms since individual preferences always depend on their social context. Insecurity with reference to correct evaluation or assessment is today of specific importance when it comes to investments as the success or failure of any investment is extremely sensible to future preferences. The third source of uncertainty is one determined by coordination processes. The consequences and implications of any action taken by individuals or an institution not only depend solely on the state of the environment but also on everything other people do. All three sources of uncertainty are caused in the first instance by a lack of information. The gathering, distribution and processing of knowledge is therefore a central aspect to any planning strategy (Knight 1991; Bayrische Rückversicherung 1987).

A third possibility how uncertainty can be defined refers to the probability of uncontrollable events. Three types of uncertainty are differentiated: First, is there an objective probability that such an event will occur (risk / prediction is possible)? Second, is there a subjective likelihood (uncertainty / anticipation possible) or third, is it not possible to predict an uncontrollable event at all (doubt and ignorance). This differentiation bases on the traditional concept of insecurity in the theory of decision processes, formulated by Frank Knight as early as 1921 (Kleindorfer et al. 1993; Werlen 1987). Once again we are dealing with a concept of Newtonian character. It believes in science being objective and mankind being able to predict future events.

Risk then has been defined as a situation in which the behaviour of a system is widely known and estimates are made as to the probability that a certain event or chain of events occurs. These estimates principally concern the future. All sorts of consequences and implications of what is done, be it of positive or negative effect, can be calculated. The term risk is traditionally linked to damage which is to be avoided, for example the failure of a business schedule or disruptive events in large technical plants (Fisch and Boos 1990; Jungermann 1990). Risks can therefore be reduced to a simple formula: risk = damage x probability. Statistic and prognostic methods are at hand to calculate risks. Mitchell (1995) explains this as follows: "An example would be estimating the risk of a specified magnitude of flood based on a long-term record of water flow in a river system. The risk could be calculated with long-term information. There would always be a possibility of error, but there could be considerable confidence in the estimate" (Mitchell 1995, p. 3). The risk of flood becomes a calculable factor since it can be determined with objective facts and (long-term) data.

In contrast, uncertainty refers to a situation in which we do not know how the system might react. There are no objective facts and parameters which allow us to calculate a likelihood. We can estimate what might happen, but it remains uncertain how likely one or the other event is (Hanf 1986). The future of the system itself is unknown but sometimes the central factors of influence (the variables) are known. Mitchell (1995) once again refers to the above example: "In the example of estimating the magnitude of a flood, one could well know the key variables that are significant for a flood, but if inadequate data were available, then it would be difficult if not impossible to estimate their probability with any confidence" (Mitchell 1995, p. 3). Uncertainty is the result of a lack of information about the system, and this in turn hinders us to define or quantify probabilities.

Ignorance as another kind of uncertainty refers to the perceptive faculty of man. There are situations in which a problem is not perceived as such and therefore not dealt with. Such a situation normally occurs because people are not aware of certain phenomena, even though they might be quite important. Not knowing of their existence at the same time implies that these phenomena are not taken into consideration, i.e. ignored. A good example is any science which is limited to the simple Newtonian cause-effect-model and completely overlooks non-linear processes. Finally, the fourth type of uncertainty is generally referred to as indeterminacy. In this situation humans completely fail to imagine what might happen. They lack the capacity to recognise interactions within and between systems, to even grasp the structure and functions of a system. We have now reached exactly the point science of complexity is foremost concerned with. Wynne (1992) defined indeterminacy as follow: "causal chains or networks are open, understanding [is] not possible" (Wynne 1992, p. 114).

The traditional differentiation between risk, uncertainty, ignorance and indeterminacy bases on man's capacity to make rational decisions, an assumption which still to a large extent dominates the decision and risk theory. Probabilities are focal point of the research and can be proven by statistical calculations. The current trend in social sciences dissociates itself from this assumption. Rather, the objective nature of probabilities is questioned and there is also a tendency to refer to the theories about complexity which root in the fundamental thought that information is always incomplete. There is no objective calculation of probabilities since each calculation is subject to the butterfly effect, to the sensitivity to initial conditions as Lorenz' example illustrates. Each estimate or prediction is subjective since it depends entirely on the individual person. The time frame chosen, the data selected, the model opted for and the individual mistakes made all determine the result. Probabilities are not a characteristic of an event, but of the person itself. This implies that different people can allocate completely different probabilities to one and the same event. It, furthermore, implies that events might not even have any probabilities at all, effectively when no-one thinks about them. Probabilities are the result of a subjective process in which information is digested. Therefore, terms such as risk, uncertainty and indeterminacy are nowadays often used synonymously. Solely the term ignorance is distinguished from the others (Eisenführ and Weber 1993; Herbold and Wienken 1993; Fisch and Boos 1990; Bayrische Rückversicherung 1987; Hanf 1986).

Current social science has, based on above reflections, today a different approach to certainty and uncertainty. Both aspects are nothing more or less than social evaluation modes for our future. We believe that something is not endangered, neither now nor in future, when it is called certain. Is something called uncertain then, we believe that it is endangered. What is of specific relevance in this context is the framework of all three temporal horizons, i.e. of the past, the present and the future. Uncertainty is caused by incomplete knowledge about the past. And since especially the past is often understood as the determinant factor for the present and the future, we normally try to gather as much information about it as possible. Our individual mode of information processing often is the obstacle. We might forget things, suppress all memory of something or just not perceive an event. There is, of course, also uncertainty in the present, especially with regard to our awareness of certain threats and their possible consequences. A number of threats, such as the global climatic change, the polluted environment or infectious diseases like AIDS are real already. There are many threats, though, which are projected into the future. They might evolve to such a state that any reaction could seem futile. This is where we have to realise that threats call for an immediate response. Existing information should be selected and interpreted in order to actively design the future. At the end of the day, future is always something uncertain because it does not yet exist, i.e. it is fictitious and imaginary.

Uncertainty, therefore, is primarily a subjectively determined human experience. It is a social construct which is affected by the individual and collective processes of our anticipation of the future (Beck 1991; Bayrische Rückversicherung 1987; Hanf 1980). As Rayner (1993) expressed it, uncertainty is a function of how human beings individually and in groups perceive their place in the world and the things that threaten it. Uncertainty as an experience manifests itself as a state of feeling and consciousness. A state of fear, helplessness or lost orientation can be enforced by permanent changes in the living conditions. However, humans do not like to live with uncertainty. They try to either ignore it or – like complexity – reduce it. The scientists Tversky and Kahnemann (1974) illustrate that people always live with a simplified model of their environment. They never have all information available and are not capable of sorting it out or assessing it fully. Their capacity to process information is limited. In the sense of Luhmann's pattern of selection they simplify any problem calling for a decision, and they use strategies of incremental analysis (working one's way through) just as well as heuristic methods (rough estimates, tricks etc). Tversky und Kahnemann have, above all, analysed the heuristic strategies, i.e. the individual perception of the probability and frequency of unexpected events. This perception depends decisively on the personal experience with such events. Individuals are generally very convinced of their own perceptions. Slovic et al. (1980) supplemented the work on this topic. He and his colleagues asked for the impact values have on the perception of insecurity. The actual frequency of 41 deaths was compared with the estimates the subjects of the test had given. The result was that the estimates of the experts differed considerably from those of the laymen. The laymen tended to be strongly affected by their individual feelings, such as fear and apprehension. This type of feeling was generated to a large extend by the media. Sensational and dramatic events suited to the media, such as floods or volcanic eruptions, were perceived far greater threats than commonplace and well-known, but less spectacular events, such as accidents or diseases.

Depending on their individual perception, there are different strategies of how humans cope with uncertainty. Dovers and Handmer (1992) have elaborated three kinds of typical behaviour. The first type of person is characterised by ignorance. He is strongly opposed to any change. Change is a threat to his personal security and has to be avoided at all cost: "Where proper reaction would threaten the status-quo, appeals to ignorance are common: these are often expressed through calls for more information, and an insistence upon inaction because of uncertainty" (Dovers and Handmer 1992, p. 270). The second type of person allows certain changes in selected areas, always provided the societal basis is not endangered: "This is characterized by incremental change, change which does not challenge the basis of our societies, but which may lead to changes in emphasis at the margins" (Dovers and Handmer 1992, p. 270). The third type is open, flexible and adaptable. He does not object to changes which affect basic values or institutional structures. Dovers and Handmer point out, though, that this type of person is rather rare: "Of course, through history this has happened, but usually only in a slow and painful way. An adaptable society would be open to the possibility of moving in a new direction quickly and relatively painlessly" (Dovers and Handmer 1992, S. 270).

Let us have a look at a good example. The political discussions about the greenhouse effect are well known to everybody and clearly illustrate the different behavioural patterns. There is one group which is still calling for further scientific research. This group is only willing to implement measures for the reduction of gases when the phenomenon has been scientifically verified

100 %. Another group is of the opinion that there is sufficient knowledge, giving cause to thoroughly change our current structures of production and consumption. These are the two extremes and there are, naturally, a number of opinions inbetween. These represent the type of person who supports some measures, but only those which do not affect any basic structures (Dovers and Handmer 1992).

Wildavsky (1988) differentiates even further and comes to the conclusion that there are at last two rivalling patterns of behaviour when dealing with uncertainty. The first is based on the anticipation of possible risks in order to avoid them. Hypothetical risks are foreseen in order to eliminate them (as far as possible). Should they, nevertheless, occur, they then have to be controlled so that any possible negative effects can be reduced. The choice will always be for the most secure and predictable activities in order to prevent the worst. Preventive measures can be the selection of the safest technology, risk variation, diversification, insurances or the exclusion of new, untested production processes. It is common standard in our world that technologies, for example, are only introduced when we believe that we can deal with their possible negative effects (Evers and Nowotny 1988).

Opposed to anticipation as one option for dealing with uncertainty is elasticity or flexibility. People who act according to this pattern consciously accept risks. Wildavsky calls this characteristic "resilience", i.e. the ability to accept the unexpected or the uncertain and to respond flexibly to changing conditions. Research and management currently pay much attention to resilience. Endeavours are made to either use reliable strategies or to develop new ones, in order to encourage the flexibility and adaptability of social systems. Geographic research in this context concentrates on two traditional areas: resource management being the first and hazard management being the other. Both disciplines have seen a transformation from control and prevention to adaptive management strategies. These changes and the current challenges facing both theories will be examined in the next chapters.

2.2 Resources and Uncertainty

Neo-classical economics has for a long time dominated the theoretical preoccupation with resources. Neo-classicists also developed the term *resource management theory* which has, however, not found acceptance in other disciplines. It has to be admitted that the term as such is unsuitably restrictive. The only relevant aspect seems to be the economic management of resources, i.e. mainly cost-benefit factors. Cultural, social, political, legal or perceptional aspects which are, above all, examined by geographers, sociologists, ethnologists and political scientists, were hardly considered by neo-classical economists. Emel and Peet remark: "Typically, socialized knowledge, ideology, and contradictions in economic and political rationality have been neglected by the dominant resource management paradigms rooted in neoclassical economics and pluralistic political science. In addition, management implies practise but not necessarily theory or explanation. Although many argue that theory is a precondition of practise, this has not been a pre-eminent theme in resource geography" (Emel and Peet 1989, p. 50). In Anglo-American science there are two terms readily available, the first being resource analysis and the second being resource management. In this study a combination of the two seems sensible and it will in the following be referred to as *resource research*. This term is to be understood in a very broad sense, ranging from the theoretical discourse about resources to the practical application of new results to, finally, the management of resources.

Again, we have to start off with a question. What exactly are resources? Resources are always the product of human perceptions and needs. Natural factors are neutral until man perceives them, recognises their potential for the fulfilment of his needs and develops methods for their utilisation. Resources are as such the result of an interactive process between man, or society, and the environment. These reflections go back to Zimmerman, his definition of resources up to now being the most widely accepted: "... resources are not, they become, they are not static but expand and contract in response to human wants and human actions" (Zimmerman 1951, p. 15).

Any resource concept is subject to change since the perceptions, needs and activities of mankind also change in the course of time, space and cultural transformation. Rees (1991) points this out in her specific definition of resources: "Natural resources are those products or properties of the physical environment which human beings are technically capable of utilising and which provide desired goods and services. Both these criteria must be satisfied before a particular part of the physical world can acquire a value as a resource. Technological innovation and improved knowledge only create the opportunities for utilisation. Whether these opportunities are taken up depends upon economic, social and political demands. Resources are, therefore, defined by human desires, needs and capacity. They are phenomena which depend greatly on the prevailing culture. (...) Since the value of resources is determined by the culture that uses them, it follows that the value of particular resources varies not only over time but also over space" (Rees 1991, pp. 6–7). Resources are, in summary, subjective, dynamic and culturally defined. They are a factor of the natural environment, suitable for human utilisation within a certain cultural system. Iron ore, cotton, a national park or the environment as a whole might all be considered resources by one culture whereas another culture is not the least interested in them (Simmons 1993; Mitchell 1989).

Burton and Kates go a step further and apply the attributes allocated to resources to natural dangers: "According to Zimmerman's view, the physical environment is neutral stuff, but it is human culture which determines which elements are considered to be resources or resistances" (Burton and Kates 1964, p. 413). The boundary between a resource and a hazard as a type of natural threat is therefore blurred. This goes quite far – a natural threat often represents a variation of a natural processes in its extreme and can in another context be considered a resource. There are some simple examples at hand: Lightning can cause a forest fire which is necessary for the preservation of an ecosystem, but it can, on the other hand, destroy someone's house. Floods can increase the fertility of cultivated land but at the same time ruin a farmer's crop. Snow is (considered) a resource as long as it falls onto ski-runs but not when it blocks off the roads of access a few hundreds metres away. The tropical climate attracts sun-seeking tourists but also is a cause of the yearly hurricane season (Smith 1994; Burton et al. 1993; Geipel 1992; Rubin et al. 1987).

The following model (Figure 2.2) depicts the differentiation between resources and hazards which is a result of the interaction between humans and their environment. The hatched area shows the acceptable range of variation, i.e. natural processes can within this spectrum be evaluated as resources. The hatched lines represent the so-called damage threshold, the point where a resource becomes a hazard or, to put it in other words, a negative resource. Smith himself describes his model with the following words: "As long as the variation of the environmental elements remains fairly close to the expected performance, insignificant damage occurs and the element will be perceived as beneficial. However, when the variability exceeds some threshold beyond the normal band of tolerance, the same variable starts to impose a stress on society and becomes a hazard" (Smith 1994, p. 10). Smith chooses the example of rain. Rain is generally perceived as a resource, but it can also cause either floods or droughts. The threshold has then been passed in one or the other direction and the formerly neutral, or rather beneficial, resource turns into a threatening hazard.

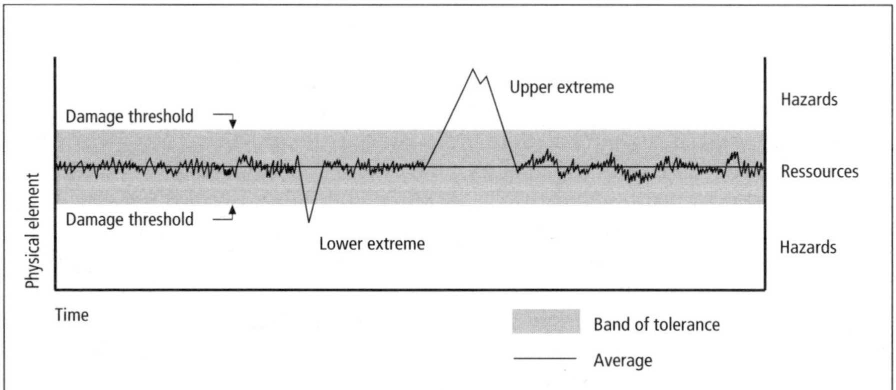

Figure 2.2. Resources and hazards. (After Smith 1992, p. 11)

These correlations seem logical and obvious. Nevertheless, resource research has for a long time purely considered the positive aspects of natural factors and events, i.e. the resources. Natural threats were dealt with in a separate scientific discipline, namely in *hazard research* which was established shortly after World War II (see Chapter 2.3). The dominance of neo-classical economics caused this division. As explained, it believes in our ability to control the environment and in the predictability of events. Simple cause-and-effect models served as tools. The basic question was always that of an optimised utilisation of resources and the necessary measures. Society was to be supplied and at the same time control over the environment maximised. These growth-oriented methods were for a long time unchallenged. They were then eclipsed by anticipative concepts which were primarily concerned with the prevention of possible hazards. Management strategies were designed to avoid activities which had been recognised as being of negative effect.

Since the mid of the 1980s scientific research of systems gained importance. Main aspect was the management of the dynamic interactions between the human system and its environment. Blaikie's and Brookfield's (1987) as well as Redclift's (1987) *political ecology approach* had a formative influence on this concept which is still dominating. Complex interactions of social, economic and political processes against the background of growing ecological problems on the local, regional and global scale are since that time receiving increasing attention. Complex interactions are recognised step by step and it has become apparent that our knowledge about complex systems will always be incomplete: "... there has been a growing agreement among organizations in the public and private sectors that they operate in increasingly complex environments, as well as under conditions of rapid change, fewer resources and greater uncertainty. These aspects have to be addressed if organizations are to be effective and efficient in realizing goals and objectives" (Mitchell 1997, p. 138). The management of resources has, as a consequence, to be flexible and adaptable. Above all, in contrast to previous approaches, the starting point of all considerations has to be uncertainty. Controlling the environment has become insignificant, instead ways of dealing with unexpected events come to the fore. Last, but not least, must any approach consider the population with its perceptions of the environment and its resources and with its preferences for certain management strategies. This is depicted in the following illustration about the development of resource management:

Development of Resource Management

If resources are defined by perceptions, needs and activities of mankind, the same has to be valid for their management. Management generally is a process of defining objectives and implementing decisions in the scope of planning, coordinating and controlling. Resource management is a decision process which naturally focuses on resources, their utilisation and allocation. O'Riordan (1971) describes resource management as follows: "... a process of decision-making whereby resources are allocated over space and time according to the needs, aspirations, and desires of man within the framework of his technological inventiveness, his political and social institutions, and his legal and administrative arrangements. Resource management should be visualized as a conscious process of decision involving judgement, preference and commitment ..." (O'Riordan 1971, p. 3.). The objective of resource management therefore has to be the search for suitable resources and the management of their utilisation and distribution. The resource manager has to undertake a sequence of decisions and actions in order to supply his goods to society.

According to Holling (1994) one objective dominated resource management over decades: a certain stability and consistency, in short, a balance of the ecosystems realised by regulatory measures. However, exactly the balancing efforts have often effected unwanted negative changes of systems. Holling gives us some examples. Fighting the spruce budworm population in the pine woods of eastern Canada with pesticides was in the short term successful. Long-term analysis has shown, however, that the forests are now sensitive to such an extent that a return of the worm would cause serious damage. Another example is the very successful battle against forest fires in American national parks. Long-term result, however, was the growth of easily flammable wood which has, in turn, caused extremely large and costly fires. A final example is the fight against malaria which has, in the long run, also caused changes in the system. Many types of mosquitoes are nowadays resistant to the pesticide DDT and our immunity against malaria is diminishing.

All examples quoted show that mankind has in each case interfered in order to keep a balance and every time exactly the opposite happened. The biophysical environment responded to human management in such a way that the fragility of the system was increased and discontinuous processes of change took place. This brings us back to Prigogine's (1980) theory of self-organisation. Ecosystems operate in a state of imbalance. They are complex adaptive systems which use self-organisatory processes in order to develop new structures. Schneider and Kay (1994) therefore urge us: "We must stop managing ecosystems for some fixed state, whether it be an idealistic pristine climax forest or a corn farm. Ecosystems are not static things, they are dynamic entities made up of self-organizing processes. Management goals that involve maintaining some fixed state in an ecosystem or maximizing some function or minimizing will always lead to disaster at some point, no matter how well meaning we are. We must instead recognize that ecosystems represent a balance, an optimum point of operation, and that this balance is constantly changing to suit a changing environment" (Schneider and Kay 1994; p. 645). Resource management which aims to produce a balance in a complex far-from-equilibrium system can, logically, not make sense. Resource management must, to the contrary, be flexible and adaptable in order to be able to keep up with a continuously transforming environment.

The geographers Burton and Kates (1964) in an earlier study examined how man copes with a dynamic and uncertain environment. Their research focused on the perception and dealing with hazards and the resulting implications for resource management: "It may be argued that the uncertainties of natural hazards in resource management are only a special case of the more general problem of risk in any economic activity. Certainly, there are many similarities, but it is only when man seeks to wrest from nature that which he perceives as useful to him that he is strongly challenged by the vagaries of natural phenomena acting over and above the usual uncertainties of economic activity. In other words, the management of resource use brings men into a closer contact with nature where the extreme variations of the environment exercise a much more profound effect than in other economic activities" (Burton and Kates 1964, p. 413). They argue that the recovery from hazards and the utilisation of resources depends on the general attitude humans have on their physical environment. Basis of these reflections is, amongst others, a cultural-anthropological study of Kluckholm and Strodt-beck (1961). They interviewed about 100 people with different cultural background in New Mexico. As suggested by Table 2.1, three different types of perception of the environment, the nature here used synonymously, were distinguished and with it the resultant resource management strategy.

Table 2.1. Different perceptions of nature according to cultural groups. (After Kluckholm and Strodtbeck 1961)

Nature perception/ cultural groups	Man subject to nature	Man with nature	Man over nature	Number of persons interviewed
Hispano-Americans	71.7	10.9	17.4	23
Texans	30.0	22.5	47.5	20
Mormons	25.0	55.0	20.0	20
Zuni-Indians	19.0	62.0	19.0	21
Navajo-Indians	18.2	68.2	13.6	22

"Man subject to nature" implies that the person believes in his complete dependence upon nature. He refrains from any type of control and accepts anything that may happen as his fate. We are here dealing with an ecocentric concept: "My people have never controlled the rain, wind, and other natural conditions, and probably never will. There have always been good years and bad years. That is the way it is, and if you are wise you will take it as it comes and do the best you can" (Burton and Kates 1964, p. 431). The second category "man with nature" consists of people who see themselves in a unifying partnership with nature. "My people help conditions and keep things by working to keep in close touch with all the forces which make the rain, the snow, and other conditions. It is when we do the right things – live in the proper way – and keep all that we have – the land, the stock and the water – in good conditions, that all goes along well" (Burton and Kates 1964, p. 431). The person belonging to the category "man over nature" dominates and controls his environment. Nature is seen as purely a supplier of the necessary resources. This anthropocentric concept is based on the belief in technical progress and control over nature: "My people believe that it is man's job to find ways to overcome weather and other conditions just as they have overcome so many things. They believe they will one day succeed in doing this and may even overcome droughts and floods" (Burton and Kates 1964, p. 432).

The hypothesis that resource management depends on the respective perception of the environment is still relevant today. Or, rather, we are today recognising that it plays an important role in resource management. Current studies devote considerable attention to cultural aspects. Culture does not only determine the perception and evaluation of resources, but also the management of the dynamic relationship man-environment (Mitchell 1997; Ratter 1997; Rees 1991). Mary Douglas' ethnological work had considerable influence on resource research. She and her colleagues, the political scientist Aaron Wildavsky, the ethnologists Michael Thompson and Steve Rayner and the psychologist Karl Dake, jointly developed the cultural theory. This theory relates closely to Kluckholm and Strodtbeck (1961), saying that dealing with a dynamic and uncertain world can best be explained by a cultural approach: "Cultural theory is a general sociological theory, constructed according to the tenets of positivistic reasoning, by way of deduction from a limited number of basic axioms, regarding human social beings and their interactions. A basic distinction is made between cultural bias, defined as shared values and beliefs, on the one hand, and social relations, defined as patterns of interpersonal relations on the other. The crucial explanatory concept, way of life, is defined as a combination of social relations and cultural bias" (Boholm 1996, p. 67). This theory might simplify, but it is an interesting and legitimate attempt of differentiation which shall briefly be explained below.

Cultural theory distinguishes four different risk cultures on the basis of their cultural shaping and social interactions. They are bureaucrats, egalitarians, individualists and fatalists, each group being characterised by a specific environmental perception and decision-making behaviour. Bureaucrats believe in being able to control and regulate the environment with technological progress and institutional structures. The social elite determines how resources are used, the remaining majority being left out of this decision-making process. Egalitarians see the environment as extremely fragile and full of surprises. They believe that resources are limited and therefore consider a careful utilisation inevitable, not last to serve the public at large and also future generations. Individualists perceive the complex and uncertain environment similarly to bureaucrats as a positive challenge. However, individualists, in contrast to the latter, see the control mechanisms not as institutional regulations but believe in the laws of the market, letting each person decide and act as he thinks it to be appropriate. This, in practice, means that the individual with the best strategy can be sure of the best advantage in the fight for resources. This concept sees the future as a time when humans have to be flexible and adapt to changing conditions. Fatalists, finally, accept an uncertain environment. To them it is dominant and can by no means be controlled. They believe in an unlimited variety of resources but their positive implications depend solely on fate (Thompson et al. 1994; Merriman and Browitt 1993; Rayner 1993; Wildavsky 1993; Douglas 1992).

Van Asselt and Rotmans (1996) have continued these thoughts. They use the cultural theory to distinguish between the respective management styles, excluding the passive behaviour of fatalists: "Cultural theory enables us to ascribe different management styles to the various perspectives. Hierarchism is associated with a preference for bureaucratic management, while the egalitarian preference can be described as communal anti-managerialism. Individualism in its extreme manifestations advocates an anti-intervention laissez-faire attitude. (...) Following the cultural theory, the hierarchistic management style can be characterized as control, while the egalitarian management style is characterized as preventive. The management style associated with the individualistic world view is characterized as adaptation" (van Asselt and Rotmans 1996, p. 132). Bureaucrats try to solve problems through a carefully monitored management. Their central objective is the preservation of the system and the rational allocation of resources, achieved by an institutional structure.

Laws, regulations and financial incentives are the preferred strategies, as for example land utilisation plans or ecological taxes. Egalitarian management style tries to avoid risks altogether. In order to accomplish this objective it relies on the anticipation of possible negative effects any human activity might have on society itself or on its environment. Sustainable management concepts are another central point of this preventative style which preferably applies communicative and educating strategies, tests as to the ecological tolerance and prognoses. Individualistic management style in turn relies on the natural market forces to regulate all activities. Possible negative effects are rectified by technological processes. Since the economic development shall not be interrupted by political interference, regulation is almost completely blended out. Preferred strategies are again financial incentives, technological progress and development.

Van Asselt and Rotmans (1996) are well aware that cultural theory depicts the situation in a simplified way, not actually doing justice to reality. However, it is the only attempt so far to differentiate culturally determined perceptions of the environment. These perceptions offer us, at least to a certain degree, important facts for resource management. The two scientists are currently trying to prepare an explanatory model of the interactions between time, space, demographic, social and economic processes. This so-called target model will emphasise individual perceptions, decisions and activities, as opposed to the entire groups defined in the scope of cultural theory. Asa Boholm (1996) is of a similar opinion. He, too, considers the defined groups too limited and also opposes to treating the groups as separate entities without any exchange between them. Boholm criticises that cultural theory suggests a world in which bureaucrats, individualists, egalitarians and fatalists live completely isolated of each other and are not capable of exchanging information, learning from or understanding each other. Thus the theory can in no way do justice to the complexity and dynamics of social systems.

The cultural theorists themselves have accepted that their theory cannot explain fully how to deal with uncertainty in resource management. Individual perceptions and patterns of behaviour have to be taken into consideration. Current research therefore analyses how individuals evaluate and change their own perceptions and how they consequently adapt their decisions and activities: "A person might very well be a hierarchist at work, an egalitarian at home and a fatalist in his spare time. The person is portrayed as a mosaic of life, each enacted in its proper context. The self is compartmentalized so that individuals are able to cope with multiple social contexts that place different demands on them. This version of cultural theory also invokes the principle of self-interested maximization and argues that the individual might find it beneficial to pursue different ways of life in different realms" (Boholm 1996, p. 77). Cultural theory is one of many which have all come to realise something very important: interaction, structure and development of a system can only be understood when all factors are equally considered, i.e. individuals and society, respectively single agents and the whole system's structure.

It has, in summary, become clear that resource management has experienced a considerable transformation. At first it was characterised by such terms as control, certainty, balance and a best possible allocation of the resources. The worst was to be avoided and to this end preventive measures taken. It then became apparent that despite these attempts unexpected events could not be avoided. This has led to the conclusion that resource management always has to imply a critical preoccupation with extremely complicated problems, always being aware that unexpected events and developments are not an exception but the rule (Mitchell 1997). This fact has for a long time been neglected, but brought to the light by the political ecology approach. Resource problems can, according to this approach, only be tackled by an analysis of the dynamic interaction between society and environment, as this includes unexpected events: "Social order and environment are not static; conflict and contradiction in the spheres of production, consumption and nature are fundamental, also fundamental is the unity of environment and development. Resource managers are encouraged to see socio-economic development and environmental quality as unantagonistic concepts, to generate a wide range of alternatives for accomplishing a management objective, acknowledge uncertainty, expect surprises, be sceptical of what we think we know, avoid doing the irreversible, integrate the environmental with economic and social understanding through every phase of the policy design process ..." (Emel and Peet 1989, p. 56).

As a result of the intensive research we are today concerned with flexible management methods and continuous learning and reflecting. This facilitates the acceptance of inherent uncertainty just as much as the adaptation to sudden and unexpected events. Such events are to an increasing extent even seen as an extraordinary opportunity to learn and adapt. This is exactly the point where we encounter an acute need for research and action. Interesting is that, in the sense of Burton's and Kates' work (1964), scientists and managers alike once again refer to the neighbour discipline of resource research – hazard research.

2.3 Hazard Research and Complexity

Hazard research is based on one of the fundamental questions of this paper: How do humans confront uncertainty? Or expressed another way: is mankind able to adapt to sudden and unexpected events, and, if so, how? The geographer Kates developed five basic questions for any hazard research. They are to be seen as paradigms which seek to

- assess the extent of human occupancy in hazard zones;
- identify the full range of possible human adjustments to the hazard;
- study how man perceives and evaluates a hazard;
- describe the process of adopting damage-reducing adjustments in their social context;
- estimate the optional set of adjustments in terms of anticipated social consequences (cf. Kates 1970, p. 438).

The questions show that hazard research is practically oriented and that it addresses not only the question of adjustments but also the psychological, economic, political and social consequences a hazard may have. Hazard research is therefore primarily concerned with the adaptability of man to the uncertainty implicit to natural processes, especially sudden and often unexpected events. It also examines the implications of the results and strategies for planning and management (Geipel 1992).

Whereas resource research completely neglected unexpected events for a long time, hazard research failed to notice their complexity for quite some time. It is especially the dynamic and uncertain situation of a hazard that is extremely complex. Comfort concludes: "Complexity is the product of uncertainty and interaction which together compound the possible issues for public action involved in emergency planning nearly beyond human calculation" (Comfort 1988, p. 6). The spectrum of uncertainties is large – how does each individual realise a hazard, how long will it last and what will be its socio-economic consequences? Once a hazard occurs, the decision-makers and the public have to react very quickly in order to save their lives and their belongings. The pressure of time usually limits the search for alternative activities. As a consequence conflicts often arise between different groups of interest. Experience has shown that these conflicts worsen because of a general ignorance about the resultant consequences of a hazard and often also because of a complete breakdown of the communication structures. The methods applied by the classical decision theory, explained in Chapter 2.1.2, are naturally not effective in such an extreme situation.

Modern hazard research therefore increasingly concentrates on complexity as the starting point of any consideration. Individual and collective perceptions, decisions and reactions of the people affected form the focal point. Hazard scientist Palm emphasises that hazard research will only be able to offer any help when it does justice to the complexity of the interaction between humans and their environment: "... we must interpret the relationships between people and their environment as an interplay among the physical setting, the political-economic context, and the sometimes unpredictable influence of agents or individuals. (...) The ultimate goal is an understanding of the actual response of real people in real place. (...) A complex world demands the ability to provide an explanation tracing complex lines of relationships" (Palm 1990, pp. 158–159). Before going into detail, we have to first define the terms hazard and disaster more precisely than it has been necessary for the earlier, more general reflections.

What are Hazards and Disasters?

The term hazard implies the state of a still existing threat and that of a past interaction between man and his environment, i.e. the realisation of a hazard (Steuer 1979). This ambiguity and also the attempts to equate certain German terms, such as risk, threat or danger, with hazard have often led to misunderstandings, especially in German definitions. (Incorrect) translations also pose a serious pitfall to any German researcher. We will at this point, therefore, distinguish between the above terms in order to clearly define the term hazard[3]. Hazard can be defined as a potential threat to humans and their belongings caused by a sudden event. Risk, on the other hand, already implies a probability for the occurrence of such an event and also refers to the possible damage that might be incurred. Smith (1994) illustrates the difference between hazard and risk with a very vivid example: Two people cross the ocean, one of them on board of a passenger steamer, the other in a rowing boat. The hazard, i.e. death by drowning, faces both of them alike. The risk, i.e. the probability of such a death, is, of course, far greater for the person in the rowing boat.

It would therefore be preferable to use the American-English term hazard even in German literature. This paper concentrates especially on the so-called natural hazards, i.e. those which refer to extreme meteorological, geotectonical or biological processes. These types of hazard were for a long time interpreted as a stroke of fate, an act of God, something on which man could not exert any influence. He was consequently completely at the mercy of a natural hazard. Our horizon has widened by now. There are extreme natural events which do not have to become a natural hazard. This is the case when they take place in an unoccupied area. They will not be defined as a natural hazard unless an occupied region is concerned (Figure 2.3). Floods, hurricanes, avalanches, volcanic eruptions and earthquakes will in occupied regions often affect unprepared individuals, groups or whole societies. They cause injuries or deaths, disrupt and destroy spheres of life, cause damage and force us to confront the situation, to adapt and to develop countermeasures (Smith 1994; Geipel 1992).

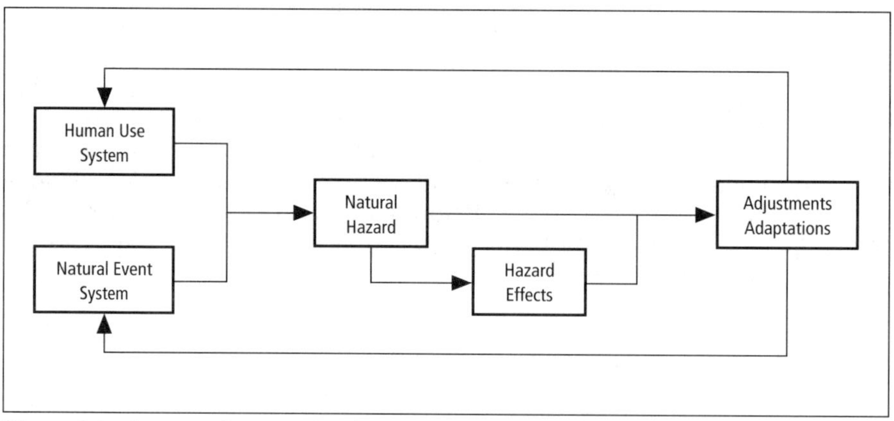

Figure 2.3. System of terms related to natural hazard research according to Kates. (After Kates 1970, p. 137)

[3] Even though the terms might in the Anglo-American language be clearly defined, the author has because of the international readership considered it important to keep these definitions in the English translation.

Kates' (1970) model illustrates that a natural hazard is, just like a resource, the result of an interaction between two systems, i.e. the systems natural environment and society. In the case of a hazard the interaction shows a rather extreme, eventful character. A hazard in most cases demands too much of a society's capacities so that dramatic consequences are often enough the result. The society tries to develop countermeasures.

Kates distinguishes between *adjustments* and *adaptations*, both being able to exert influence through feedback. Adjustments are undertaken from time to time, serve a certain purpose and also include short-term reactions and measures in a hazard situation. Adaptations, on the other hand, are long-term processes which are characterised less by detailed planning than by spontaneous action. They are in almost all cases culturally motivated. Their purpose is to facilitate coping with future hazards and preventing damage as early as possible (Burton et al. 1993; Geipel 1992; Geipel et al. 1988; Dobler 1980; Kates 1970). Kates summarises these correlations as follows: "... a natural hazard is an interaction of man and nature, governed by the coexistent state of adjustments in the human use system and the state of nature in the natural events system. In this context, it is those extreme events of nature that exceed the capabilities of the system to reflect, absorb or buffer that lead to the harmful effects, ofttimes dramatic, that characterize our image of natural hazards. However, it is also the continuous process of adjustment that enables men to survive and indeed benefit from the natural world. Therefore, the burden of hazard is twofold: a continuing effort to make the human use system less vulnerable to the vagaries of nature, and specific impact on man and his works arising from natural events that exceed the adjustments incorporated into the system" (Kates 1970, p. 438).

Human beings and their perceptions, decisions and activities play a far more important role than originally thought since natural hazards are the result of the interaction between the natural environment and its utilisation by society. The perception of a potential extreme natural event is the basis for its turning into an actual threat to man. This threat can last a long or a short time and become reality at any time. Each individual evaluates the threat differently and, therefore, reacts in his personally appropriate way. The threats we are talking about are, as we have seen, natural phenomena of different characteristics. Let us go back to the example of the two people crossing the ocean. The natural event could in this particular case be a storm, characterised by its violence and duration. If the dreaded situation, death by drowning, actually occurs, we are then talking about the realisation of a hazard (Smith 1994). This takes us to the next important term – disaster.

The term disaster is usually used in a very broad sense. It generally describes a sudden event with fatal consequences for man, as for example a war, a nuclear accident or an earthquake. The United Nations Disaster Reduction Organization (UNDRO) proposes a standard definition according to which a disaster is: "... an event concentrated in space in time, in which a community undergoes severe danger and incurs losses to its members and physical appurtenances that the social structure is disrupted and the fulfilment of all or some of the essential functions of the society is prevented" (UNDRO 1984, p. 4). A disaster usually causes a breakdown of the existing social structures and functions which can only be rebuilt with the extreme endeavours of the affected society. Traditionally, a difference is made between technological and natural disasters. The former are understood as man made disasters, whereas the latter are linked to natural events. Today this distinction is hardly ever referred to since most disasters have both human and natural causes. The same naturally goes for hazards, as shown by Kates' (1970) model. They, too, are neither of a purely technical nor a purely human nature: "Hazards (...) may be seen as existing along a continuum from technological to natural, where every hazard has both natural and technological components" (Mossler 1996, p. 88). There is therefore a growing tendency to refer to them as environmental hazards.

To the sociologists Clausen and Dombrowsky (1993) the standards of the affected society are the decisive aspect. An extreme, abrupt event can in one culture constitute a disaster, whereas another culture might not (yet) define it as such. Clausen and Dombrowsky use the term cultural disaster. It implies that a disaster is on the one hand the result of a risk-distributing dealing with resources, i.e. a culturally conditioned behaviour. On the other hand does a disaster as a radical change force the society to a new interpretation of its structures in the scope of collective processes. Based on these reflections, Clausen and Dombrowsky have developed a theoretical model of these processes. The basic idea is that disasters develop over a longer period of time and are characterised by the social interactions. A disaster is therefore a dynamic figuration, the case of a change which has become extreme (Clausen and Dombrowsky 1993; 1987).

The definitions of the terms hazard and disaster presented above are part of an ongoing discussion about their meaning in general and in a growing spectrum of systems and standards. A wide range of disciplines have by now turned their attention to hazards and disasters. This includes geography, sociology, medicine, psychology, ethnology and political science just as well as geology, meteorology, volcanology, hydrology and associated engineering sciences. Not all disciplines and their development can, of course, be paid attention to in the scope of this work. It will concentrate on the evolution of hazard management in geographical terms and at the same time continue reviewing the understanding of hazard and disaster.

Different Approaches to Hazard Research

Hazards have always played a significant role in the history of mankind. Archaeological studies show that ancient cultures protected their buildings against earthquakes as early as 2,000 years ago. A systematic scientific treatment of the subject did not commence until after World War II. Pioneer of this research was the geographer Gilbert White (1945) of the University of Chicago. He investigated floods in the United States and recognised the apparently rather paradox phenomenon that the damage was growing even though the government had increased its efforts (Flood Control Act 1936) and spendings (more than US$ 5 milliards) to control the floods. White came to the result that exactly the extensive structural measures (dams, dykes etc.)

prompted people to feel secure and settle in areas prone to flooding (White et al. 1958; White 1945). This investigation showed that purely technical measures are obviously unsuitable to solve such problems. Similar studies were applied to other hazard situations, such as volcanic eruptions, droughts, earthquakes and hurricanes. A collaborate programme was to this end started in 1967 by the University of Chicago, the Clark University (later Brown University) and the University of Toronto. White was supported by his former students Burton and Kates, already mentioned above. The so-called White–Burton–Kates School of Hazard Research was established, later on also involving a considerable number of other scientists.

These *traditional approaches* to hazard research have their origin in human ecology. For this reason their focal point of interest is the interaction between extreme natural events and human utilisation of resources. Physical geographers in this context aim at the possible prediction of extreme natural events in order to be able to minimise the socio-economic consequences. Anthropological geographic approaches, on the other hand, examine the perceptions of hazards in the population and the countermeasures developed. Kates (1962) made a valuable contribution to the latter issue. He used Simon's theory of rational decision-making (1956), and examined the decision process for the development of countermeasures. This process is, according to Kates, determined by the following factors: personality, role-specific behaviour, information available and the precise circumstances at the time. These factors, in effect, account for the different perceptions of a hazard and different patterns of behaviour when adapting to a hazard situation. Resource managers for example distinguish different behavioural patterns which have already been outlined earlier, but shall briefly be recalled at this point:

1. The hazard is placed in a cyclical pattern so that the past might reveal facts for the future;
2. The hazard is completely denied;
3. The hazard is evaluated as a once-in-a-lifetime stroke of fate;
4. The hazard is seen as a rare and extreme event which will not repeat itself for a long time (Burton and Kates 1964).

The *International Geographical Union's Commission on Man and Environment* has since 1965 completed more than 40 field studies in 17 countries in order to obtain comparative data about the adaptation to different types of hazards (Burton et al. 1993). The findings suggest that the individual perception of a hazard and, as a consequence, the behaviour at the time largely depend on the personal time horizons, cognitive constraints to confronting uncertainty and the evaluation of the probability and magnitude of a potential event. Thus, maladaptations to different hazards are, in the scope of traditional approaches, to a large extent the result of a wrong perception, incomplete information, inflexibility in the decision-making process or personal interests. Burton, Kates and White cite a vivid example for a case that is typical for the predominance of personal interest. They interviewed an estate agent who planned to construct a house near the Missouri on a site which had been flooded just a few years ago. Having been asked, the estate agent's first reaction was to deny the risk of flooding. On persistent questioning, he revealed his personal interests and his personal time horizon, saying: "There isn't any risk. I expect to sell this building before the next flood season" (Burton et al. 1993, p. 109).

Traditional hazard research faced growing criticism in the 1970s. Especially its basic assumption that a disaster is a direct result and therefore characteristic of a hazard was challenged. The conclusion of the White–Burton–Kates School that an increasing number of people lived in endangered areas because of a simple ignorance about the threat was also criticised. This assumption led, in the opinion of the critics, to an insufficient consideration of greater socio-cultural and political-economic correlations in hazard and disaster research. The growing criticism to the traditional approaches finally cumulated in Hewitt's book 'Interpretations of Calamity from the Viewpoint of Human Ecology' (Hewitt 1983). This still significant book examines the social, political and economic structures and processes which exert an important influence on the effects hazards have on individuals, social groups and whole societies. The book has been considered the break-through to the political-economic approach of hazard research, extending the traditional approaches by a number of decisive factors and relating closely to the political-ecological approach of resource research.

The political-economic approaches contain a vast range of paradigms, concentrating on the economic and political balance of power and the resulting social consequences of hazards and disasters (O'Keefe et al. 1976; Hewitt 1983; Susman et al. 1983; Cuny 1983). Hazards and disasters are, in this sense, no longer understood as simple cause-and-effect models where natural events exert influence on society, but as non-linear systems in which social processes play an important role. The thoughts go even further because hazards and disasters are now considered the result of inequalities in the prevailing social structure. They are characteristic of the existing social, economic and political processes: "In other words, it is not simply a cause and effect (linear) relationship between an extreme event and a vulnerable population, but rather that hazards may be created as particular economic development trends erode or distort a population's framework for decision-making regarding disastrous events. It is not only that disasters occur and affect particular populations but, more importantly, that hazardous conditions exist in the day-to-day context of a population and that extreme events accelerate the crisis" (Marston 1983, pp. 342–343).

This understanding prepared the ground for the introduction of economic development and political power relations in hazard research. Development and marginalisation theories were applied and, as a result, hazards and disasters were embedded in the context of comprehensive processes of social transformation and underdevelopment. The problematic of developing countries received more attention in hazard research since especially in the poorest and smallest states of our world the tendency of death, injury or damage following a disaster has grown for decades (Mossler 1996; Blaikie et al. 1994; Smith 1994; Varley 1994; Marston 1983; Haas et al. 1977). One of the main reasons is that marginal social groups have to live in hazardous regions. They often do not live there because they are ignorant of the danger but because of a lack of alternatives. Their access to resources is just as limited as is their possibility to exert influence. The threats posed by nature are often peripheral in comparison to their daily fight for survival: "Vulnerability to hazard may be chosen as a lesser evil, because the hazard is infrequent, compared to the more pressing day to day problems of homelessness, lack of income or inaccessibility. Poor people are vulnerable to hazards as a result of processes that have deprived them of any power to affect their own physical, social or economic environment. Their only freedom of choice is to choose between different disasters" (Maskrey 1989, p. 25).

Hazards and disasters do not concern all people in the same way. The poor, the elder and marginalised groups of a society are much more vulnerable than its richer and more influential members. Vulnerability therefore is the key concept of political-economic approaches. It seeks to combine the risk which people and societies are exposed to and their social and economic abilities to cope with the damage incurred. Vulnerability is no static variable but reflects a dynamic process in which technological innovations, economic development, increase in population, scarcity of land and other pressure on resources play an important part. Blaikie et al. therefore emphasise the dimension of time which is implicit to their definition of vulnerability. This refers mainly to the capability of mankind to rebuild their life structures following a disaster. This confronts marginal groups of any society with special difficulty and their vulnerability to subsequent disasters is also likely to increase.

This immanent worsening effect has been illustrated by many authors, amongst them Susman et al. (1983), Maskrey (1989) and Blaikie (1994). Hazards and disasters are time and time again experienced by vulnerable parts of the population, but the instant assistance offered and the subsequent reconstruction hardly ever manage to reduce the vulnerability of the affected people: "Marginal people remain marginal. In fact, disasters produce more marginal people – people who have survived but are unable to recover their livelihoods, who are destitute and forced to live in even more vulnerable situations" (Wisner 1993, p. 130). One of the central objectives of political-economic approaches in hazard research is therefore the reduction of this vulnerability through structural transformations in society. However, such a one-dimensional concentration on purely social aspects is also subject to criticism: "Drawbacks on this paradigm lie in its deemphasis of natural mechanisms in the hazard equation – the view that yes, we can make social adjustments, but we also have imperfect knowledge of geophysical mechanisms of hazards" (Mossler 1996, p. 88). Sayer (1983) puts it even more clearly: "... we may be interested in the causes of flooding and this will often include social as well as natural events. But although floods may be the effects of social actions this does not make the floods social. (...) Understanding the social character of the actions which caused the floods (...) would not be essential for understanding the latter" (Sayer 1983, pp. 55–56).

Integrative approaches have therefore gained importance for hazard research, at the latest since the 1980s. System-oriented analyses have in this context received more attention than before. The integrative approaches offered are rather different, but do all intend to link traditional and political-economic approaches and other relevant paradigms (Blaikie et al. 1994; Smith 1994; Geipel 1992; Vermeiren 1991; Lewis 1990; Palm 1990;

Mitchell et al. 1989). Maybe the most important aspect is that all of them understand hazards as open and complex systems for which a multitude of elements is just as characteristic as cyclic and instable transformation processes: "The central concern of each of these paradigms is on the total system involved in the environmental hazard or disaster, the numerous variables that compound the hazards, creating a state of dynamic equilibrium, the physical and social orders in the system and/or the institutional frameworks which produce preparedness, response and adaptation to hazards" (Mossler 1996, p. 88).

Mitchell et al. (1989) have therefore proposed a new model to depict the correlations between the system natural event and the system society (Figure 2.4). To them a hazard consists of two sub-systems, one of these containing the components of the hazard itself and the other being the context in which these components become effective. The first sub-system comprises four major components: natural processes, the population, adaptive reactions and net losses which can be expressed by factors such as risk, vulnerability and costs. The model suggests that all components interact via seven feedback loops. Natural processes affect society (1) and vice versa (2). The resultant adaptations of the society on the one hand again affect the natural environment (5) and can, on the other hand, furthermore change the vulnerability of the population (4). The latter usually incurs costs (6) which are evaluated by the society (7) and lastly result in new adaptations (3). This correlation which is internal to the first sub-system is exogenously affected by the second subsystem, the time context with its respective factors. Amongst these factors are the ongoing globalisation, environmental problems, population growth or new knowledge and technologies, all of them subject to a dynamic and hardly predictable process of change. This is the reason why Mitchell et al. call for an intensive research into the exogenous factors and their interaction with the hazard components (also Geipel 1992).

The US-American hazard researcher Palm (1990) also explicitly refers to the necessity of integrative approaches in order to do justice to the complexity of hazards. She criticises that previous approaches have concentrated too

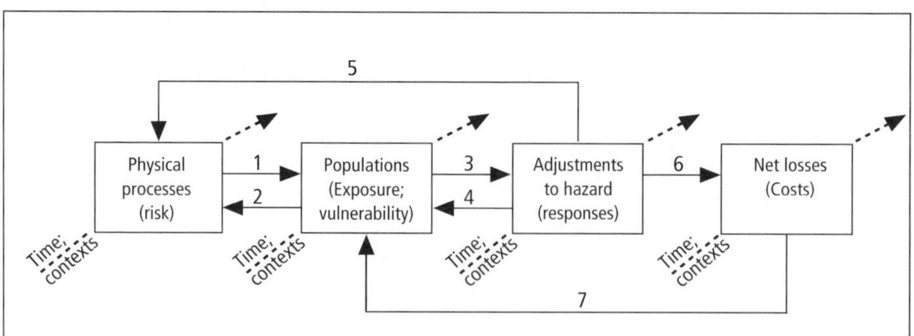

Figure 2.4. Natural hazard system according to Mitchell. (After Mitchell et al. 1989, p. 404)

much on the individual decision-making behaviour relative to uncertainty. In her much noticed book 'Natural Hazards – An Integrative Framework for Research and Planning' she presents her structuralistic approach with a model in which the individual is not seen as something isolated but dependent on the prevailing social structures (Figure 2.5). The model in the first instance addresses the individual, his experiences, aims, knowledge and resources. Usually, this individual plays a certain role in a family or a household and the allocation of the resources is subject to joint decisions. Each household is in turn embedded in society, the determinant factors at this level being of political, economic and cultural nature. Society is then again embedded in its environment, containing resources and hazards. So-called gatekeepers regulate the exchange between the micro level (individual, household) and the macro level (society). Gatekeepers can be planners, managers, emergency staff, financiers, businessmen, estate agents or land owners. They are, in summary, individuals who can dispose of important resources and have decision-making powers. They put social norms into practice and thereby affect the activities of any individual.

Both Mitchell's et al. (1989) contextual approach and Palm's (1990) structuralistic approach make clear that integrative approaches intend to do more justice to the complexity of hazards. They examine as many factors as

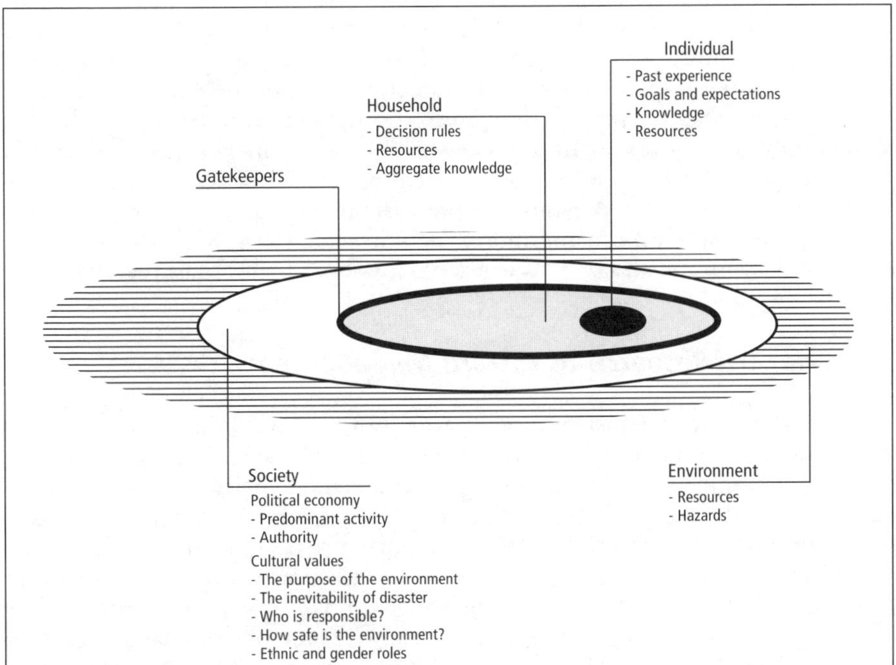

Figure 2.5. The structuralistic model according to Palm. (After Palm 1990, p. 81)

possible and their respective interactions. To Palm the theory of complexity has the following advantages: "What does this science of complexity have to contribute to the understanding of environmental hazards? At the very least, it should caution us about assuming that we can develop quasi-deterministic models of relationship of individuals and societies and their environment. If very slight changes in energy can result in very different evolutionary tracks of simple physical systems, then how much more likely is that the inevitable differences in circumstance that characterize societal response to environment will result in different evolutionary paths in human systems" (Palm 1990, p. 61). The theory of complexity offers hazard research the opportunity to examine hazards from different perspectives, to analyse development processes and to consider the interactions between individuals and society. It has often been seen, for example, that self-initiated processes of interhuman communication and action emerge naturally in situations of disaster – processes of self-organisation. These processes have, regrettably, to date not been analysed systematically so that we are not yet in a position to encourage them in a supportive way.

This deficiency brought the hazard researcher Comfort (1994) to apply the theory of complexity to the analysis of different self-organisatory processes in situations of disaster. She derived the following implications for hazard research: "These indigenous and self-initiating networks in disaster response and recovery operations, as in other domains, exhibit high levels of intragroup communications, highly similar sets of values and goals, and often unarticulated but effective systems for feedback and evaluation of performance. Such self-evident natural networks frequently based on geography or self-expressed membership in group, are found to be effective mechanisms for disseminating information, catalysts for untried innovations, and important vehicles for socialization and group learning processes among members" (Comfort 1994). The necessity of such research in concrete disaster situations, analysing the structure, function and dynamic of the systems through an integrative approach, is recognised more and more. (Blaikie et al. 1994; Smith 1994; Berke et al. 1993; Rubin et al. 1993; Wisner 1993; Geipel 1992; Maskrey 1989).

From Hazard Research to Hazard Management

The theoretical results just outlined naturally find their practical application in hazard management. We are dealing with a multilayered decision-making process. It comprises relief actions in acute emergencies directly after an earthquake, a hurricane or a flood. It also refers to strategies for a long-term reconstruction of communities and to anticipative measures such as the avoidance or reduction of potential hazards and disasters by controlling and limiting vulnerability. All these challenges are linked to the temporal phases of a hazard (Figure 2.6). First, there is the pre-event phase, characterised by the dimension of the threat. The second phase is the impact and trans-event phase during which the natural disaster comes into force and continues to occur. This acute emergency is finally followed by the post-event phase. Following this pattern, hazard management itself is also divided in characteristic

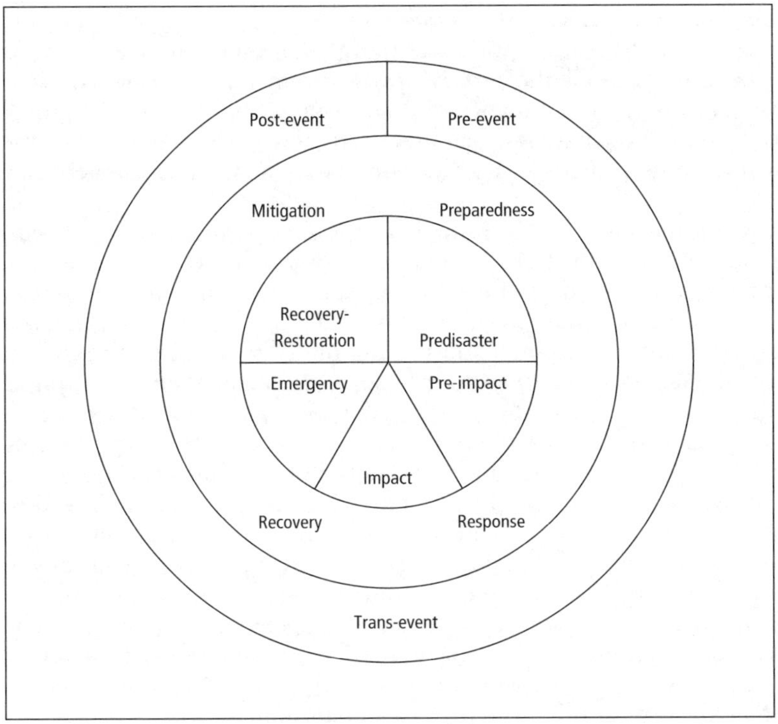

Figure 2.6. The phases of hazard management. (After Comfort 1988, p. 165)

phases – mitigation, preparedness, emergency and response and, finally, restoration and reconstruction (Smith 1994; Comfort 1994).

The *mitigation phase* or *disaster prevention* comprises a wide range of activities which all serve the purpose to avoid or reduce the negative consequences of potential hazards: "Mitigation can be defined as a management strategy to balance current action and expenditures with potential losses from future disaster occurrences" (Comfort 1988, p. 67). The strategy was for a long time limited to structural measures, such as the construction of dams, dikes, earthquake-proof buildings. Another important preventative factor are the development and establishment of disaster prediction and warning systems. These have been used for a relatively long time and are of growing importance. The theoretical discussions of the 1980s furthermore led to the realisation that socio-economic and political measures are equally important in order to reduce vulnerability. Such measures can be educating or informative campaigns which heighten the awareness of both the population and the decision-makers. Analyses of the prevailing social, political and economic systems can also show possible weaknesses and lead to remedies in the scope of long-term planning strategies.

The *preparedness phase* is the time when people have been put on the alert, expecting the actual occurrence of the disaster. Important protective measures are undertaken without delay. This might be the provision of shelter, warning and even evacuation of the local population, organisation of emergency teams etc.: "... preparedness entails those activities closest to the onset of a disaster which minimize damage and enhance disaster response operations" (Comfort 1988, p. 113). The model shows that these activities have their roots in the mitigation phase and also flow into the boundaries of the acute emergency phase.

The e*mergency* and *response phase*, in effect the time of direct crisis, reflects the relief directly before and after the occurrence of the disaster. The relief can comprise spontaneous measures of the special teams, the search for and rescue of the injured and buried, the recovery of the dead, food supply and medical care, the provision of shelter in schools, tents or halls, the first clearing-up operation on the access roads or other measures deemed suitable The phase following this immediate response phase is the so-called *restoration phase.* It attempts to provisionally restore the basic functions of society, e.g. the supply of water, electricity, gas and telephone, the repair of buildings and houses or at least the provision of adequate shelter. This phase is again closely linked to the final *phase of reconstruction.* The activities at this point all aim at the resumption of the predisaster state and the return to 'normal' social relations in the long-term. The economy is revived, infrastructure and housing rebuilt, on the whole, the vital functions and activities of a society re-established. The phase of reconstruction after a disaster is hardly distinguishable from other development efforts. At the end of the day the objective remains the same: to give the people the opportunity to return to an autonomous life.

As we have seen, hazard management comprises preventative phases (mitigation and preparedness) and phases of reactivating (emergency, response and reconstruction). These phases do not necessarily have to follow one another. To the contrary, they generally overlap and affect each other. However, over years they have been examined as isolated matters. It was not until a few years ago that researchers began to realise that hazard management is a never-ending process during which all phases have to be considered equally and at the same time. These thoughts lead to the necessity of a purposeful and sustainable planning strategy which integrates preventive measures just as well as long-term reconstruction strategies.

Putting this theory into practice is more than difficult and has hardly been realised so far (Anderson and Woodrow 1989; Geipel et al. 1988; Rubin 1987). The reasons are on the one hand the complexity of a hazard system and the limited cognitive capacities of humans to deal with this complexity. The main reason is linked to the perception of hazards. Despite all the research results of the past few years, the majority of people still sees a hazard as an isolated event which occurs very rarely and does not have much in common with any 'normal' development. The threatening dimensions and the possibly sudden realisation of a hazard are in most cases forgotten, just as the necessary preventative measures for the reduction of their vulnerability are. In the disaster situation itself people tend to react *ad hoc,* trying to satisfy short-term needs whilst disregarding all general objectives of a long-term planning strategy. Furthermore, in a disaster situation other potential hazards are normally neglected. This, in effect, means that hazards themselves are viewed as separate entities: "Many programmes respond to a particular hazard type in a specific and limited time period. As such, they cannot address vulnerability, which is a complex relationship between people and their social, physical and economic environment" (Varley 1994, p. 109).

The tendency towards more and more disasters of increasingly serious consequence forces us to consider alternatives and to react (Figure 2.7). This is especially true for the developing countries which are extremely vulnerable, due to their steadily growing population, their economic and political dependence on other states and extreme social differences. The damage caused by a hazard in developing countries is up to 30 times higher than in industrial nations (Fenner 1996; Blaikie et al. 1994; Varley 1994). It should not be forgotten that pressure is not only exerted on the local (or national) level but also on a global scale, with the climatic disaster surely representing the greatest challenge to mankind today. A sustainable reduction of such a global hazard can, if at all, only be realised through an intensified fight against poverty in the developing countries and at the same time an ecological reform in the industrial nations. This would, of course, be a century's programme (Deutsches IDNDR-Komitee 1996; Burton et al. 1993).

In 1989 the so-called Tokio Declaration (Resolution 44/236) was passed by the General Meeting of the United Nations. It brought the *International Decade for Natural Disaster Reduction* (IDNDR) into being which attempts to deal with the problems above mentioned. Its priority is to diminish the losses of human lives and property and the social and economic damages following a disaster. Unfortunately, the IDNDR has for a long time concentrated on

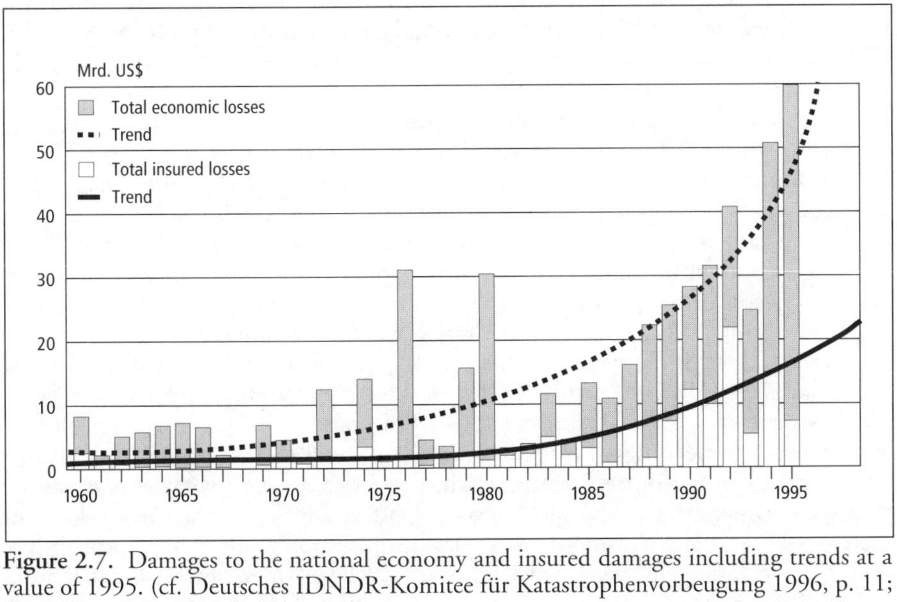

Figure 2.7. Damages to the national economy and insured damages including trends at a value of 1995. (cf. Deutsches IDNDR-Komitee für Katastrophenvorbeugung 1996, p. 11; after Münchener Rückversicherung)

technical preventive measures. The assumptions made also based on rather optimistic evaluations of the capacities of natural and engineering sciences in view of the reduction of hazards. The committee thus repeated known mistakes. Many hazard researchers criticise this ongoing neglection of complex interactions between humans and their environment. Blaikie et al. (1994) for instance fault the dominance of old paradigms in the scope of the IDNDR. The authors have written a book titled *At risk – Natural hazards, people's vulnerability and disasters* in which they appeal to the IDNDR to focus on the reduction of vulnerability: "... the book (...) arrives in the context of this decade in the hope that it will establish the vital importance of understanding vulnerability in the context of its political, social and economic origins. It tries to move beyond technocratic management to a notion of disaster mitigation that is rooted in the potential that humans have to unite, to persevere, to understand what afflicts them, and to take common action" (Blaikie et al. 1994, p. XIV).

The criticism uttered and the results of the *United Nations Conference for Environment and Development* (UNCED) in Rio de Janeiro (1992) and the *United Nations Conference on Natural Disaster Reduction* (UNCDR) in Yokohama (1994) have towards the end of the decade finally led to an increased resonance to integrative approaches. One of the prerequisites for this development was a changed perception of hazards and disasters, seeing them no longer as isolated events of an extreme nature but as integral part of the development endeavours of the countries. After all, hazards and disasters are a consequence of vulnerability, adaptability and flexibility of humans, in short, a consequence of their daily decisions and activities (Blaikie and Brookfield 1987). These correlations have been verified quite some time ago. They are depicted in the illustration (Figure 2.8) developed by the *Pan American Health Organization* (1994) and now finally taken into consideration by the IDNDR. The committee itself recently called for an integration of hazards and disasters in sustainable development plans.

The reduction of vulnerability has thus become an important matter of concern to the IDNDR. To this end just access to resources and their sustainable utilisation are necessary. Social structures have to be enforced so that management strategies can assist to better absorb the shock caused by a disaster and to initiate a quick and constructive reconstruction process. All this requires sustainable management approaches with following objectives: 1. mobilisation of the people affected by a natural disaster, 2. self-reflection of a society about its basic values and objectives, 3. self-identification of problems and 4. development and discussion of alternative problem-solving strategies (Blaikie et al. 1994; Smith 1994; Berke et al. 1993; Cuny 1993; Geipel 1992; Comfort 1988; Maskrey 1987). This has to be the way for hazard management to become an integral part of sustainable development planning: "... countries must journey along a winding path where progress is made as countries recognize that hazard management is more than a simple logistic exercise. It is a development and planning responsibility, a responsibility calling for multidisciplinary collaboration ..." (Pan American Health Organization 1994, p. 4).

Figure 2.8. The correlation between natural disasters and development. (After Pan American Health Organization 1994, p. 36)

2.4 Challenges to Geography

We have addressed a wide range of aspects in the above chapters – ranging from complexity and uncertainty to resources and hazards. We shall now briefly reflect on the results and then consider the implications these theories have for the modern geographer. Complexity, as we remember, means emerging, developing, evolving of new structures and patterns – "from being to becoming", as Prigogine (1980) appropriately put it. The theory of complexity is therefore concerned with dynamic processes, sudden changes and disruptions which are not the exception but the rule in the continuous becoming of a non-linear system. Dynamics, transformations and disruptions are felt as a threat by the majority of people. Growing complexity therefore implies a growing feeling of uncertainty. Humans do not like to live with such a feeling, though. They try to either repress it or to even diminish it by striving to control complex systems. The reflection about resource management has shown that most of the strategies developed in the past decades aimed at preserving the balance of the respective systems on a long-term basis. These strategies completely neglected the dynamics of the interaction between humans and their environment, i.e. unexpected and sudden events. Hazard management strategies then concentrated strongly on sudden events, but neglected the consequences of long-term measures and the complexity of the interaction human-environment. The current trend in resource management therefore aims at developing adaptive strategies, enabling us to obtain the necessary flexibility and adaptability for changing conditions. Hazard management, on the other hand, currently places more emphasis on the integration of its strategies into sustainable development planning.

The logical conclusion for research seems obvious: we have to find ways to generally accept and creatively cope with complexity and uncertainty. We have to learn to adapt flexibly to complex systems without loosing sight of our long-term objectives and interests. We need certain knowledge in order to orientate ourselves, where we are, where we come from and where we are heading to. And this knowledge has to be acquired again and again in a continuous learning process, thereby maintaining our capability to adapt whenever necessary. Robert Bennett and Robert Estall of the Geographic Institute at the London School of Economics have in this context spoken of a "management of change". Similarly to Funtowicz and Ravetz, this implies that such an orientating knowledge can only develop in a cooperative and iterative practice in which we learn with and of each other. This is the only way subjective reality can be grasped in a world of change, since it is finally related to individual and collective perceptions, objectives and interests.

But what are at last the implications for geography? This appears a useless question, if one considers the following: understanding and managing resources and hazards, analysing the interaction human-environment, the combination of individual and collective perceptions, interests and activities and the orientation in a complex world – all these are focal matters of geography. Geography should, in the light of the developments outlined above, be facing a very exciting future with plenty of potentials and challenges. Nevertheless, the debate about what geography is and what it might contribute to a world of change carries on. This discussion was mainly led by Anglo–Saxon scientists, above all by Johnston (1993), Harvey (1990; 1989), Bennett (1989), Bird (1989) and Stoddart (1987). Despite specific differences they all pinpointed the central challenges geography faces. These do not solely consist in describing this world of change but, above all, in understanding the world as it really is in order to then be able to actively shape it as best as possible. This, of course, requires that geography changes from a purely descriptive or even prescriptive discipline to an emancipating and liberating one which helps mankind to help itself (Johnston 1993).

This idea of an emancipating geography is also gaining acceptance in Germany as the following quotation shows: "A renewal of geography starts off where it distances itself from its monopoly of world descriptions and interpretations and rather emphasises the possibility to subjectively experience the world. In such a geography people tell each other their experiences with the world and this exchange is in itself an experience. Total chaos will, of course, not determine our future. Perceptions happen in an orderly manner, just that the respective principles of order now at the same time parallel the respective arrangement of personal interests in life. Such a changed geography distances itself from its obsession with lecture and supports individuals in a process, the naturalness of which we are no longer aware of: We perceive the world and we grasp it in an endless autobiographic activiy[4]" (Daum 1993, p. 46). Ratter (1992) offered us an interesting attempt in this direction in her book "Karibische Netze" (Caribbean nets). She illustrates how subjective experience can be integrated into geographical analysis. To this end Ratter guides us across the horizons of different lebenswelten in an interesting interplay of self-reflecting and narrative elements, originating in the author herself or the inhabitants of the examined regions. In her analysis Ratter links two classic elements, i.e. theory and empirics, with narration and illustrates an emancipating way to approach a complex system.

Rhode–Jüchtern (1996) takes these thoughts a step further. In his book titled "Den Raum lesen lernen" (*Learning to read a place)* he postulates the change of perspectives as an important concept for the geography of the 21st century. 'Learning to read a place' is to Rhode–Jüchtern an old and, in times of pensiveness, restlessness and change of paradigms, again a new task of geography. To find orientation in a complex world requires the readiness and willingness to take advantage of a number of perspectives, to cross borders, to constantly reflect and see things in relative terms and to transmit the own and the strange. It is about letting humans tell their story, grasping their picture of reality, understanding their motives, at the same time being conscious that their perceptions are subjective and might have been overtaken by development the next day. Rhode–Jüchtern shares Johnston's (1993) view: *"Geography as an emancipatory discipline has to base on an understanding of the difference which makes the mosaic of the world"* (Rhode–Jüchtern 1996, p. 23).

Understanding the difference between the natural and the social, between the individual and the collective, between the expert and the layman and lastly between the researcher and the research object is an element which has kept on appearing during our reflections so far. This understanding can be realised better by an integrative and interdisciplinary science like geography than by a highly specialist one which limits itself to certain aspects. This is the reason why the current tendency towards fragmentation poses a threat to geography, even attacks its growing competence and potential. Especially the research into complex systems is already announcing a new type of synthesis between natural sciences and humanities (Mittelstaedt 1993). Many geographers therefore call for a renaissance of the entity between physical and hu-

[4] Quotations which are in italics are originally German and therefore freely translated by the author.

man geography (Abler 1993; Johnston 1993; Unwin 1993; Stoddart 1987). Unwin for example has made this very clear: "There is a long tradition of such geographical research, and even though many geographers continue to advocate separate human and physical dimensions of the discipline, it would seem that if geography did not already exist, then there would be a need to institutionalize a new discipline concerned with such issues" (Unwin 1993, p. 210).

This unity in geography is of great relevance in order to be able to, firstly, understand the interaction human-environment better and to, secondly, find the required solutions to problems. Integrative and interdisciplinary does not mean that everything must be recorded and analysed. This would in view of complexity be impossible. Geography is one science amongst many which will always lag behind complexity. Integrative and interdisciplinary is rather the preparedness and readiness to cross borders, to consider multi-perspectives and to mediate between humanities and natural sciences. Geography will have to make complexity comprehensible by examining central factors and interactions – human perceptions, behaviour and activities in a certain place.

A place is always the result of the interaction between humans and their environment. A place is created by humans, both by those who live there and by those who live somewhere else. Johnston (1993) illustrates this as follows: "People learn who and what they are in a place, being influenced both directly and indirectly by the others who live there. The nature of these influences partly reflects the physical and the created environments which provide the enveloping context within which lives are structured. As people learn and act, so they in turn affect the nature of the physical environment and restructure the created. Life is a continuing dialectic between peoples and places, which proceeds over a range of spatial scales form the global to the very local" (Johnston 1993, p. 23). The task of geography is to observe, understand and structure these complex interactions, but by doing so it reduces complexity. This is the paradox situation this dissertation is also facing and which can be summarised with the following question: *"How is orientation possible when the realities are plural, everything almost endlessly complex and often non-linear and when science, too, is producing no more than its own reality?"* (Rhode–Jüchtern 1995, p. 166).

To Rhode–Jüchtern (1996) the answer is clear: geography needs a three-dimensional perspective:

1. Seeing what is visible, i.e. the perspective of the present.
2. Seeing the genesis of the things that exist and why this has survived and nothing else, i.e. the genetic perspective and, lastly,
3. Seeing the predictable, possible, potential developments, i.e. the constructive perspective

These theoretical thoughts would in practice lead to a reflexive and constructive geography which would try to not only explain complex systems but to also actively manage them. Next to describing and explaining, geography's task must be finding and discovering. Discovery is in this context not to be understood in the traditional sense as the discovery of new places, but as the discovery of new future places. Place is in this sense a horizon of possibilities, decisions and activities which can be structured by a geography willing and capable to ask for alternatives. The constructive perspective therefore leads to research about the future, still an aspect which geography neglects. It is now, against the background of sustainable development, gaining importance. The future of our lebensraum is a great challenge to geography and, as Stiens (1989) has put it, it is a call to a discipline with tradition, to open up to a stronger consideration of the time category future (cf. Stiens 1989, p. 187).

However, the future is always uncertain. This is exactly the reason why the three-dimensional perspective has to be supplemented by a fourth: the inclusion of sudden and unexpected events. A change of perspective is therefore not limited to different perceptions, interests and objectives or different scopes of space and time when analysing a complex system. It also demands the examination of complex systems in all possible states of chaos and order. This is the only way to avoid a typical mistake of many geographical analyses – focusing on the apparently safe, orderly and expected, thereby cutting uncertainty out. Wilson and Clarke (1985) have expressed this phenomenon: "Many geographical analysis of dynamical systems have focused on static spatial patterns or structures, representing these as equilibrium or steady-state situations within the ongoing dynamic process. Change is then handled by forecasting the independent variables associated with the system and then calculating the new equilibrium or steady state" (Clarke and Wilson 1985, p. 429). Such a strategy has limited effect since it hinders us to include the perspective for potential hazards, for uncertain, not predictable and chaotic transitional phases, all of them being so characteristic of complex systems. Reasonably successful prognoses therefore have to include the non-linearity of social, economic and political systems. Mittelstaedt (1993) even calls for prognoses which are self-regulating. They will not lead to wrong decisions and misplanning which might be deeply regretted a few months later but cannot be corrected.

Avoiding misplanning and keeping options open for future generations is a *leitmotif* of sustainable development. This theory is not last a reaction to the growing sense of risk societies feel. It also suffers from a lack of perspectives and concentrates too much on what is supposed to be secure, in order and in balance. Sustainable development endeavours to support green technologies, eco-tourism, a balanced economic and social world with positive, fair and secure standards and values. It, nevertheless, still neglects the creative handling of uncertainty and the necessary inclusion of potential hazards. This, in effect, must be its future task (Bayrische Rückversicherung 1993; Evers and Nowotny 1988). It has been suggested to distinguish sustainable development from sustainability. Dovers and Handmer (1992) for example said: "Sustainability is the ability of human, natural or mixed systems to withstand or adapt to endogenous or exogenous change indefinitely. Sustainable development is therefore the pathway or deliberate change and improvement which maintains or enhances this attribute of the system, while answering the needs of the present generation" (Dovers and Handmer 1992, p. 275). Sustainability in this sense refers to the capability of complex systems, especially complex social systems, to cope with changing conditions, to permanently adapt and, nevertheless, satisfy present needs. Resilience thus becomes the normative objective of sustainability. Sustainable development in this context is the process which leads there and which can be very variable.

Presupposing that the normative objective of sustainability is resilience, then complexity, uncertainty and the structuring of the future have to be linked. This first of all requires the acceptance of complexity and uncertainty. Living with complexity and uncertainty also requires that flexibility is learned. Mankind has to be able to adapt to leaps and bounds in the future development. This can only be realised if we ourselves change. This is why more and more systems theorists, futurologists and geographers are searching for methods which increase the capability of resilience, i.e. the capability to adapt to natural systems. There is general consensus about the need for other forms of planning which no longer consider chaos, change and variation as threats which have to be avoided. To the contrary, these elements should be considered essential for the dynamic and creative self-organisatory processes of societies. Planning, therefore, has to be understood as a process during which humans not only learn to cope with complexity and uncertainty, but manage and utilise these factors according to their interests and objectives: "Planning (...) will instead be an adaptive, organic process that seeks to influence change and take advantage of unexpected opportunities, rather than prevent change. This process recognizes constraints, is reflective, reflexive, and creative (...) it will be a planning that is complex, opportunistic, and anticipative, yet better suited to a non-equilibrium, self-organizing world of increasing complexity, connectedness, change, regionalisation, and pursued sustainability" (Slocombe 1990, pp. 22–23). But what does such planning precisely look like? And how can it be put into practice? These questions will be the focal point of the following chapter.

One thing has, in summary, become clear: a reflexive, constructive and emancipating geography with a four-fold perspective has a high status in a world of change. Moving between perspectives and different spatial and temporal levels, between perceptions and interests and between order and chaos is absolutely vital for the understanding, describing and structuring of the interaction human-environment. Geography's task therefore is to make the world aware of complexity and uncertainty, of different live worlds and to analyse the potentials we have to structure and influence the space around us. This challenge is not only one for experts. It is, to the contrary, a social challenge, concerned with a systematic openness for options, the search for alternatives, permanent possibilities for self-transformation, capability to learn and tolerance to mistakes, in short a challenge concerned with the creative and flexible conduct of a society under the conditions of complexity and uncertainty.

It is less important to foresee the future than to be prepared for it (Perikles).

3 Strategic Planning as Methodical Consequence

Planning can generally be defined as a well-ordered process using information to develop a projective framework (plan) which conveys development objectives (Götze 1991). An important element of planning is the application of alternative plots so that the various possible and necessary decisions are recognised at an early stage. This ensures that the planner is able to integrate all sorts of conceivable future environmental conditions and to opt for suitable ways to realise his goals. This is the point where planning turns into management. Management is a well-ordered process for the organisation and directing of complex systems. It is central to any management strategy to select certain steps of action and to bring them into force, in order to be able to influence the evolution of the system as much as possible. During an ongoing management process it is often necessary to question the planning objectives set earlier and, in some cases, to define them again. At this point management returns to the planning stage. It therefore appears that the theoretical distinction between planning as an anticipating decision process (determining planning objectives) and management as organised implementation (realising of the planning objectives) can hardly be sustained in practice. Planning and management generally tend to blend into another. For exactly this reason the two processes have so far not been differentiated.

Chapter 2 has shown that planning and management traditionally strive for a state of security and stability. Systems with a high degree of complexity do not allow for this, though. Management strategies which strive for a balanced stabilisation can in the case of disruptive events even result in a diminishing regeneration capacity of a complex system. Even detailed plans which take complexity of a system into consideration, often do not serve the purpose since they can only be brought into line with the changing conditions at rather infrequent intervals. Unexpected events and disasters lead to short-term reactions, external to the original planning and management strategy. Furthermore, the long-term interests are often lost sight of (Ganser et al. 1991, Slocombe 1990). A legitimate question arises in this context: Can planning and management in the light of complexity and uncertainty of systems meet the necessary requirements with a view to the design and application of future-oriented concepts? In other words: What is the benefit and importance of planning and management strategies if the concepts are not capable of influencing and guiding the development of complex systems?

There are at present a good many systems theorists and futurologists addressing exactly these questions. The number of those who still believe in the possibility of actively being able to influence the future grows constantly, despite or because of complexity and uncertainty (Flechtheim 1995; Gell–Mann 1995; Merry 1995; Jungk 1990). New and positive ideas for the future are given a strong status in order to increase our adaptability to constantly changing conditions. These scientists call on the independence, capabilities and creativity of mankind to ensure a continuous development within their specific environment. Banathy (1993) in this context refers to the evolutionary competence as an indispensable factor of complexity and uncertainty. In order to develop this competence individuals, organisations and societies as a first step have to describe their perception of the present as a basis for the subsequent definition of their future objectives. Banathy calls the latter the "guided image of the future", the model around which central values revolve and which motivates any action taken. Merry puts it as follows: "Having a positive image of the future built on values, that transcend self-interest, is the bedrock of being able to deal with problems of the present" (Merry 1995, p. 188).

Funtowicz and Ravetz (1994) as well as Mittelstaedt (1993) also emphasise the importance of alternative visions of the future when trying to handle complexity and uncertainty. Alternative visions form the basis for the development and discussion of completely new values which, in the context of the sensitivity to initial conditions, can lead to tremendous consequences in such a social system. New values can set off processes of self-organisation and as such be the moving force behind changes in the formation of social systems. But maybe global crises and disasters are the impulse for such changes. After all, the systems theorists suspect that complex systems on the verge of chaos show the greatest capacity to process information, the best performance and the greatest choice between the various development options. Chaos, complexity and uncertainty in this sense do not represent a threat but extremely vital elements, enabling the initiation of dynamic and innovative processes in a non-linear world: "Nonlinearity and uncertainty open the gates to choice. Nonlinear dynamics are far more widespread and much more open to choice among uncertain futures. For many human goals and purposes varying degrees of nonlinearity are helpful" (Merry 1995, p. 144).

These considerations clearly show the necessity for another sort of planning and management which accepts chaos, complexity and uncertainty as an opportunity for the organisation of our future. Some of the latest challenges have already been mentioned in Chapter 2. In the following planning shall be focused on: The relevance of planning processes has, amongst others, already been mentioned. It guarantees a continuous exchange of information and flexible communication and determination mechanisms to find solutions and come to better decisions. At the same time the permanent reflection and modification of the values, objectives and actions is ensured. The necessary skills are resilience, creativity, a certain tolerance towards mistakes and a range of perspectives. But how can these skills be supported? Which planning strategies can assist in the constant process of acquiring and questioning the necessary knowledge about the past, the present and the future? When addressing these questions, the focus shall be on available strategies which, having been put into practice, offer a fund of practical experience.

3.1 What is Strategic Planning?

Strategic planning has been used with conditions of complexity and uncertainty since the 70s. It is a decision-orientated process during which the analysis of the present situation combined with the future opportunities and risks leads to the definition of the relevant objectives and strategies. According to Götze (1991), the characteristics of strategic planning are a long-term planning horizon, the dominance of complicated problems, imprecise information, the investigation of key variables and the development of rather simple plans. As the name implies, strategic planning focuses on the development of strategies. A strategy[1] is traditionally a complete instruction which defines a certain action for each point of decision. The strategy is thus characterised by certain objectives and the timing of the necessary steps of actions. Central aim of the strategy is the best possible utilisation of the existing resources to make the most of given opportunities and avert threats (Hoffmann 1993; Meyer–Schönherr 1992; Götze 1991).

At first sight, the development of such strategies seems to be important when dealing with complexity and uncertainty. Nevertheless, this traditional concept is faced with growing criticism. Lane and Maxfield for example comment: "What is a strategy? Once upon a time, everybody knew the answer to this question. A strategy specified a precommitment to a particular course of action. Moreover, choosing a strategy meant optimizing among a set of specified alternatives, on the basis of an evaluation of the value and probability of their possible consequences. (...) For a while, strategy as optimizing precommitment was a growth industry. And it still flourishes in some environments. But it is fair to say that this notion of strategy is falling into increasing disfavor ..." (Lane and Maxfield 1995, p. 1). The increasing criticism of the concept is quite justified, taking the basic assumptions into account. This concept is again based on the rational decisions of humans and the attempt to predict the consequences of any action taken with the aim to foresee and control complex systems.

The theoretical considerations have shown, though, that probabilities are a rather subjective assessment and that our knowledge about complex systems is never complete. Alternatives and their effects cannot be predicted, especially in social systems, since there are too many agents involved. The result is an incomprehensible and unpredictable number of possible developments of the whole system. This makes it impossible to choose the optimum or safest strat-

[1] The term *strategy* originates in the Greek *strategos*, composed of two elements. The element *stratos* refers to something that comprises all other things, and the element *igo* refers to acting (Götze 1991).

egy. Furthermore, changing conditions can lead to opportunities or risks which might not have been taken into account, and overlooking such transformations can result in adherence to what is apparently the best strategy but no longer meets the requirements. For this reason strategic planning nowadays emphasises a process-orientated character. This is in effect the only way to make sure that the agents constantly re-interpret and adapt their identity, their values, their visions, their relationship and not last their strategies. Defining the most probable development and the optimised course of action taken is thus subordinated whereas a constant and active reviewing of the environment becomes priority and allows a fast and effective reaction in case of unpredicted events (Lane and Maxfield 1995).

Lane and Maxfield (1995) therefore suggest another form of strategic planning which is characterised by two interlinked steps. The first step is of cognitive nature. The agents first of all determine in a discursive practice their own values, visions and objectives as well as their assumptions of the interests and objectives of those agents which have influence on their scope of action. Those agents can be competitors, politicians or consumers. This step is consequently a specification of the mental constructions, ideas or *weltanschauungen*, factors which naturally influence the identity and actions to a considerable degree. The second step is of a structural nature. During this second step the identification and support of productive and innovative relationships with other agents take place. Central to this step is the joint and innovative development of products, projects and strategies. Both steps have, of course, to be repeated constantly since interpretations, identities and interactions are subject to a continuous process of change: "The world in which you must act does not sit passively out there waiting to yield up its secrets. Instead, your world is under active construction and you are part of the construction crew – and there isn't any blueprint. Not only are the identities of the agents and artifacts that currently make up your world undergoing rapid change, but also are the interpretations that all the agents make about who and what these things are" (Lane and Maxfield 1995, p. 2).

Collingridge (1983) defines effective strategic planning similar to Lane and Maxfield (1995) as the organisation of investigation and adaptation processes. The aspect of adaptation, which Collingridge calls *flexing*, gains further emphasis. Flexing focuses on the wide range of options which are to be kept available. The decision makers do not opt for the apparently safest strategy but for that which offers the best opportunities. At the same time alternative strategies are kept on stand-by. A continuous monitoring of the environment allows a flexible handling of the alternatives available in case of need: "Flexing (...) suggests trying for the best, risking the worst and being ready to reverse a decision should the worst occur (...) flexing opens up the future, promotes flexibility, and eases the adaption of policy to information (...) flexing insists that insurance is taken out in terms of monitoring and enhancing flexibility, so that if the worst happens, the damage it does is limited ..." (Collingridge 1983, pp. 163–166). This strategy avoids getting into a routine or simply adhering to a given strategy. Also, innovative development opportunities will not be neglected as risks are deliberately taken. The success of flexing depends on the ability of the planner to observe the developments, to interpret them and to react flexibly, whereby intuition and creativity play a considerable part.

Both approaches are characterised by a continuous reflection and discussion of the existing values and objectives as well as the constant observation of the environment so that a fast and flexible adaptation of the existing strategies to changing conditions is always possible. At the same time the anticipation of future potentials, problems, situations and possible force to act are taken into consideration. Such situations can be met more effectively and successfully when at least considered beforehand. In this context the above mentioned positive visions of the future are also of great importance since they increase our adaptability and challenge our ability. They also challenge our creativity and ability to act. Summarising all these approaches, successful strategic planning should comprise the following elements:

1. The acceptance of complexity and uncertainty and the creative handling of these factors.
2. The constant reflection, discussion and modification of values, interests, objectives and visions.
3. The active observation of the environment.
4. Keeping alternative development strategies open.
5. Encouraging the evolutionary competence.

Scenario planning is such a strategic planning approach which aims at fulfilling all these requirements. It is a systematic approach which is employed to help overcome the barriers of traditional thinking, to illustrate even complicated developments and create flexible strategies.

3.2 What are Scenarios?

The word *scenario* is of Latin origin (*scaenarium*) and signifies as much as a plot outline or a sequence of events which form the thematic frame of a theatre play or a film. The word is thus used mainly in the theatre. The US-American futurologist Hermann Kahn was the first one to introduce the term scenario into strategic planning. At the beginning of the 1950s he designed strategic military games for *Rand Corporation* which he called scenarios. These scenarios were visionary descriptions of changing situations which were designed to teach the military strategists to emerge victorious. Following this Kahn and Anthony Wiener about ten years later began to develop alternative future visions of the world. In contrast to the rather visionary character of the military scenarios, the options shown here were based on the actual political and military conditions. In this sense, according to Kahn and Wiener a scenario is the illustration of a possible logical sequence of events which can be followed up from a certain starting point to a possible situation in the future (Kahn and Wiener 1968).

This definition proposed by Kahn and Wiener (1968) is today the most widely used and accepted one for a scenario. There is a tendency towards concentrating more than before on complicated, mainly qualitative interrelations which scientists nowadays try to comprehend as fully as possible. Systems scientists therefore define scenarios as follows: a scenario is a description and mental construction of possible developments and events in a complex system, which also takes the interactions between the system's elements into account. Thus the mathematical probability is replaced by a plausible and credible argumentation which integrates unexpected events and trend reversals in the assumptions (Hansel and Lambrecht 1993). Schoemaker, too, emphasises the role of the system elements and the relevant interactions: "Each scenario tells a story of how various elements might interact with each other under certain conditions" (Schoemaker 1995, p. 26). Schoemaker speaks of *stories* and scenarios are in fact stories which describe a hypothetical sequence of events and points of decision. The description is in most cases literary and should be as vivid and visual as possible. The target groups of the scenario, be it the citizens, the politicians or managers, have to be able to visualise how they would live in the future presented to them, how they would decide and act. This is the way a scenario can trigger off the imaginary journey into the future (Ministerium für Arbeit, Gesundheit und Soziales des Landes Nordrhein–Westfalen 1988).

The theoretical considerations have shown that there cannot be only one future, a theory which traditional projections like to transmit. The future is non-existent, it is imaginary and subjective and for that always multidimensional. The futurologists therefore talk about many different *futures*. Flechtheim, for example, has emphasised that it is without question that we cannot today predict the world of tomorrow. Therefore it is justified to use future in its plural form (cf. Flechtheim 1987, p. 200). For this reason, the design of just one scenario is relatively unreasonable. To the contrary, different scenarios have to be designed and be compared in order to allow a comparative analysis of the alternatives. Scenarios are as such sketches of alternative, conceivable futures of complex systems. They illustrate how these futures could develop as a consequence of hypothetical events and interactions (Stiens 1996; Georgantzas and Acar 1995; Schoemaker 1995; von Reibnitz 1992).

The main difference between the traditional strategies of prognosis and scenario planning is their different understanding of a system (see Table 3.1). The first method tries to make a prognosis, if possible even the one and only correct prognosis, for the future. To this end it attempts to quantitatively register the actual status of the system. Scenario planning, on the other hand, tries to identify and analyse the basic interactions that are immanent to the system, its key variables and possible chains of reactions. Scenarios are thus first of all process-orientated: "*Scenarios are always only an image of what could be. They are neither desires nor reality, but only possibility, without any judgement about the actual probability. Their importance is mainly to illustrate to those agents who have to take decisions which factors influence the further development and to which extent this may happen. It is far more important to investigate the basic interactions than to ask at the end of the day how this or that figure came about*" (von Ilsemann 1980, p. 118).

Complexity is not transferred to computer aided simulation models. On the contrary, it is presented as a picture, sketches of conceivable developments, and as such comprehensible and a basis for discussions. Basic

Table 3.1. Comparison of the different understandings of systems. (After Hoffmann 1993, p. 552)

Criterion	Traditional prognosis	Scenario planning
Basic approach	The future is predictable	The future is not predictable
	Patterns and interdependences of the past do not change	Patterns and interdependences of the past can change
	Search for the best answer	Illustration of various possible visions of the future
	Creating a pseudo-security	Creating insecurity
	Eliminating doubts	Evoking doubts
Function	Part of the decision process	Aid for the preparation of decisions
Methodical principles	Quantitative methods	Quantitative and qualitative methods
	Analysis of "hard facts"	Analysis of "hard and soft facts"
	Focusing on the result	The result is of subordinated importance
	Assumptions and interdependences are usually not considered	Emphasis lies on assumptions and interdependences
Test criteria	Accuracy / exactness	Comprehensiveness
	Objectiveness	Relevance of the contents
	Theoretically proven	Thoroughness, overall picture

assumptions about important agents and interactions are taken whereby it is important to always show different alternative developments. Uncertainty is neither denied nor reduced, but consciously produced. It is the doubt which allows the escape of the own *weltanschauung*, opens the horizon for new and innovative perspectives with which the existing structures can be questioned. Arras has summarised the importance of scenarios as something which offers us the opportunity to think forward in a structured way. Scenarios ensure that we do not attempt to grasp the future as a certainty but to remain uncertain or even uneasy about the variety of futures on offer. Scenarios can prompt us to discuss the futures and to then draw our conclusions about what we have to do at the given moment in time (Arras 1986, p. 42).

The comparison illustrates that scenarios can broaden the possibilities of the quantitative prognosis in a wide range of contexts.

1. Scenarios are often designed for a longer period of time.
2. Scenarios take a large number of variables, interdependences and possible developments into account.
3. Scenarios are able to work with quantitative and qualitative factors, i.e. they are capable of considering imprecise factors such as values, interests and objectives.
4. Scenarios allow a change of perspective since basic structures are always analysed and questioned.
5. Scenarios consciously integrate unexpected events and discontinuities into their illustrations.
6. Scenarios are comprehensible and can be discussed on a broader basis since the various visions of the future are presented in a literary and pictorial

form. (Clemons 1995; Schoemaker 1995; Hansel and Lembrecht 1993; von Reibnitz 1992).

Von Reibnitz (1992) illustrates the handling of complexity and uncertainty with the so-called scenario-funnel (see Figure 3.1) The funnel is the symbol for the complexity and uncertainty of a system in the future. The present is located at the narrow end of the funnel. We are at this point still able to determine some more or less reliable factors and structures. However, if we try to project the development of these factors into the future, we quickly reach the point where complexity and uncertainty have reached such an extent that we are unable to reliably predict the development of the system. The influence present structures have constantly diminishes on the time axis, and the range of possible *futures* opens up like a funnel. New agents and factors can for example take up important parts or alternative measures and projects can be introduced. Also, sudden and unexpected events, i.e. disruptive events can appear which might turn the evolution of the system into a completely new direction. A projection of the present structures of a system into the future is for these reasons neither possible nor useful.

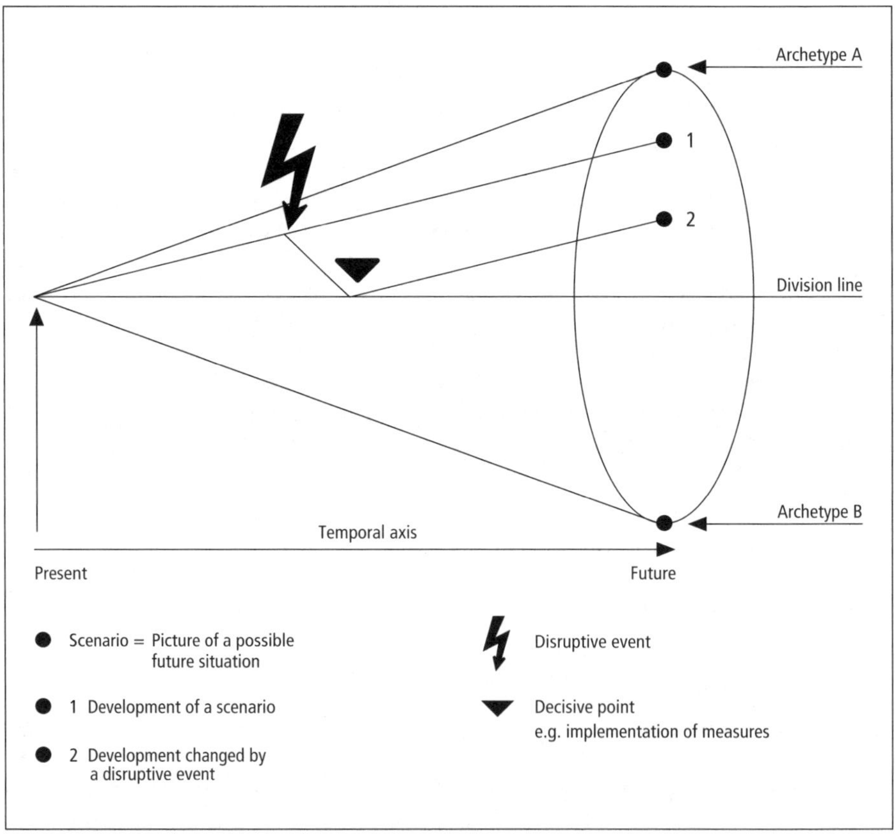

Figure 3.1. The scenario-funnel. (After von Reibnitz 1992, p. 27)

All conceivable future visions or scenarios can be found on the sectional plane of the funnel, if one now draws an intersection through the funnel at a random point of the future. But not all futures can be taken into consideration and this tempts planners to opt for one of the following two, very typical patterns of behaviour: either the planner resigns because of the uncertainty and complexity of the system and consequently sticks to the traditional quantitative strategy of prognosis with all its shortcomings. Or else, the planner consciously faces uncertainty and works out alternative images of the future. Von Reibnitz (1992) points out that it is in most cases sufficient to create two rather different but in themselves consistent images of the future. These can be seen as scenario archetypes. They are the results of different developments which should be as extreme as possible so that they are to be found on the edges of the funnel.

3.3 Procedures of Scenario Planning

Quite a number of different approaches to scenario planning have come up since the 70s[2], most of them closely related to the systems sciences and futurology. Until now it has not been possible to settle on one common and standardised procedure for scenario planning. However, there is agreement on the most important elements. Götze (1991) summarises that all approaches to scenario planning have the following in common:

- a careful analysis of the present situation and perhaps the past in order to gain an insight in the structures of the – generally complex – area of research;
- making plausible assumptions for factors with an uncertain future development;
- proposing a number of possible, consistent visions of the future and related paths of development on the basis of the results obtained;
- processing of qualitative and quantitative data;
- integrating the elements of analysis, prognosis and the system itself;
- proceeding systematically and comprehensively (Götze 1991, p. 76).

Furthermore, all approaches are characterised by the systematic order of each working step. Nevertheless, the exact order of the steps, their contents and objectives as well as the instruments and strategies applied do vary in the different approaches. One criterion of differentiation for example is the percentage of quantitative and qualitative methods used. This gives Meyer-Schönherr (1991), amongst others, cause to distinguish between the *hard* and the *soft approaches.* He classifies those approaches as hard which are purely quantitative world models and do not constitute scenarios in the narrow sense. Soft approaches, on the other hand, are those which are based on

[2] Relatively comprehensive information about the various scenario techniques can, amongst others, be found in Götze (1991) in the field of management planning, and Bieberstein (1979) in the field of scientific consultancy services to politicians.

common sense and use quantitative models – if at all – only partially as an aid for the processing of information (cf. Meyer–Schönherr 1991, p. 24). The soft approaches are further differentiated by Meyer–Schönherr between purely intuitive, not formalised procedures and those which are systematic, formalised procedures. The work of Kahn and Wiener (1968) for example belong to the former type. They aimed at approaching the matter in its entirety and therefore refused any systematic procedures. The handling is consequently very simple and determined by two steps. In a first step the main tendencies of a problem are identified and projected into the future. Thus a scenario is obtained which does not yet account for any unexpected events. Some projections are then varied in the second step in order to be able to develop alternative visions of the future. Attempts towards a more formalised procedure since Kahn and Wiener have been enforced, though. One of the reasons being that the purely intuitive approach has not been accepted as a method (Meyer–Schönherr 1991).

The formalised procedures are characterised by a systematic process during which qualitative and quantitative factors are equally taken into consideration. These types meet the requirements of a methodical approach better. Götze (1991) at this point distinguishes between the deductive and the inductive procedure. The former relates to determining the framework of each scenario. This outline is in most cases defined by the combination of a few factors. It also has to be determined how many scenarios are to be developed. Assumptions are then made with regards to the possible developments of those factors. A deductive approach is generally only useful when dealing with less complex systems. The inductive approach is based on the possible development of the various factors which are then combined in so-called raw scenarios. It is common practice to include a large number of factors in the analysis. Later on the number has to be reduced to a practicable scale. This reduction is in most cases done by intuition. These types of approach have the advantage that the interrelations within the system are analysed thoroughly and a broad basis is given for possible development alternatives. Both approaches are often combined in practice.

A further differentiating criterion between the various approaches of scenario planning is the temporal order of the assumptions the scenario is based on (see Figure 3.2). *Anticipatory procedures* imply assumptions about conditions at the end of the period of planning. Potential events at the beginning of the period can thus be deduced. This type of procedure is often used to develop new, innovative visions, instruments and strategies. Scenarios which have been developed with this method are called *normative scenarios.* They base on a future vision which is determined beforehand and naturally influenced by wishes, values, objectives and hopes. Taking the vision as a starting point, the temporal path of development is traced back to the present. This is necessary in order to be able to realise the desired vision of the future. The so-called *contrast scenarios* are related to the normative scenarios. Here the present structures are contrasted just as the possible visions of the future which might develop from these structures are. The advantages of the anticipatory procedures are their orientation towards the objectives and the better op-

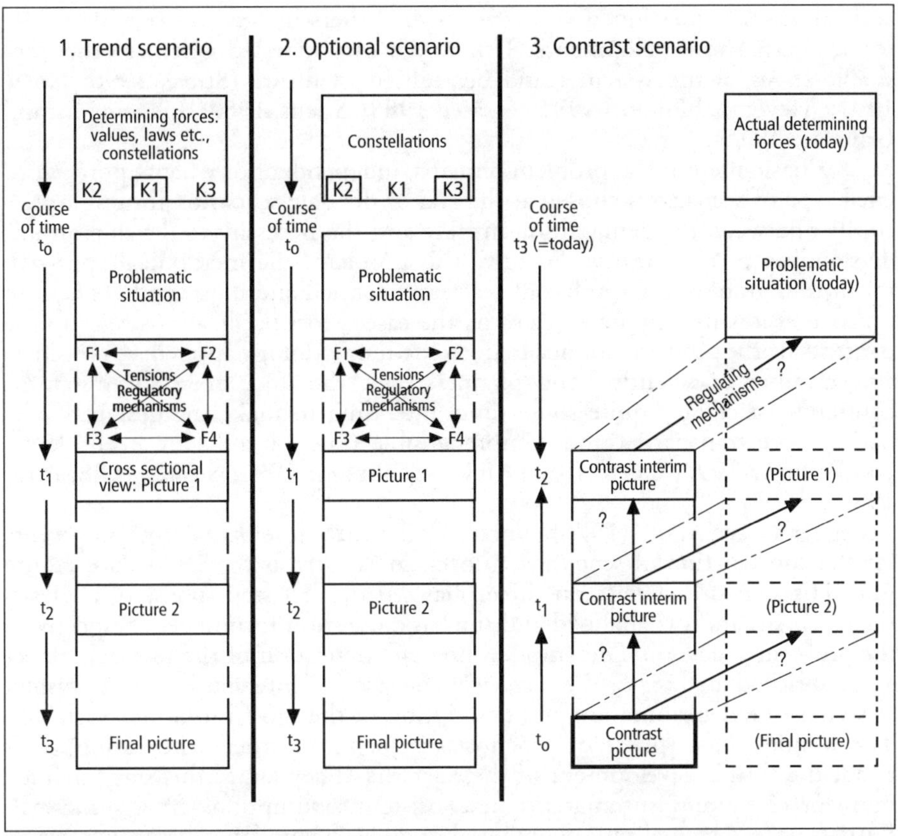

Figure 3.2. Operating schedule of the three different scenario methods. (After Lienemann and Unholzer 1975, p. 243)

portunity to be able to identify revolutionary developments (Stiens 1996; Götze 1991; Sträter 1988; Stiens 1982).

In contrast *exploratory procedures* are based on the realities of the present, to then deduce the resulting consequences for the future. These types of procedure strive at ascertaining and sounding out the future. They question positive and negative developments. The relative scenarios are always deduced from the present. The *trend* and *optional* scenarios are related to the exploratory procedures. Trend scenarios start at a determined point and follow the system's path of development up to a specific point of time in the future. Neither an objective nor a certain vision may be prepared. The dynamic of the system is explicitly determined by the currently dominating actions of the agents. Trend scenarios therefore project the present structures into the future and attempt to describe the most likely development, assuming that the conditions remain unchanged. This type of scenario is often used as a comparative basis for other types of scenarios. Optional scenarios, on the other hand, consciously assume different starting situations. Existing values, objectives and actions are questioned and replaced by others in order to be able to describe alternative developments. This way it can be tested whether other, desirable states of the system could be realised in future (Stiens 1996; Götze 1991; Meyer–Schönherr 1991; Sträter 1988; Stiens 1982; Lienemann and Unholzer 1975).

One basic methodical problem remains, independent of which approach or what type of scenario is chosen at the end of the day: scenarios always imply a conflict between the demand for entirety and the necessity to develop an understanding of complexity. To solve this problem, the methodical approach for the design of a scenario has to be clearly defined and illustrated. This is, in practice, unfortunately not very often the case. Most publications refer to the contents of the scenarios rather than to the methodology applied. A considerable number of scientists and planners have therefore been striving for a common pattern for quite some time, intending to make the design of scenarios more transparent and comprehensible (Hansel and Lambrecht 1993; von Reibnitz 1992; Götze 1991; Meyer–Schönherr 1991; Sträter 1988; Arras 1987).

Segner's endeavours (1976), amongst others, to develop a common process for the construction of scenarios go back to as early as the 1970s (see Figure 3.3). His suggestions base on three phases. In a first step (phase of analysis) the given situation is analysed and the basic question formulated, respectively the problem specified. This implies the exact definition of the research object and a detailed analysis of the historical and present interactions in the system as well as the determination and description of the most important variables. The second phase (phase of prognosis) is then concerned with assumptions about the future development of these factors. They are at this stage defined more precisely and summarised in consistent assumptions (raw scenarios). Furthermore, an analysis of potential sudden events, the disruptive events referred to earlier, is undertaken. The third phase is that of implementation. Scenarios, as different as possible but internally consistent, have to be created, formulated and interpreted. This includes the analysis of the possible effects disruptive events may have. Measures are developed and whether they are suitable to help with the realisation of the anticipated future visions has to be tested. Each of the three phases is divided into specific steps of work which are undertaken in parallel, subsequently or even repeatedly. The whole process works on an iterative basis, supported by different methods.

The use of three characteristic phases has by now widely been accepted. Almost all approaches are based on this method. As mentioned earlier, certain variations are available when constructing scenarios, particularly in view of the sequence of the steps undertaken. The *Battelle–Institute* has certainly proposed one of the most mature approaches for strategic planning. This also includes the variations of the approach which have been worked out mainly by former members of the institute, such as von Reibnitz. The work on the *Battelle* approach started in 1977 in cooperation with the *Battelle* institutes in Frankfurt and Geneva as well as the *Battelle Memorial Institute* in Ohio (USA). Since then it has permanently been refined. A sequence of eight steps is characteristic for the concept. These steps are structured even more strictly

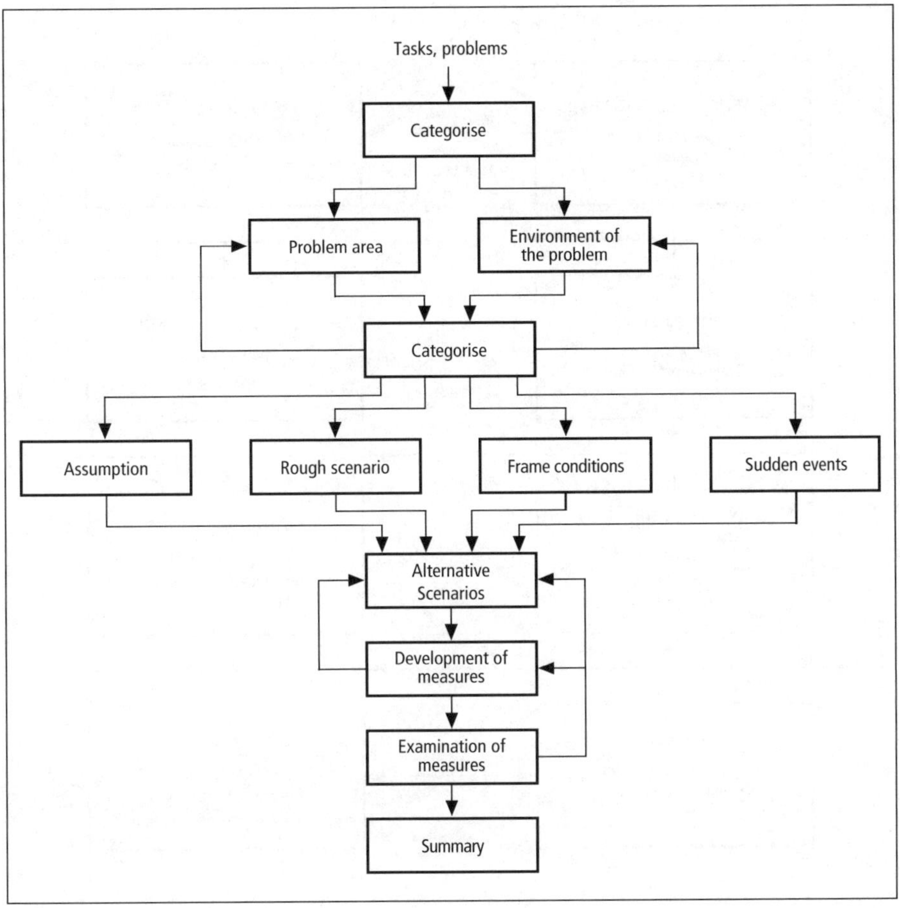

Figure 3.3. Logic sequence when creating a scenario according to Segner. (After Segner 1976, p. 31)

than those proposed by Segner (1976). They are shown only briefly in Figure 3.4 since a closer examination will follow in Chapter 6.

The various *tools* represent a further alternative when designing a scenario. Some of them are: creative techniques or model-aided processes such as interactive and cross-impact analyses or prognoses (see Chapter 6). Which type of tool is used depends on the objectives and whether the scenario is in fact worked out by a single person or a larger group of people. The more people are involved, the lesser the risk to make implausible assumptions whereas the range of experience, interests and information naturally widens considerably. It is, for this reason, useful to create a scenario with a team in which everyone learns together and from each other (Götze 1991). Von Reibnitz (1992) considers the size and the line-up of the team a decisive factor. When discussions are planned during the process, the group should not comprise more than 16 persons. The team members should be as heterogeneous as possible, i.e. of

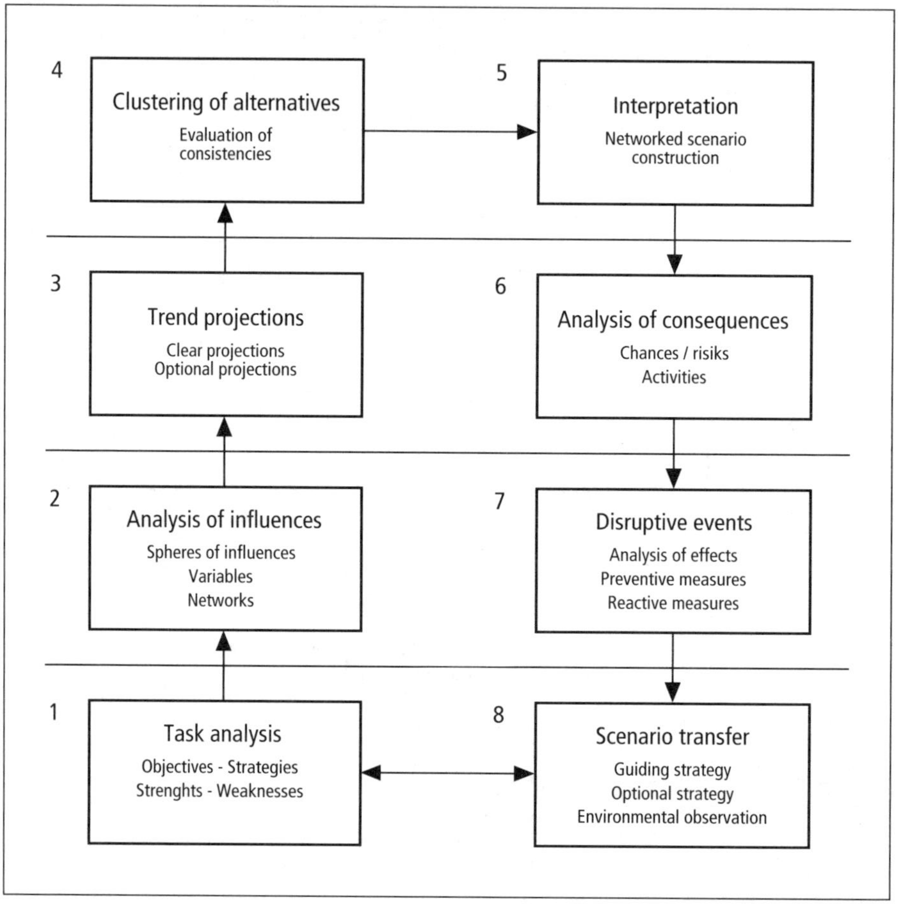

Figure 3.4. The eight steps of scenario planning according to von Reibnitz. (After von Reibnitz 1992, p. 30)

different age, sex and with different technical know-how, experience, competence and authority. Strategic planning in a company will therefore usually involve representatives of different management levels. In the sector of public planning and politics scenarios are generally prepared by a team of experts and then presented to the public for discussion.

One further alternative for the creation of scenarios should be mentioned at this point. The *time* allocated to the project. It can range from a single day to some weeks or months, depending on the objectives and, of course, the budget available. Scenario planning for a larger company or for the development of a town or region will naturally only be possible in the scope of a project taking some months. A scenario project comprises preparatory discussions, phases of analysis and a number of workshops, each taking about two to three days. The workshops serve for team discussions of each step, whereas the gathering of information and their analysis, based on model-aided processes, take place during the phase of analysis. Following this the results are presented. When scenarios planning is used for systems with lesser complexity, the eight steps for developing them can be managed during a scenario workshop of two or three days. This is equally valid for managers, planners, politicians and citizens who wish to learn about scenario planning and train for it (von Reibnitz 1992; Götze 1991).

3.4 Potentials and Problems of Scenario Planning

The model of the eight processing steps described above illustrates the potentials of scenario planning very clearly. These potentials are summarised in the following. One of the potentials is its function as a tool for dealing with complexity. Complexity is systematically made comprehensible when analysing important variables and interactions. The whole complex system is broken down into its individual constituents; the chains of activity between the variables being of special importance. The reflection on the conditions of the system and its environment can at this stage lead to a better understanding of the system itself. The next step is the various assumptions about the possible dynamics of the variables. The so-called trend projections are then grouped into internally consistent pictures of the future, naturally taking the analysis of the variables and its results into consideration. Stiens, in this context, talks about decomposing and afterwards composing the future. Central to these steps is the cognitive process. According to Stiens, it is not until decomposing has taken place that strategic sketches can be developed, i.e. specific compositions of single and separately tested solutions are made possible. This constitutes the basic difference between this particular method and the traditional simulation method which is like a black box to the planner, giving him, who is not the operator of the model, the finished or prefabricated results (Stiens 1996, p. 90).

Scenario planning cannot only be a tool for grasping complexity, but also a means of encouraging the constructive coping with uncertainties. A systematical preparation for alternative futures is the objective, rather than the prediction of the future. Continuous reflection plays an important role in existing values, interests, objectives and visions. This ensures that our mental constructs, which considerably affect our decisions and actions, can be specified and generally questioned. Reflecting encourages an openness for other argumentation, experimenting with new, innovative ideas and the consideration of alternative development options and strategies. We are dealing with alternatives which open up the horizon, increase flexibility and, above all, create the confidence that changing conditions can be met. The analysis of possible disruptive events takes place earlier on and describes the conditions under which sudden unexpected events can come about and what implications they may have for the future development of the system. Thus patterns of thought can be sensitised with a view to the chances and risks any future development may bring. Leeways are thus considered when deciding, acting or structuring (Stiens 1996; Georgantzas and Acar 1995; Stiens 1982).

When looking at the potential of scenario planning, it becomes evident that communication is a key factor when dealing with complexity and uncertainty. This is valid for all phases of scenario planning, i.e. for analysis, prognosis and implementation. When scenarios are planned by a group, all decisions and actions depend on the communicative processes between the team members. They have to pose questions and find answers. They have to illustrate their mental constructs. They have to analyse complexity of the system in question, define variables and discuss possible paths of development for the system. Communication remains a key factor up to the final use of the scenarios in practice. After all, complicated interactions have to be presented and made comprehensible. They have to be translated into literary, pictorial stories, readily understood by everyone. Arras (1989) therefore understands scenarios as a methodical tool for communication, preparing complicated interactions for discussion via illustrating depiction. Scenarios in this context even have a didactic value, since they facilitate the collective thinking and learning of scientists, politicians, planners and citizens.

Scenario planning not only offers potentials, but also bears risks. The criticism mainly refers to the lack of theoretical ground and the subjectiveness of the method. The supporters of the traditional method of prognosis in particular find fault with the lack of comprehensible laws and conditions for the assumptions made. They consider it a great risk to make only striking verbal statements, the internal consistency and logic of which cannot be checked due to their mainly qualitative premises. These critics consider scenarios unspecific and not scientifically founded. It has to be remembered, though, that scenario planning is deliberately based on the assumption that there is no such thing as an objective truth and thus subjective interpretations are given a particularly important status. The critics are right in saying that any scenario is created *ad libitum*. Scenarios are, as Trischler (1991) puts it, as much within the planners discretion as they have to be when it is scientifically impossible to find empirically measurable and intersubjectively checkable facts for future developments.

The criticism of subjectivity can thus be rejected. Sträter argues as follows: "The lack of scientific approach is criticised against the background of a scientific understanding and research practice which unilaterally pushes quantifying strategies to the fore whilst on the whole cutting other non-measurable phenomena out. This is done even though the modelling of the social reality with mathematical functions is not even suitable for grasping the surface of society's manifestations. Quantifying strategies are also specifically unsuitable when trying to depict the social interactions concealed behind the surface. (...) The reproach of a lack of scientific character therefore cannot offend scenario planning at its core. It does rather reflect the intention to describe and depict the present and future realities more accurately, with more subtle differentiation and with a greater complexity – even if on the whole more speculatively – as opposed to the traditional quantifying strategies of analysis and prognosis" (Sträter 1988, p. 419).

The criticism which refers to the actual process of constructing scenarios has on the other hand to be taken more seriously. Well-founded scenarios require the analysis of a wide range of quantitative and qualitative variables and an interdisciplinary cooperation. The input of personnel, material and time can be relatively high. Sträter (1988) gives this as a reason why scenarios have not yet reached the status of other prognosis strategies in political and administrative areas. Furthermore, Sträter sees a possible risk of misuse of scenario planning in exactly these areas since the application of primarily qualitative information increases the risk that an excessive influence may be exerted by subjective and not verifiable expert verdicts, which can, in the extreme, consciously manipulate the results of a scenario (cf. Sträter 1988, p. 431). This criticism is in fact justified, as for example the analysis of various scenarios for urban planning has shown (Fellner and Gestring 1988). The practice of municipal planning often gives a specific black-and-white picture, presenting an extremely positive and an extremely pessimistic picture of the future. The negative alternative is so black that basically only one alternative remains – that which has been envisaged from the beginning. According to Häußermann and Siebel (1989), the actual protagonists of scenarios designed for urban development are trends and factual constraints, above all the one and only objective of economic growth. The core objectives of scenario planning, i.e. the analysis of complexity and the depiction of alternative leeways are thus completely undermined (Fellner and Gestring 1989; Institut für Landes- und Stadtentwicklungsforschung 1989).

Häußermann and Siebel (1989) are of the opinion that scenarios should also have political and planning leeways and their conditions as a central theme. They suggest that scenarios are generated competitively by different, interdisciplinarily composed groups. This would, on the one hand, have the advantage that a wide range of different evaluations would contribute to the scenario planning process. On the other hand, the group would be invited to develop future visions which would pose a real and innovative alternative to existing patterns of development. The potentials of scenario planning could in this way surely be far better exploited than has been the case so far. This constructive suggestion has until now unfortunately not been tested in practice. A further problem still remains, i.e. putting the contents of the scenarios into concrete planning practice. The risk is evident: politicians and planners would possibly not pay enough attention to the ideas and strategies, with the result that they stick to their present practice. The more incredible and the less realistic the scenarios are, so might this risk further increase. (Institut für Landes- und Stadtentwicklungsforschung 1989).

Arras (1989), too, is preoccupied with the reproach that real alternatives are not worked through when using scenario planning. To him, this criticism concentrates too much on the aspects as regards contents. The most important characteristics of scenario planning are lost, i.e. process and recognition, the joint discovery, understanding and learning: "*These thoughts do not intend to reject any criticism of the strategy or critical observations as regards the contents. They intend to make clear that criticism which refers explicitly to the text and follows common scientific criteria, disregards the process of assimilation and thinking ahead, and thus does not do justice to this method. This has to be made perfectly clear, otherwise a method whose potentials are by far not yet exhausted would, by deficient and undifferentiated criticism, be shown in the wrong – and in my opinion an undeservedly negative – light"* (Arras 1989, p. 29).

Arras therefore calls for a critical assessment of the method which should consider the contents of the scenarios as well as the communication processes. This has so far not been realised in the debate on scenario planning. Scientific analyses have up to now either focused on the theory, the method or contents or on the practice of scenario planning in specific regions.

3.5 Methodical Consequences for this Paper

This paper takes up the above mentioned suggestions of Häußermann and Siebel (1989) and Arras (1989). The potentials and the problems of scenario planning will be critically examined. Both aspects, the results as regards content and the communication processes will be considered equally. To this end the separate steps of scenario planning will be undertaken in the scope of a group process and afterwards assessed according to the following questions:

1. What is the contribution scenario planning can make when dealing with complexity and uncertainty?
2. Do scenarios facilitate the joint thinking, learning and acting of decision makers and persons concerned and do they facilitate the decision for certain planning strategies?
3. Can the potentials of scenario planning be used more effectively when scenarios are designed competitively by different groups?

The relevance of scenario planning can in this context only be determined by its social practice in the prevailing *lebenswelt.* After all, scenarios are a tool to contribute to a new consciousness about the constitution of the lebenswelt and the possibilities and limits of its shaping. The question is whether this method still functions when the *lebenswelt* is subject to chaos, i.e. in a concrete disaster situation? Can scenarios help to divert the insecurity we face when trying to recover from a natural disaster by presenting an option to actively restructure our future? The current endeavours of hazard research and management confirm that these two questions are particularly relevant. The development and application of methods which allow a constructive handling of natural disasters in all phases of hazard management gain more and more importance. The main objective is the integration of hazard management into a sustainable development planning strategy. The advantages of scenarios have in this context been pointed out by a number of scientists (Geipel et al. 1988; Rubin et al. 1987; Dobler 1984; Foster 1980). So far, though, in the majority of cases the use of scenarios in hazard management has been limited to the simulation of possible natural disasters. As yet, their potential for planning and management during the time of a natural disaster has not been assessed (refer to Chapter 5). Scenario planning is therefore, in the scope of this book, applied to a concrete disaster situation and also analysed in view of the above mentioned unsolved questions.

The actual *lebenswelt* which is the focal point of this work is the Caribbean island Montserrat. The island is characterised by its extremely sensible eco-system, limited resources and recurrent natural disasters like floods, hurricanes and earthquakes. In short, Montserrat is a very vulnerable insular system which is permanently exposed to unexpected events. The sudden awakening of the volcano Soufrière in the south of the island was probably the event of most serious consequence in the history of the island. The volcano has erupted several times since July 1995 (see Chapter 4.2). My work is based on a number of field work visits to Montserrat, before and during the disaster. In the sense of the above considerations the field work served to gain the best possible understanding of Montserrat as a complex system in order to later on be able to transfer this understanding to appropriate and flexible strategies. The already stated classic working phases of scenario planning have been undertaken: the phase of analysis, the phase of prognosis and the phase of implementation.

The following chapter refers to the phase of analysis. Sub systems are defined and described for the complex Montserrat system. Focal points are the identification of important events, actions, agents and strategies, both in the past and the present. A classic analysis of historical documents, current newspapers, literature, statistics, development plans and maps cannot be considered sufficient. Montserrat as a system of which one is ignorant can only be understood when the local protagonists are consulted and their perceptions and opinions digested in the learning process. In line with the endeavours of qualitative social research it is particularly important to learn about previously unknown perceptions of Montserrat from the perspective of the people living and working on the island. Also, my personal ideas as an outsider have to be constantly reflected on and, if need be, revised. To this end more than 100 conversations took place with experts, and a further 140 intensive interviews were held with the local population. The first talks took place during a period of apparent security, that is before the eruption of the volcano. The second talks took place during a period of insecurity and disorder since there was a constant threat of the volcano erupting with the resulting air of impending disaster.

Once the complex system Montserrat had been approached in this mainly qualitative way, a pre-test was launched. The further steps of scenario planning were applied in the scope of a scenario project of several months. Members and students of the Department of Economic Geography in Hamburg and representatives of the private sector were involved in the project. The project primarily served the purpose of critically examining each working step of the method and developing an appropriate strategy for the concrete case. The resulting strategy was then applied during a two-day scenario workshop on Montserrat at the time of the volcanic crisis. This scenario workshop was organised in cooperation with the local planning authorities. Altogether 23 representatives of the public and private sector, schools and non-governmental organisations (NGOs) of Montserrat attended the workshop. In order to meet the von Reibnitz (1992) requirements as to the members of the team, two completely different, interdisciplinarily composed groups were selected. They competitively developed scenarios and strategies for Montserrat in the scope of the scenario project and the workshop. In this way, not only the results of the insider and outsider perspectives can be compared but also the individual communication processes. And furthermore, scenario planning's possible contribution to coping with a natural disaster can be examined.

No island in these seas is bolder in its general aspect, more picturesque and I think I may add without exaggeration, more beautiful in the detail of its scenery – indeed might be tempted to say considering its fortunes, that it has the fatal gift of beauty (Davy).

4 Montserrat – A Complex System

Montserrat is part of the Lesser Antilles in the Eastern Caribbean, an archipelago extending from Puerto Rico to Trinidad (Figure 4.1). The Antilles were formed by the subduction of the Atlantic plate beneath the Caribbean plate. Since this process continuously expands from the Northeast to the Southwest, the islands are of different geological ages and origins. Montserrat itself is of volcanic origin. The island is composed of pyroclastics, agglomerates and lava flows and is geologically young, its formation process having commenced less than 50 million years ago. It is positioned 43 km (about 26 mi) southwest of Antigua and 65 km (about 40 mi) northwest of Guadeloupe. Montserrat is about 18 km (roughly 11 mi) long and 11 km (almost 7 mi) wide, with its total surface extending to no more than 102 sq km (a good 39 sq mi) (Figure 4.2).

Archaeological research has shown that Montserrat was settled by American Indians, the Arawaks, probably as early as the first century AD. Excavations indicate that the first settlements were located at the eastern coastal flats (Trant's) and the rivermouth of the White River at the southwestern coast. Oral history says that the Arawaks were later driven off or even killed by the Caribs. Archaeologists believe that the Caribs did not settle permanently on the island, but rather used it as an interim location for their invasion of other neighbouring islands. The Caribs were the first to name the island – Alliouagana, the land of the prickly bush (Fergus 1994).

Christopher Columbus discovered Montserrat on 11 November 1493 during his second expedition to the Caribbean. The island with its rough relief and the lush vegetation reminded Columbus of the abbey Montserrat (*Santa Maria de Montserrate*) near Barcelona. He named the island after the abbey, but it was too small to attract his genuine interest. Columbus was at anchor for a short time and then carried on his journey to Hispaniola, hoping to finally find gold. It took another 138 years until, in the year 1631, a small group of Irish and English settlers landed at Carr's Bay and started to colonise the island. A decisive century began for Montserrat. The news of the

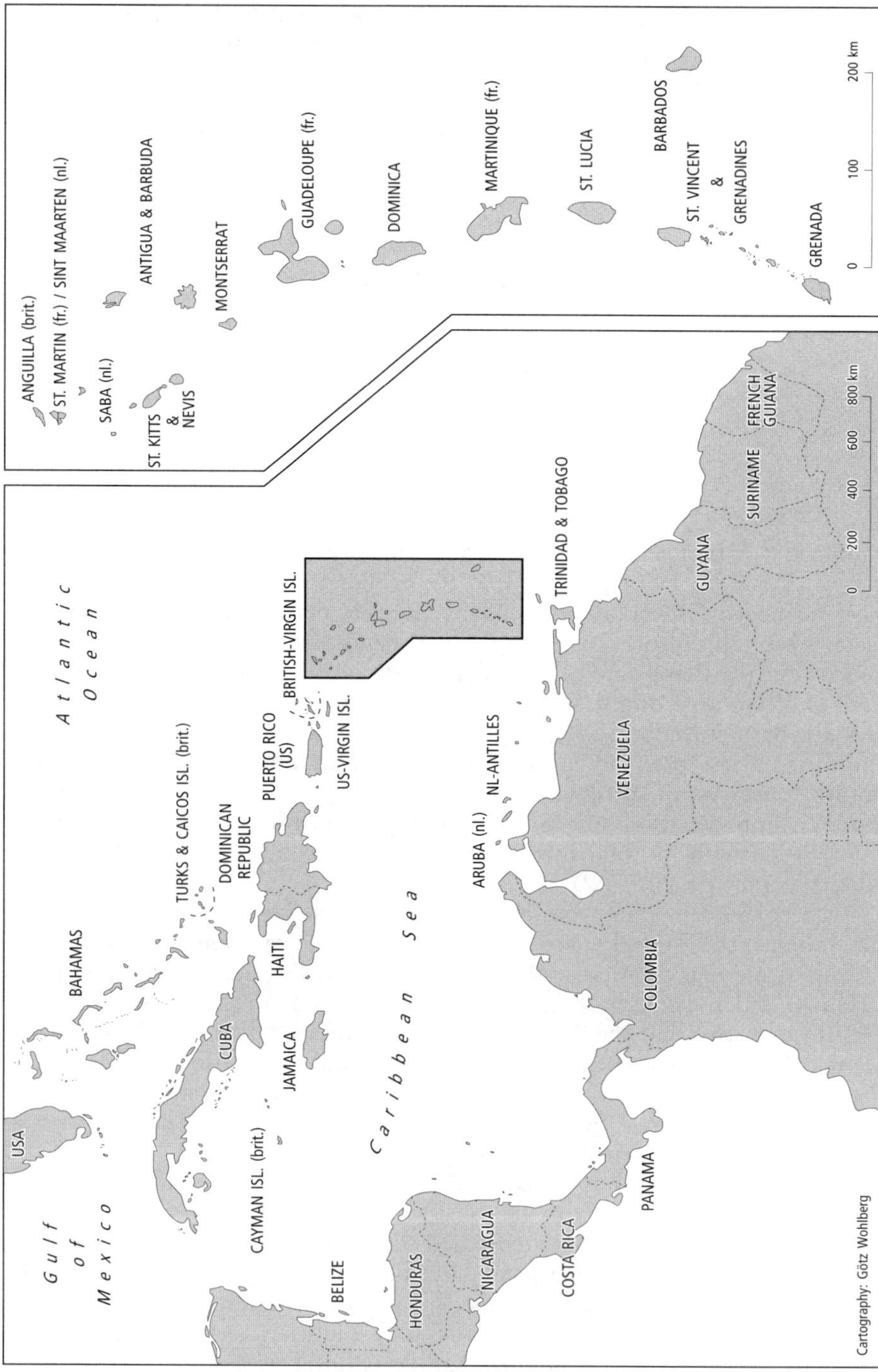

Figure 4.1. The Caribbean region

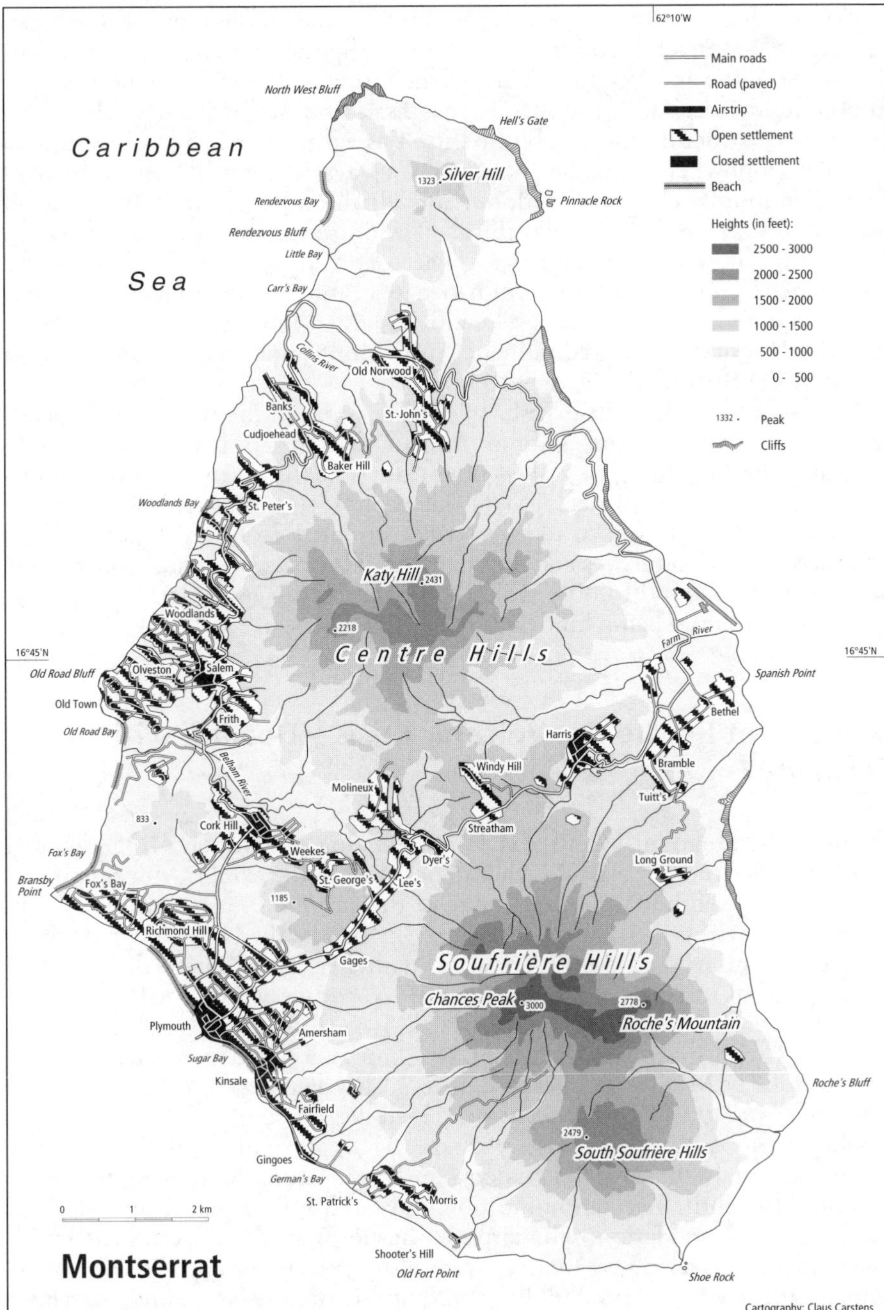

Figure 4.2. Topographical map of Montserrat

Roman-Catholic colony spread very quickly and thousands of Catholics headed for Montserrat, taking their refuge in flight of the Protestant persecution on St Kitt's and Virginia. Most of the refugees settled in Plymouth, later to become the capital of the island, or Kinsale and St Patrick's. A classic colonial social structure had established itself as early as in 1654. There were a few rich landowners, e.g. the Blakes of Galways who owned entire belts of land. The middle class was predominantly Irish and there were hundreds of black slaves on the island. Two thirds of Montserrat were deforested within 50 years to allow the cultivation of tobacco and sugar cane. The British and the French struggled for predominance over the sugar-producing islands, Montserrat being no exception. These fights lasted more than a century until in 1783 the question of predominance over Montserrat was finally decided in favour of the British side (Fergus 1994).

Montserrat is today still British. In 1967 it received the status of a dependent territory with internal self-government (see details below). The 1960s were also the beginning of a slow but steady and relatively successful transformation from the former agricultural economy, during the 20th century predominated by cotton, to today's service-oriented economy which is based on tourism, light industries and construction. However, the endeavours of the population of about 11,000 people (1991 census) for an economic development are constantly being hindered by natural disasters.

4.1 Montserrat Before the Volcanic Crisis – Dealing with Certainty?

Hurricane Hugo hit Montserrat on the evening of Saturday, 16 September 1989. The destruction was sudden and of catastrophic extent. Montserrat was for more than ten hours subject to one of the strongest hurricanes the Caribbean has experienced in the 20th century[1]. 11 people died that night, 40 were injured and more than 3,000 left without shelter. Since Hugo destroyed all meteorological instruments on the island, its average wind speed could only be guessed. The estimates range from 225 to 240 km/h (roughly 140 to 150 mi/h) with maximum speeds of 380 km/h (about 235 mi/hr). The following Sunday Montserrat, because of its lush vegetation also known as *The Emerald Island of the Caribbean*, was declared disaster area.

More than 98 % of the houses were damaged, 50 % of them heavily, 20 % being completely destroyed. The docks in Plymouth harbour were destroyed, as well as the entire telecommunication and radio infrastructure and 80 % of the hotel accommodation. The airport, the hospital, the government buildings and the diesel power station were also severely damaged. The overall damage incurred just on buildings and infrastructure amounted to about US$ 400 million. Furthermore, hundreds of animals, most of the forest areas and a large part of the harvest were victims of the hurricane. Alone 60 % of

[1] According to the Simpson-Saffir scale, hurricane Hugo reached category 4 (speeds up to 210 to 248 km/h). This is only exceeded by category 5 (speeds of more than 248 km/h).

the forest were defoliated and the damage in agriculture and fishery amounted to more than US$ 4 million (Ministry of Agriculture, Trade and Environment 1994). Hugo also destroyed the *International Air Studios*, established in the 1960s by the former manager of the Beatles, Sir George Martin. The studios had previously been a magnet for many international musicians and bands who produced their albums on Montserrat. The musicians spent over US$ 30,000 monthly, the studios were thus an ideal source of foreign currency, combined with the advantage of hardly any adverse ecological or social effects. However, the studios have never been rebuilt.

The extent of the natural disaster was incomprehensible to the Montserratians. Despite all prior warnings nobody had believed that Hugo would hit Montserrat. After all, the weather had been particularly nice and quiet in the days preceding the hurricane. Since the last one had been some 61 years ago, only a few had ever experienced a hurricane. Since then only two minor hurricanes had reached Montserrat (1966 and 1933). Therefore, nobody was prepared sufficiently for the impact of Hugo. The complete breakdown of the communication facilities and the impassable condition of the roads led to serious problems and temporal constraints in the coordination and performance of emergency measures. These were carried out completely *ad hoc* since most of the people in charge were not adequately trained. The national emergency management was not functioning until 48 hours after the disaster had struck the island. There were neither cooking nor sanitary facilities in the shelters which were even damaged in parts. A responsible management team had not been set up, and there were numerous complications with regard to food supply, medical care and assessment of the damage. However, there were some individuals who worked very effectively and made a number of important decisions. One of them was the then Governor Christopher Turner. In his career he had already experienced five other hurricanes in the South Pacific and took charge of coordinating the emergency measures. There were also numerous members of the local ministries and authorities who collaborated with teams from the British frigate H.M.S. Alarcity, anchoring only a few hours after the incident and assisting with the clearing up operation (Berke and Wenger 1991).

The ensuing phase of restoration and reconstruction also happened *ad* hoc, without planning and was very slow to start. For example, the water supply for the whole island could not be guaranteed until the end of October, i.e. one and a half months later. Restoring the electricity supply in some areas took until February 1990, and some telephone connections were not fully re-established until July 1990. The reconstruction phase was dominated by a "just put it back" mentality. Few endeavours were undertaken to weigh up short-term needs and long-term interests. The dependence on non-local and foreign assistance, especially of the British mother country and non-governmental organisations (NGOs), such as the *Red Cross* and the *US Peace Corps,* posed a serious problem. The reason being that on the one hand the local government was momentarily incapable of managing the situation and took on too little responsibility, and, on the other hand, a lack of integration between the local government and the population regarding the reconstruction activities. The regional assistance provided by the *Caribbean Economic*

Community (Caricom) also caused problems. An *ad hoc* established group, the *Caricom Disaster Relief Unit* (CDRU), coordinated the transport of the relief goods given by the neighbouring islands, but the work of the committee caused confusion to those responsible on Montserrat. It took six months after the disaster had occurred until the Montserratian government established a special organisation, the *National Housing and Rehabilitation Unit,* and with it re-established its control over the reconstruction activities (Berke and Wenger 1991).

This brief summary of the hazard management comprises just the most important reactions and adjustments in the first 12 months following Hugo. Naturally, the reconstruction process was not finished after such a relatively short time and, in any case, can only be reviewed many years later. It became apparent, though, that the national hazard management contained considerable shortfalls, particularly regarding the reactive phases. There were some endeavours to improve hazard management, but these were not realised. One example is the proposed actualisation of the national disaster plan. The new version of the plan was rejected by the government shortly after its presentation in 1990. A full-time post for a national disaster coordinator was not set up and the budget for hazard management was only marginally increased from EC$ 5,000 to EC$ 8,000[2] (Berke and Wenger 1991). The government and the majority of the local population both focused on the full return to the pre-disaster state. Hugo and other hazards fell more and more into oblivion, whilst the mobilisation of all available resources came to the fore.

If one now has a look at the temporal phases and takes the basic theoretical assumptions into consideration, then the following hypotheses can be made regarding the human adjustments to Hugo and the evolution of the system Montserrat: the *lebenswelt* of the Montserratians was completely turned upside down by this event suddenly overtaking them. This disorderly state is felt as something very threatening. Therefore, priority is given to the re-establishment of the previous order. A state of stability and security is striven for and the objective of all activities is the *status quo ante* Hugo, even if this requires enormous efforts. People show a strong resistance towards any change which they accept only in selected contexts, if at all. Montserrat, however, is a complex system *far-from-equilibrium.* It is subject to a continuous process of change and will therefore never return to its previous state. Hugo is within this system the case of extreme change. The hurricane set off internal fluctuations which through processes of self-organisation result in a new order, more complex than the previous one.

The complexity of the system Montserrat, i.e. its structure, function and development, shall in the following be examined and the hypotheses mentioned be reviewed. The focal point shall be the interaction between the human beings and their environment or, to put it another way, the understanding of real people in their real world. In the sense of the previous reflections the interaction human-environment is always an interplay between (1) the natural environment, (2) the political-economic and socio-cultural context and (3) the partly unpredictable influences of agents or single indi-

---

[2] EC$ 1 = US$ 0.27

viduals. This interplay can only be understood properly through an integrative approach which equally considers different perspectives and different levels of standards. Palm's (1990) approach which has been presented in Chapter 2.3 is particularly suitable for this task. Her differentiation between the three levels of analysis, i.e. the macro level (basic structures), the meso level (the gatekeepers) and the micro level (the individual), will be used for the following approximation to the complex system Montserrat. The interactions between the different levels will within the scope of the analysis also be illustrated.

Such an analysis, however differentiated it might be, can, due to the prevailing complexity, never represent more than a small extract of the continuous alterations such a system is subject to. Palm also emphasises this: "The web of relationships of the individual, society, and the environment is always changing, modified by the nature of the structure that binds them together, and by the incomplete articulation existing within the structure at any given time. Any empirical research, then, is a snapshot in time of this set of relationships, and should be seen as an approximation of their nature" (Palm 1990, p. 92). The research approaches the complex system Montserrat in two different situations – five years after Hugo, i.e. in the summer of 1994 and during the volcanic crisis in the autumn of 1996. The intention is to compare the three levels of analysis in these two completely different states of the system.

4.1.1 The Macro Level – Basic Structures

The macro level, according to Palm (1990), represents the basic forces and structures which influence the interaction human-environment. Central questions, amongst others, are: how environmentally sensitive are the predominant economic activities? How does the political organisation function? How are decisions made regarding the utilisation of land and resources? Which are the dominating structures in the society? In order to be able to answer these questions the complexity of the system has to be understood. In the scope of this work the differentiation into sub-systems which is usually put to use in systems sciences will be applied. The most important subsystems identified are then analysed with reference to their most important components, structures and events. Following sub-systems are determined for Montserrat: the environment, the social, the political and the economic system. Each of these sub-systems will in the following be briefly characterised, taking the institutional and legal organisations and potential constraints into account.

The Environment

The topography of Montserrat is dominated by four massifs, ascending from North to South. These are the Silver Hills, the Centre Hills, the Sourfrière Hills and the South Soufrière Hills (Figure 4.3). The Silver Hills (403 m / 1,323 ft) and the Centre Hills (914 m / 3,000 ft) are very old volcanoes which were active about 4.4 to 1.5 million years ago. They have ceased volcanic

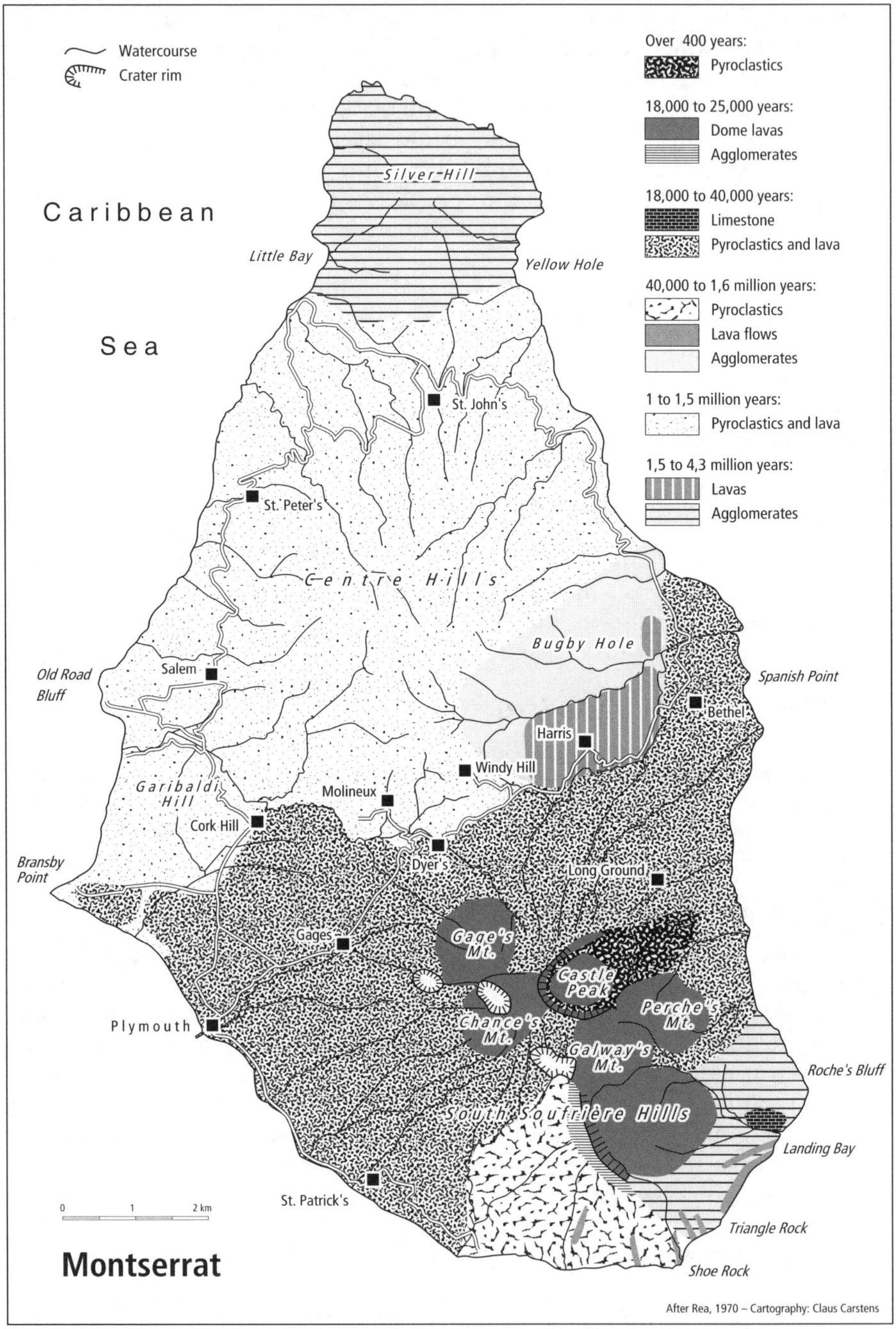

Figure 4.3. Geological map of Montserrat

activity and are strongly eroded. The southern part of Montserrat consists of younger volcanoes. The South Soufrière Hills showed their last activity around 1.8 to 0.9 million years ago. The youngest volcanic centre on the island are the Soufrière Hills. It is a comparatively small, but still active volcano. The steep topography of the island continues below the marine floor. The 45 km (28 mi) coast line are thus characterised predominantly by steep cliffs (70 %). The remaining part consists of a few coastal wetlands and altogether 15 smaller sandy beaches, all of them with the exception of Rendezvous Bay being of black volcanic sand.

Montserrat's soils are of primarily volcanic origin. They consist of clay, loam and sand; the clayey soils predominating in the north and the sandy and loamy ones in the south. The soil classification describes about 40 % as severely eroded and not recoverable soils. 30 % of the sandy soils are in a deteriorated condition. Only 14 % of the soils are to be rated relatively stable. They are to be found around the headwaters of the rivers or situated on the higher slopes. According to Corker (1986) about 40 % of the soils are suitable for agricultural production, however, 10 % of them located in settlement areas. The most fertile areas can be found along the southwest coast and the east-western corridor between the Centre and the Soufrière Hills.

The climate of the island is maritime, subtropical with constant north-east trade winds. The temperatures vary only slightly over the seasons (by max. 5° C / 9° F). The average temperature is around 26° C (79° F) (Corker 1986). The average yearly precipitation on Montserrat varies between 1,000 mm (39 in) at sea level and more than 2,500 mm (97.5 in) in the altitudes. A dry season (February to June) and a wet season (July to January) can roughly be determined, however, there can be considerable yearly variations. The spatial and temporal variations of the rainfall facilitate an extensive variety of different microclimates, ranging from dry woodland and cactus scrubs in the coastal flats to rain and cloud forests in the altitudes. The central western slopes of the Centre Hills and the saddle region between the Centre and the Soufrière Hills are also very humid. The north and the south of the island are due to the constant onshore winds relatively dry and often even exposed to droughts (Figure 4.4).

The indigenous forest landscape of the island has nearly disappeared in favour of standing timber and agriculture. Montserrat's present vegetation is mainly secondary forest, accounting to about 71 % of the entire island's surface. The most important types of vegetation are rain, cloud or moist forests, dry woodland and cactus scrub. There are two mangroves, one in Fox's Bay and the other in Carr's Bay. Altogether 132 types of trees have been identified on Montserrat, 17 of them being endemic. The evolution of the fauna is closely related to the isolated island position. It is extremely adapted to the local conditions and as such very sensitive to external influences. Birds and reptiles are the most common animals. Blankenship (1990) identified 11 species of reptiles and 59 species of birds, the only endemic one being the Montserratian Oriole (*Icterus oberi*). This national bird is quite well examined. 1,000 birds were counted in 1984, but hurricane Hugo reduced the stock to half the number (Island Resources Foundation 1993). The list of the otherland species is short: 13 species of mammal, alone nine being bats, and

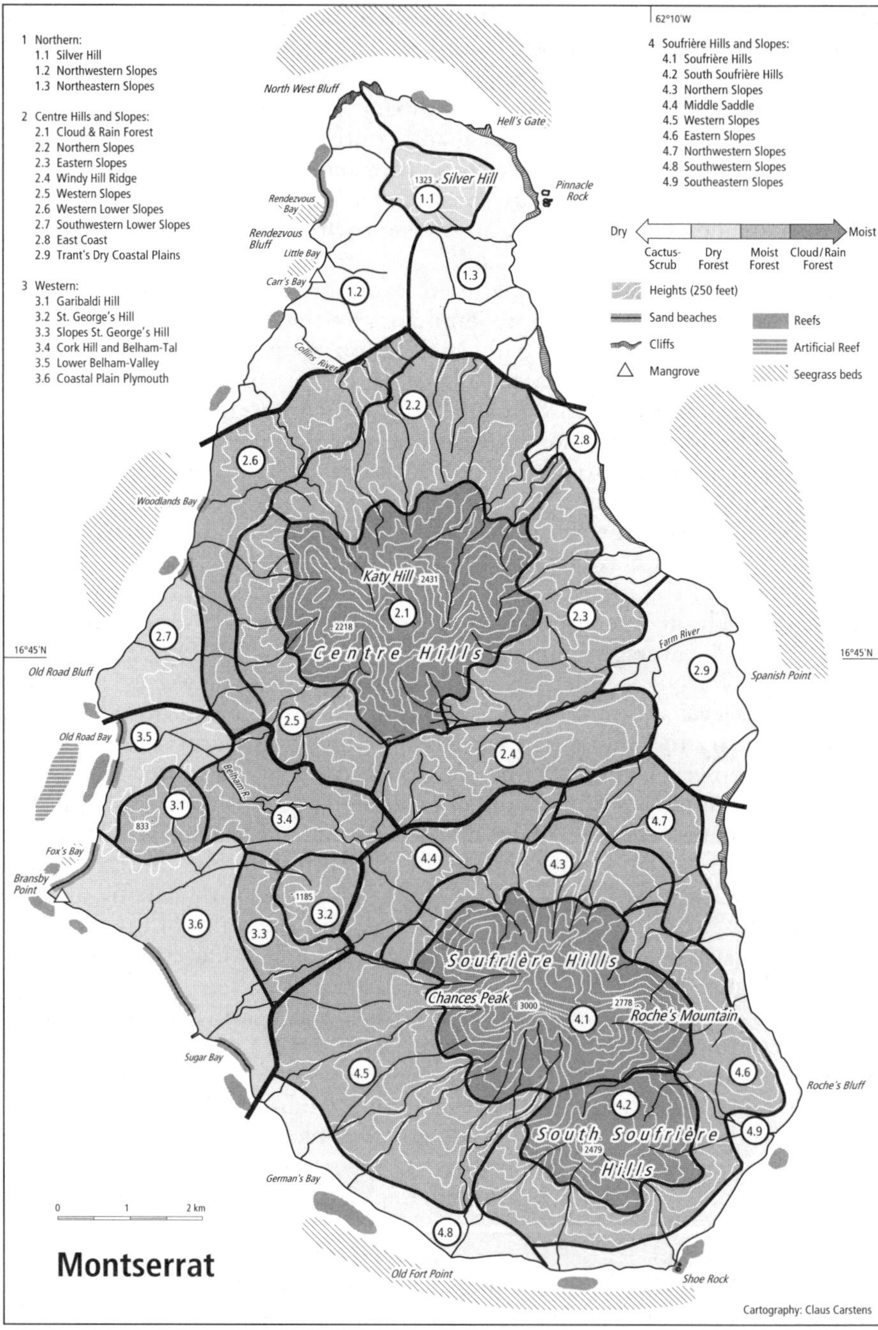

Figure 4.4. Natural regions

three types of amphibian. One of the latter, the toads called Mountain Chicken (*Leptodactylus fallax)* are of special interest. They inhabit just Montserrat and Dominica in the East-Caribbean. Even though the *International Union for Conservation and Nature* (IUCN) has listed them as threatened with extinction, they are known as delicacy. The fish stock around Montserrat is limited because of the narrow and steep coastal shelf and the few coral reefs.

The relatively early and active integration of conservation into the political issues is characteristic of Montserrat. The *Forestry Ordinance* (Chap. 95, 1956) for the protection of the forest areas was for example passed as early as 1956. The main reason for the law was the important protective function the forests have for the watersheds and soils of the island. However, the actual implementation of the law was very slow. So far just one protected area, the *Silver Hill Forest Reserve*, has been established. The *National Parks Ordinance* was formulated in 1970. It was designed as a further law for the creation of national parks and protected areas, but until now it has not been passed since the support of the former governing party could not be obtained. As a result, Montserrat did not have an overall conservation legislation until the mid 1990s. Furthermore, the responsibility for environmental issues still remains with different public authorities (Table 4.1). This often causes coordination problems in the conservation and resource management sector. Different organisations are for example responsible for the management of the watersheds. The recent *Forestry, Wildlife, National Parks and Protected Areas Act* (No. 3, 1996) provides for three national parks and a number of protected areas as well as an effective and integrative resource management. Furthermore, a national park administration is planned, but the implementation and control of the law remains uncertain.

Table 4.1. State organisations with responsibility for environmental management

Organisation	Division	Responsibility
Ministry of Agriculture	Forestry Department	Forest management, coordination of environmental management, management of watersheds
Trade & Environment (MATE)	Forestry Board	Protection of forests on private and public land
	Fisheries Department	Management of marine resources
	Physical Planning Unit (PPU)	Land utilisation planning
	Development Control Authority (DCA)	Control of construction and development projects and distribution and sale of plots
	Land Development Authority (LDA)	Lease of agricultural land
	Pesticides Control Board	Control of imports and utilisation of pesticides
Ministry of Finance, EconomicDevelopment and Sports	Development Unit	Development and economic planning
Ministry of Education, Health and Community Services	Environmental Health	Waste and sewage disposal, control of mosquitoes
Ministry of Communications and Public Works	Montserrat Water Authority (MWA)	Water management

Apart from the public sector there is one non-governmental organisation (NGO) which has an important function in the island's environment and resource management. The *Montserrat National Trust* was established as early as 1969 and is financed by the subscriptions of its more than 100 members, donations and financial aid from the British government. The two principal tasks of the National Trust are supportive measures for the protection of Montserrat's natural and cultural resources and the environmental education of its citizens, visitors and the government. To this end the National Trust runs local campaigns like the yearly programme to clean the beaches or the *Best Village Competition.* It also maintains a museum and a library, restores historical sites and manages a bird protection area near Fox's Bay. The National Trust collaborates closely with different governmental organisations, scientists and other (inter)national groups.

Figure 4.5 shows Montserrat's land use pattern in 1995. The most important forest areas are the Centre Hills, the Soufrière Hills and the altitudes of St. George's and Garibaldi Hill. Dry woodland vegetation dominates on the Silver Hills. Agricultural production accounted for around 25 % of the island's entire area in 1960, but has diminished to around 1,200 ha (2,964 acres). The reasons being the reduced profitability of the insular agriculture and forced shifting to other land uses, such as living, working and consumption. Residential, commercial and industrial land use is concentrated around the western coastline and the east-western corridor between the Centre and the Soufrière Hills. Commercial use is found mainly in the capital Plymouth. An examination of the distribution of property has revealed four categories: (1) private estates, often left fallow, (2) private land with commercial, industrial or residential use, (3) government owned land, such as the forests of the Centre and Soufrière Hills and a few industrial plots, (4) former estates which are today also government owned. Altogether 3,115 ha (7,694 acres) of land are private property (Physical Planning Unit 1995b).

As with most of the islands in the region, Montserrat's history of development and land utilisation planning is comparatively young. The first law, the *Town and Country Planning Ordinance*, was enacted in 1975. Amongst other things it provided for the establishment of a specific administrative unit, the so-called *Development Control Authority* (DCA). The DCA has since then been in charge of the coordination and control of potential development projects. However, for a long time there was no legally binding development plan and this obstructed the progressive work of the authority. Such a plan was developed in 1976 in collaboration with the *United Nations Development Programme* (UNDP), but it had not been ratified because of a lack of local support. It was not until 1995 that the *National Physical Development Plan* (NPDP) was ratified, providing for integrated development and environmental planning for the entire island (see Chapter 4.1.2). Also, the year 1994 saw the establishment of a further important authority in this respect, the *Physical Planning Unit.* It is integral part of the Ministry of Agriculture, Trade and Environment (MATE) and functions as the executive organ of the DCA.

The pressure upon the natural environment had because of the small and stable population been relatively weak. A number of expert studies confirmed the comparatively good condition of Montserrat's environment (Island

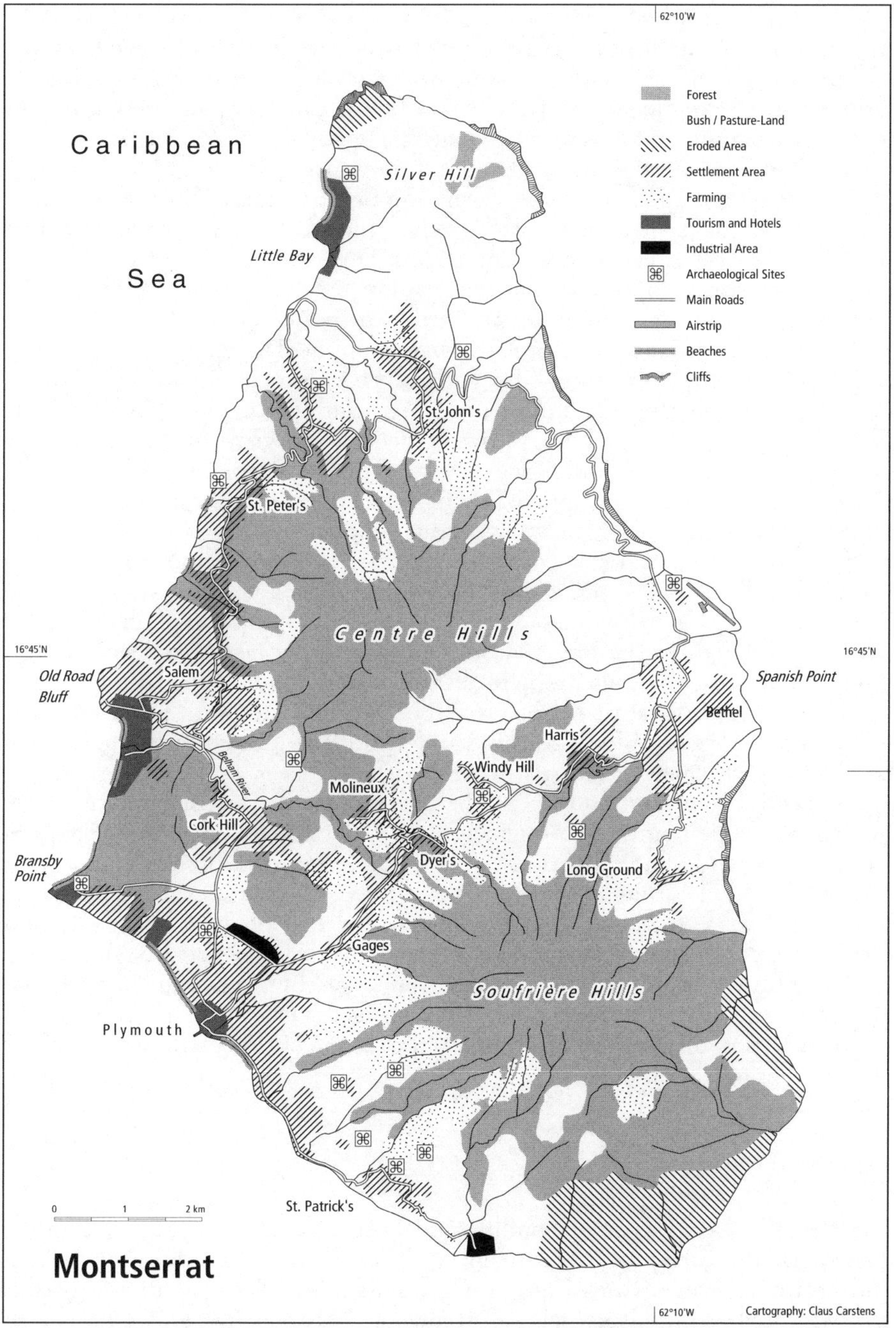

Figure 4.5. Land use on Montserrat before the volcanic crisis

Resources Foundation 1993; Corker 1986; Caribbean Conservation Association 1980). But the insular ecosystem has over the past decades shown some constraints which can mainly be put down to its extreme fragility. Rather serious is the strong erosion of the soil which is caused by deforestation and ineffective resource management practices. A good example of the effects of such unregulated agricultural and forestal usage are the Silver Hills. They are by now severely eroded and scrub vegetation predominates. The erosion process is further worsened by the steep relief and the seasonally concentrated rainfall which often even causes landslides. The same applies to the agricultural use of steep slopes and the overgrazing by free-roaming animals, like goats and donkeys, causing problems on the entire island.

A further potential constraint is the water supply. There are only two rivers which permanently run water, the *Belham River* and the *White River*. Sufficient water supply was guaranteed until the beginning of the 1990s but bottlenecks are expected for the future, especially in view of the constantly growing water consumption. Since almost all of the reliable springs have been tapped there is hardly any potential for expansion. The illegal quarrying of sand on the beaches poses the most serious problem in view of the coastal and marine resources. Beach and coastal erosion are aggravated and turtle nesting is severely affected. The wet coastal flats are another area of the island which is endangered by two major factors. One is the golf site in Belham Valley which has altered the natural environment. The other factor is illegal rubbish dumps which are a serious problem on the entire island. Many of them came up during the removal of debris after the hurricane disaster. The rubbish is for the most part not bio-degradable and also often located in the hardly accessible gorges (locally called *ghauts*) of the island.

Montserrat is not only subject to the negative effects the use of its resources by man has. The island's ecosystem is also vulnerable to a number of natural hazards. These include hurricanes, volcanic eruptions, floods, landslides, earthquakes and droughts. Table 4.2 below shows all natural disasters Montserrat has experienced between the second half of the 17th century and 1985. The potential effects any disaster may have is far greater on Montserrat with its limited resources and space than in large countries. Hurricane Hugo demonstrated this very vividly. A worsening factor is also the possibility that different types of disasters can affect the island either simultaneously or shortly after one another.

The Social System

Montserrat's society is small, mobile and lives together in relative harmony. The social infrastructure is relatively well developed. The education system and health services provided are of a good standard. The population is thus well-educated (compulsory school attendance between five and 14 years of age). The per capita income is, with US$ 6,800 p.a., by Eastern Caribbean standards comparatively high. Because of a number of reasons, the population enjoys a modern standard of living. Montserrat is dependent territory of the British mother country, relatively closely linked with the international

Table 4.2. Disasters on Montserrat from 1667 to 1985. (After Fergus 1994 and Barzey 1989)

Type of disaster	Date	Number of dead	Extent of damage and injured
Hurricane	1667	NA	NA
Flood	1672, 24.12.	NA	marginal damage
Earthquake	1672	none	marginal damage
Hurricane	1737	none	partly severe damage by floodings
Hurricane	1740	none	harvest destroyed
Hurricane	1744	none	harvest destroyed
Hurricane	1766	many dead and injured people	hundreds without shelter, half of Plymouth destroyed, streets and bridges destroyed
Flood	1767	some dead people	bridges and streets severely damaged, some houses flooded/washed away
Hurricane	1772	none	harvest and houses destroyed
Hurricane	1792	none	minor damage to the harvest
Hurricane	1816, 16.09.	NA	harvest completely destroyed
Earthquake	1843, 08.02.	6 people dead	relatively strong earthquake with a range of damage
Hurricane	1899, 07.08.	100 dead, 1,000 injured 9,000 without shelter	severe devastation on the entire island
Hurricane	1924, 28.08.	36 dead, many injured 5,000 without shelter	estimated damage to houses, infrastructure and harvest more than £100, 000.00
Hurricane	1928, 12.09.	42 dead, 100 injured 1,000 without shelter	severe devastation, especially in Plymouth severe damage to houses and infrastructure 600 houses completely destroyed, estimated damage more than £150, 000.00
Hurricane	1933	none	minor damage
Hurricane	1966	none	minor damage
Earthquake	1985, 16.03.	none	6.6 on the Richter scale, damage to houses

economy, and the great number of Montserratians living abroad tend to send their savings to their family at home. All these contribute to a comparatively modern standard of living whilst maintaining traditional structures. The *Ministry of Education, Health and Community Affairs* is, amongst other things, responsible for the administration of employment and related laws, public health and waste disposal. It is supported by various other authorities such as the *Department of Education* and the *Department of Environmental Health* (see organisation chart).

The 1991 census reported a population of 10,625. Compared with the previous 1980 census Montserrat's population declined by 0.8 % over the given period. This is the second largest decline in the past 120 years. Between 1871 and 1991 there was a constant alternation between population growth and decline. The largest growth was reported between 1921 and 1946. The authorities registered 14,333 inhabitants in 1946, this being the largest number in the island's history (Table 4.3). Many Montserratians at that time returned home from World War II. However, since then the population has on the whole been steadily decreasing.

The most significant factor determining Montserrat's population is migration. Hundreds of liberated slaves left Montserrat as early as 1838 after the abolition of slavery in order to seek work or land in neighbouring countries.

Table 4.3. Population of Montserrat, 1871 till 1991. (After Pannell Kerr Forster Associates 1994 and Physical Planning Unit 1995b)

Year	Male Population	Female Population	Total Population	Average Growth in p.a. in %
1871	3,978	4,715	8,693	–
1881	4,639	5,444	10,083	1.48
1891	5,331	6,431	11,762	1.54
1901	5,580	6,635	12,215	0.38
1911	5,245	6,635	12,196	–0.02
1921	5,094	7,026	12,120	–0.06
1946	6,362	7,971	14,333	0.67
1960	5,407	6,760	12,167	–1.17
1970	5,395	6,103	11,498	–0.57
1980	5,582	6,024	11,606	0.09
1991	5,282	5,343	10,625	–0.80

Since then Montserrat has experienced a number of migration shifts, the first lasting from 1845 till 1900. During this time a total of 6,000 people emigrated to the richer colonies Trinidad, St. Vincent and Guyana. A larger number of Montserratians was also involved in the construction of the Panama Canal around the turn of the century. Afterwards, most of them obtained US or Canadian citizenship. In the twenties, there was work in the booming sugar industry on Cuba or the Dominican Republic. There was a further exodus in the thirties when many Montserratians emigrated to Aruba and Curaçao in order to work in the oil refineries. Finally, the fifties and sixties saw one of the greatest emigration waves in the island's history. Great numbers of Montserratians emigrated to the British mother country and some to the United States or Canada. The main cause for this wave was the breakdown of the local cotton industry and the loss of jobs.

This continuous pattern of emigration over the past 150 years has had deep impact on this society. It affected not only the absolute size of population but also the age-sex structure (Table 4.4). There have been significant changes over the past 30 years in this respect. The proportion of the younger age group (under 15 years of age) has declined considerably. Whilst it accounted for 43 % of the total population in 1960, it was reduced to 26 % by 1991. The other end of the age scale shows a completely reversed picture. There were only 8,1 % elder people (over 65 years of age) recorded in 1960. This percentage has risen to 14 % in 1991. Montserrat's population has a disproportionate number of old people. Also, the sex ratio has changed in favour of the male population. The proportion of women has over the past years declined steadily, from 56 % in 1960 to 50 % in 1991.

The increase in the percentage of older people is caused by a number of factors (Figure 4.6). The mortality rate has declined and, at the same time, also the rate of births. 300 births were registered in 1975. Since then no more than an average of about 200 births p.a. is recorded. In 1991 a mere 169 births were registered. This trend is further intensified by emigration. The majority of emigrants is between 15 and 44 years of age. Many get married

Table 4.4. Determinants of the population changes, 1970–1991. (After Physical Planning Unit 1995b)

	Year	Males	Females	Total
Population	1970	5,395	6,103	11,498
	1980	5,582	5,582	11,606
	1991	5,282	5,343	10,625
Intercensal changes	1970 - 1981	197	–79	108
	1980 - 1991	300	–681	–981
Natural increase	1970 - 1980	680	549	1,229
	1980 - 1991	473	355	828
Netto migration	1970 - 1980	–493	–628	–1,121
	1980 - 1991	–773	–1,036	–1,809

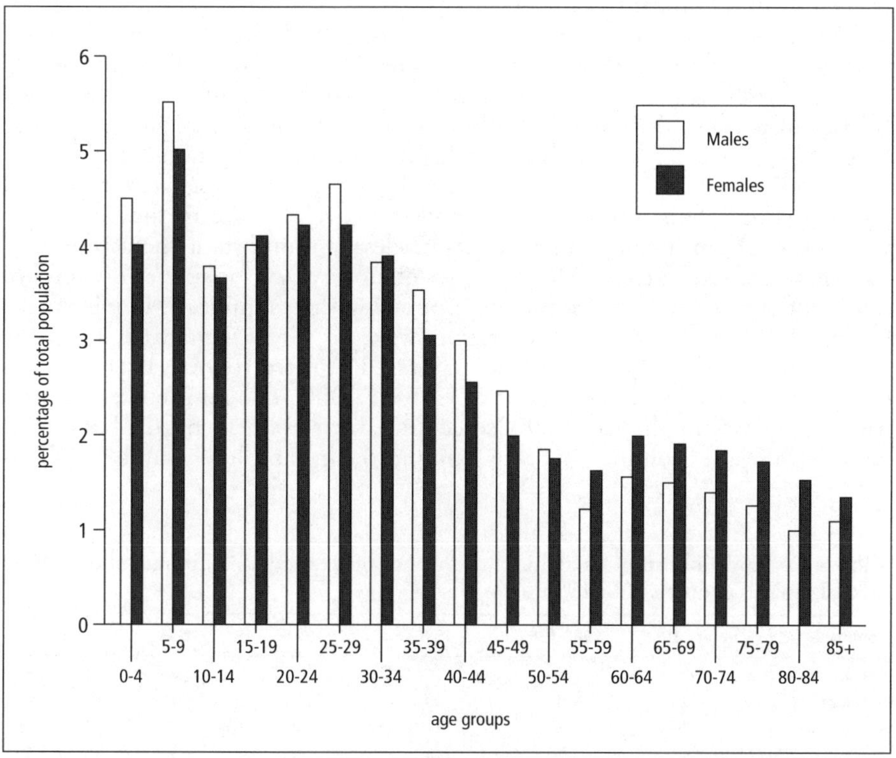

Figure 4.6. Age-sex structure, 1991. (After Pannell Kerr Forster Associates 1994)

and have families abroad. The changed sex ratio is a result of a modified pattern of emigration. During the 1960s it was mainly males who left the island. This pattern has changed and since the end of the 1980s more and more females are recorded to leave the island. Unfortunately, there are no exact details available but it is estimated that more than 12,000 Montserratians live abroad. The majority of them living in Boston, New York and London or, in the Caribbean, on Antigua, St. Kitts and St. Martin.

These trends have until the end of 1994 been alleviated by the immigration of a number of middle-aged people from other Caricom states. They often took over jobs Montserratian emigrants had left, and in this sense they had an important socio-economic function whilst at the same time stimulating the general demand. The majority of male immigrants came to Montserrat during the reconstruction phase following Hugo. They mainly worked in the construction sector. It is apparent that immigration was extremely high in 1990, i.e. one year after Hugo. 2,067 foreigners lived on Montserrat in 1991, contributing a relatively high percentage (20 %) to the total population. About 75 % of the immigrants came from neighbouring islands, alone 20 % of them from Dominica (Table 4.5).

The family structures have as a result of emigration been subject to change. A larger part of the employed male and female population lives and works abroad whilst it is the older generation that remains on Montserrat. The average size of families and households has therefore declined. In 1980 there were 3,708 households with an average of 3.1 people. The total number of households had risen to 3,878 by 1991, but only about 2.7 people were registered per household. This represents the smallest household size in the Commonwealth Caribbean. Family unity is, nevertheless, an important characteristic of the Montserratian society. This becomes obvious when looking at the level of remittances received from emigrants supporting their families at home. At the beginning of the first larger migration wave to Great Britain in the early 1950s around US$ 72,000 p.a. were registered. A mere 10 years later this sum had risen to the eight-fold amount of US$ 617,000. The remittances received from emigrants in the early 1990s accounted for about a third of the total national import volume (EC$ 81 million), in absolute numbers about

Table 4.5. Foreign born population by country of birth and year of immigration, 1991. (After Physical Planning Unit 1995b)

Country	before 1980	1980-1989	1990	1991	NS	Total
Antigua	90	48	19	6	8	171 (8.3%)
Dominica	85	181	105	27	11	409 (19.8%)
St. Kitts	111	39	10	5	10	175 (8.5%)
Guyana	4	78	210	64	1	357 (17.3%)
Other Carib.	132	147	97	29	16	421 (20.4%)
U.K.	44	64	24	22	1	155 (7.5%)
USA	36	64	51	21	–	172 (8.3%)
Rest	56	59	57	28	7	201 (9.7%)
Total	**558**	**680**	**573**	**202**	**54**	**2067**

EC$ 26 million. These financial contributions are important, particularly for many poorer households. They are a type of socio-economic security, avoiding serious social conflicts and tensions. It has been analysed that the close family ties are not the only reason for the financial support Montserratian emigrants grant to their families. Living abroad is for many emigrants only a temporary phase to earn money and support their family. The majority of Montserratians living and working abroad plans to return to the island, at the latest when they go into retirement (Pannell Kerr Forster Associates 1994).

There are different interest groups in Montserrat's social system, but these are on the whole rather small. Local groups are the trade unions, church circles, community groups, businessmen and the *National Trust.* The most important denominations are the Anglicans and the Roman-Catholics and the Methodists. Some people also belong to the Seventh Day Adventist Church or other smaller groups like Jehovah's Witnesses. A further interest group are the foreigners living on Montserrat, i.e. the expats. There are expats from various other Caribbean states who work on Montserrat, further citizens from North America or Europe who have their own business on Montserrat, and pensioners from the States and Canada who enjoy their old age on the island. There is no evidence of social tensions or feelings of resentment against the expats. To the contrary, most of the local population believe they are a benefit to the island.

One of the most significant constraints for the social system is the declining population. There are a number of implications: (1) the brain drain causes a continuous decrease of qualified employees, (2) the market shrinks and the rentability of many social and public institutions and facilities is jeopardised, (3) the high proportion of older people pressurises a number of social benefits. A further factor is the imbalanced spatial distribution of the population. In 1991 as much as 65 % of the population lived in the southwest of the island, thus pressurising the resources available in that region. Internal migration also results in the predominance of older people in rural areas whilst the younger population competes for jobs in the urban centres. Many school leavers cannot find adequate employment and, as a consequence, often leave the island, either to find jobs or to visit colleges of further education or universities.

The Political System

Montserrat is a British dependent territory. The constitution of 1898 states that the British monarch is the Head of State, locally being represented by a Governor. The Governor is responsible for external affairs, defence, judiciary, inner security, administration of the public sector and since the early 1990s for the control of the offshore financial sector. In early 1998 the British government decided to abolish the term *dependent territories.* It was deemed outmoded as it refers to dependence instead of partnership. The Foreign Minister Robin Cook stated in his speech to the Dependent Territories Association in London in February 1998 that the dependent territories shall after the enactment of the respective law in future be called *United Kingdom Overseas Territories.*

The democratically elected local government functions on the basis of the Westminster Parliamentary model. Political institutions are the *Legislative Council*, being the highest legislative body of the country, and the *Executive Council*, responsible for the control of the inner affairs. The Legislative Council is elected for five-year periods of office. The party gaining the most seats in the seven wards takes office. Head of the Government is the Chief Minister. There are three further ministers in charge of the different ministries (Figure 4.7). The Legislative Council consists of the representatives elected in the wards, the Financial Secretary, the Attorney General and two further nominated members. The members of the Legislative Council elect their speaker who functions as chairman. All formalities and procedures in the Legislative Council base on those in the British Parliament. There is public access to all sittings. The Executive Council is the cabinet, consisting of four elected Ministers of Government, the Financial Secretary and the Attorney General under the chairmanship of the Governor. The cabinet determines the political strategies and handles the daily agenda. Its members meet weekly whereas the Legislative Council is only obliged to one sitting every other month. The law on Montserrat is closely oriented to the British Common Law.

Montserrat is a member of the *Caribbean Economic Community* (Caricom) which was established in 1973. The community supports the economic cooperation and integration of its 12 members, *More Developed Countries* (MDCs) and *Less Developed Countries* (LDCs) in the Caribbean. The member nations are Antigua and Barbuda, Barbados, Belize, Dominica, Jamaica, Grenada, St Kitts-Nevis, St Lucia, St Vincent and the Grenadines, Guyana and Trinidad and Tobago. Montserrat is also part of the sub-regional group of LDCs, the *Organization of Eastern Caribbean States* (OECS) which was established in 1981. Its member states Antigua and Barbuda, St Kitts-Nevis, St. Lucia, St Vincent and the Grenadines, Dominica, Grenada and Montserrat cooperate in political, ecological and socio-economic sectors. The *Eastern Caribbean Central Bank* is an important institution of this organisation. It regulates and controls the common currency of the member states, the *Eastern Caribbean Dollar* (EC$). The exchange rate has since 1977 been brought into line with the US-Dollar.

Montserrat's political landscape is dominated by four political parties. By European standards these parties would be of hardly any significance since individual politicians dominate their field. The *National Development Party* (NDP) was established in 1984 and represents the conservative spectrum. The liberal *National Progressive Party* (NPP) is relatively young, having been established in 1991. The early 1970s saw the establishment of the two parties representing the centre politically, the *People's Liberation Movement* (PLM) and the *Progessive Democratic Party* (PDP). For one decade, between 1960 and 1970, politics on Montserrat were determined by the *Montserrat Labour Party* under W H Bramble. Bramble did not see any future in agricultural production on Montserrat and instead concentrated on tourism. Numerous plots which had previously been arable land were sold to property speculators. The construction sector and the real estate business began to flourish on the island. Villas were built and sold (mainly) to North American pensioners. So-called *residential tourism* developed and led to considerable economic growth.

However, it also caused rising costs of living and foreign enclaves emerged. Mainly the youth thus started criticising Bramble's economic concept.

In 1970 Bramble's son Austin came into office with the *Progressive Democratic Party* (PDP). He stopped the sale of land and concentrated economic policies on the promotion of trade and industry. Numerous projects were initiated, for example the agricultural production of chili, a tannery and a weaving mill. High quality national sea island cotton was processed in the mill. Public projects were accomplished by numerous private activities, as for example *Montserrat Electronics, Antilles Plastics, Allen's Lime Juice*, the *American University of the Caribbean* (see economy) and the *International Air Studios.* Austin Bramble also placed emphasis on Montserrat's agricultural production. But by the mid 1970s prices for consumer goods exploded and Montserrat incurred a considerable balance of trade deficit and the highest rate of unemployment in its history (20 % in 1978). The oil crisis and the international recession that followed were the main causes for the economic crisis. Nevertheless, the citizens criticised the political and economic strategies of their Chief Minister (Irish 1991). They had got used to a higher standard of living and were not prepared to make sacrifices.

In November 1978 the *People's Liberation Movement* (PLM) with John Osborne secured the majority of votes and came to power. Osborne's success was mainly a result of the support his party had received from the *Montserrat Allied Workers Union* (MAWU). Osborne placed his trust in free-market economy, a strategy which was absolutely contrary to that of his social-democratic predecessor. He abolished the import barriers Bramble had put up. 54 smaller companies established themselves, 200 jobs were created and the salaries of civil servants and white-collar workers rose rapidly. Osborne had quite pragmatic objectives: each Montserratian should be given the chance to build his own home, buy a car and have some savings. Opening up the market seemed to be the right way. Taxes were abolished for numerous products, especially for construction material and cars. Osborne and his government also did away with income tax and offered tax reductions to various companies. Only six years later 319 new houses had been built, 630 sites and 1,550 cars had been sold. Montserrat's economy florished, especially the sectors construction, light industry and offshore finance; environmental issues, however, were not really on the agenda.

Osborne was re-elected in 1983, but soon faced some problems. Anguilla's offshore conditions were far more attractive, and many bankers and companies withdrew their investments from Montserrat. The boom in the industrial sector terminated and the Ministers of Government could not find the necessary consensus with the Secretaries of State. External problems with the integration partners in the Caribbean and the British mother country did one more thing. The situation escalated when the newspapers reported about mismanagement and a series of corruption and drug scandals. Public opinion called for political change, even though the construction sector had recovered in the reconstruction phase following Hurricane Hugo.

The newly established *National Progressive Party* (NPP) won the election in the autumn of 1991. This team of young Montserratians managed to regain public confidence which had suffered from the various scandals during

Osborne's government. Reuben Meade, to date the youngest Chief Minister in the history of Montserrat, emphasised a closer cooperation with the integration partners in the Caribbean and with the British mother country. All political, economic and social strategies were therefore developed in close exchange with the Governor. The new government based its negotiations with the British mother country on its document *Philosophy of Development* which finally resulted in the *Country Policy Plan* (Chapter 4.1.2).

Montserrat's political administration is illustrated in Figure 4.7. The four ministries head 23 authorities with different responsibilities. The four ministers are supported by a Secretary of State each. The number of public servants has increased considerably during the past 20 years. 470 public servants were recorded in 1970 whilst this figure rose to 772 in 1990. In 1993 the government employed 786 people (cf. Table 4.6) The salaries paid to the public servants account for 60 % of the entire public spending. Statistically each public servant is responsible for a mere 13 citizens.

The close proximity of the politicians to the population is characteristic of micro states like Montserrat. Everybody knows each other and the political formation of opinion is therefore closely linked to family ties and friendships. The birth place of a politician can also be significant. A candidate from a certain village can usually be sure to receive the majority of his neighbours' votes since the villagers trust in his representing their interests. Political opinion is on Montserrat therefore far from bound to a certain party, but rather linked to individuals. Politicians vice versa often decide by the case rather than by their party's programme. In the extreme it might even happen that a minister changes to the other camp when the opposition wins an election.

The extreme costs caused by the political administration put considerable strain on the political system. The pool of potential decision-makers is small. Only few people are sufficiently qualified for their positions. Many capable heads furthermore leave the island. Individuals can with the support of their family, friends or even neighbours quite easily get into well-paid posts, even though they might not have the necessary political experience. Many of the voters are easily persuaded by the art of rhetoric. Strategic planning and consistent policies are often impeded since the voting pattern can change from election to election, possibly causing a complete turn-around in the development strategies. A further potential constraint is the necessary coordination of political strategies and measures with the colonial administration, and the financial dependence on Great Britain.

Figure 4.7. Government and administration

Table 4.6. Number of public and civil servants employed, 1993. (After Pannell Kerr Forster Associates 1994)

Ministry or department	Number of public/ civil servants
Police and fire brigade	101
Tax office	31
Ministry of Health	129
Ministry of Education	226
Ministry of Agriculture	38
Ministry of Communication Works	52
Others	209
Gesamt	**786**

Table 4.7. Public revenue and spending with balance 1989–1994 (in million EC$, the data for 1994 hab been estimated). (After Pannel Kerr Forster Associates 1994b)

Year	Revenue	Spending	Balance
1989	36.4	32.8	3.6
1990	42.7	36.2	6.5
1991	44.4	39.2	5.2
1992	39.5	40.0	–0.5
1993	38.7	40.4	–1.7
1994	42.8	42.6	0.2

The Economic System

Like any other small island, Montserrat has limited resources and a small home market. Nevertheless, its economy is diversified and there is no predominating sector. The gross domestic product (GDP) increased at an average rate of 3.2 % p.a. between 1980 and 1993 (Table 4.8). Following economic tendencies have been analysed for this period despite the impact of Hurricane Hugo: (1) the market for real estate and housing shrunk, (2) the offshore financial sector grew, (3) agriculture and the hotel and restaurant trade showed a slight growth, even though their contribution to the GDP was on the whole comparatively low, (4) public services increased, (5) the processing industries declined and (6) the construction sector also showed a tendency to decline, but was from time to time experiencing recovering phases due to public investment projects. This was particularly valid for the reconstruction phase following Hurricane Hugo (Physical Planning Unit 1995b). The economy generally recovered quite soon after Hugo, this amongst other thanks to the financial aid granted by Great Britain (£ 20 million). The average yearly growth in GDP recorded in 1992 was 4.8 % and in 1993 4.0 %. The trade deficit remained despite a considerable increase in exports from EC$ 2.78 million (1991) to EC$ 4.26 million (1993) (Montserrat Statistical Office 1994).

Table 4.8. Gross domestic product by sector, 1980 and 1993. (After Pannell Kerr Forster Associates 1994b and Montserrat Statistical Office 1994)

	1980		1993	
Sector	GDP in current prices (in million EC$)	GDP-percentage at factor cost (in %)	GDP in current prices (in million EC$)	GDP-percentage at factor cost (in %)
Agriculture, forestry and fishery	2,40	4,1	7,55	4,8
Processing industries	3,48	6,0	2,39	2,2
Construction	5,23	9,0	19,99	12,7
Wholesale and retail trade	8,92	15,3	17,95	11,4
Banking and insurance sector	2,67	4,6	14,68	9,3
Hotels and restaurants	1,75	3,0	7,14	4,5
Real estate and housing	14,28	24,4	25,35	16,1
Public services	8,7	14,9	26,57	16,9

Montserrat's labour force is in comparison with other Caribbean states well educated. About 36 % have received higher education (secondary school level). Of the total labour force available in 1991 (4.866) almost 6 % (263) were unemployed. The unemployment rate for the 15 to 19 year age group was as high as 14 %. The majority of employees was working in the construction sector (29 %) and the services sector (33 %). The primary sector declined by a further 1.5 % between 1988 and 1991 (7.5 %). The processing industries accounted for a mere 4 % in 1991, having declined from 11 % in 1988. The percentage of employees in the construction sector has in the same period risen from 19 % to 29 %, mainly as a result of the reconstruction activities following Hurricane Hugo. Employment in the other sectors has remained relatively constant and accounts for about a third.

Agriculture had traditionally been Montserrat's predominant economic sector. The production of sugar, tobacco and indigo characterised the colonial period. Later on, sugar cane was replaced by limes and cotton. Cotton was an important product until the 1960s. However, agriculture on the whole and especially cash crops for export have since the 1950s been continuously declining. This is, on the one hand, due to a lack of investments and, on the other hand, to the fact that the younger generation is no longer prepared to work in the agricultural sector. Many of the former plantations nowadays lie fallow. Agriculture has stagnated during the past decade, despite attempts of the government to revive this sector. One of the main reasons being the vast devastations by Hurricane Hugo. There are estimates that there were a maximum of 200 farmers living on the island in 1990. Their average age at the time was 50 and most of them were tenants on state-owned land. They cultivated mainly potatoes, tomatoes, onions, chillies, mangoes and limes. This ensured that the local market was more or less satisfied. Fisheries has an even lesser significance than agriculture. Only 60 Montserratians are full-time fishermen. Main fishing grounds are Carr's Bay and Plymouth. The marine resources bear further potential, particularly with respect to deep-sea fishery.

Looking at the secondary sector, we find that construction is contributing significantly to the national economy. This trend began as early as the 1960s when the government started promoting tourism and encouraged the construction of villas, apartments and hotels. In the early 1990s the construction sector went through a number of promising phases. Schools, a new harbour and a new government building had to be put up after Hurricane Hugo. The national hospital, the Glendon Hospital, was completely modernised and extended. There were numerous other smaller projects going on at the time. Manufacturing has, in comparison with construction, a rather marginal significance. An industrial park was established in Plymouth in 1977 and employed around 900 people by 1988. This has not recovered from the vast destructions it suffered from Hurricane Hugo. Only five of the companies have resumed production, but on a limited scale. The furniture industry and electronic spare parts predominate this sector. Around the mid 1990s a rice mill was built for the milling of rice from Guyana. Processing the grains in the British dependency offers the advantage of preferential import taxes for members of the European Union (EU).

Tourism has in the regional comparison experienced only a moderate development. Whilst it increased by 6 % to 10 % on the other Caribbean islands between 1980 and 1991, Montserrat recorded a mere 2 % increase. Only about 1 % of the total number of tourists in the Eastern Caribbean stay on Montserrat (1992: 17,277 overnight guests). Most of the guests are from the Caribbean, the States or Europe (cf. Figure 4.8). The number of cruise tourists has risen considerably since the new jetty was built. In 1992 a total of 5,723 cruise tourists visited Montserrat. The accommodation offered on Montserrat is rather unusual by Eastern Caribbean standards. Most of the guests are allocated in villas and apartments or private guest houses. There are only three smaller hotels with 116 rooms (230 beds), a number of guest houses with 88 rooms in total and a further 90 villas and 70 apartments. 315 villas are private property, but are, at times, also let. The occupancy rate varies between 70 % between December and April and 30 % in the summer months. The average period of stay recorded for the hotels is a week, whilst tourists in the other types of accommodation usually stay for three to four weeks (Physical Planning Unit 1995b).

We had already referred to residential tourism as a characteristic of Montserrat. The demand was for a long time mainly stimulated by North American and British pensioners spending the winter on Montserrat in rented villas or their own property. Because of their seasonal appearance they are referred to as *snowbirds*. Around 300 pensioners a year passed the winter on Montserrat. Their spending contributed to 22 % of the national income generated by tourism. This residential tourism has hardly any negative ecological

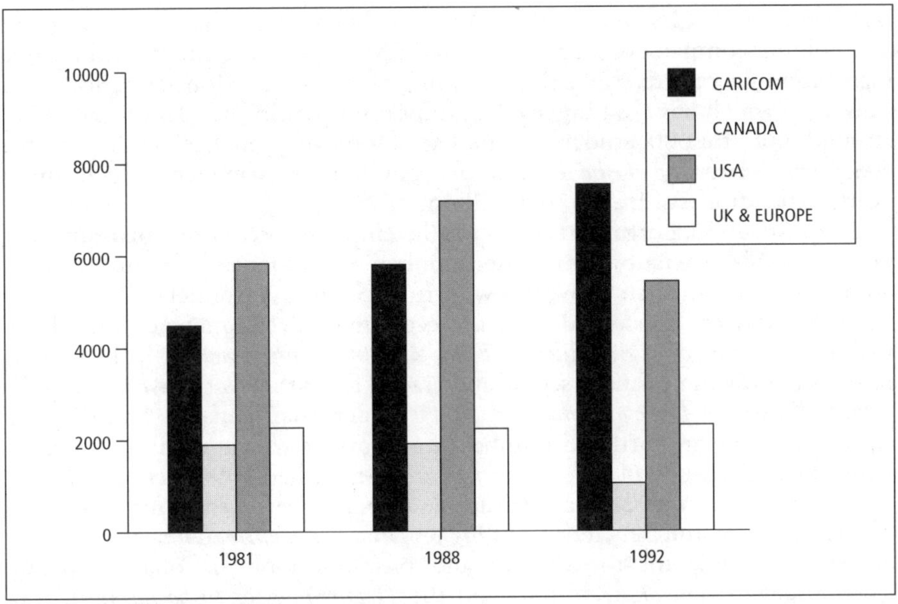

Figure 4.8. Countries of origin, tourists on Montserrat in 1981, 1988 and 1992. (After Pannell Kerr Forster Associates 1994)

or social effects and it is relatively independent of global structures. However, as most other sectors, residential tourism also suffered from the effects of Hurricane Hugo. The disaster had a considerable impact. The majority of the villas was destroyed and a great number of pensioners left the island. A slow recovery of residential tourism has been noted since the mid 1990s, but it seems that the overall structure has changed since many villas are nowadays rented at comparatively short notice (McElroy and Albuquerque 1992). Montserrat is in contrast to other Caribbean islands not a typical destination for beach holidays. It is rather the originality of the island, its nature and cultural life and the hospitality which appeal to Montserrat's guests. The island also offers relatively cheap accommodation and some peace and quiet away from the typical hectic related with mass tourism. It is not just by chance that Montserrat advertises with the slogan: "The way the Caribbean used to be".

Around the mid 1980s Montserrat became involved in the offshore financial sector. It charged low fees and had very flexible licensing formalities. Within a short time over 300 banks had been registered on Montserrat. But in 1989 the lack of control and loopholes in the legislative framework led to a scandal. Cases of corruption and money laundry were uncovered in some banks. As a consequence, almost all licences were withdrawn and the control of the offshore finance sector given to the British Governor. In the early 1990s the *Offshore Finance Centre Unit* was established. It grants licences and promotes and controls the sector. These restrictive measures have been effective. Business is recovering, mainly due to the registration of so-called *Inter-*

national Business Companies (IBC). Any type of company, be it an investment and holding company or a bank, can take advantage of the simple and flexible registration procedure. Since the beginning of the 1980s Montserrat also has a university which has always played an important part in the island's economy. Around 400 to 600 students complete a one-year medical course at the *American University of the Caribbean* (AUC). It is estimated to generate a yearly expenditure of EC$ 3 to 4 million.

The institutional organisations generally emphasise economic planning and policies. This refers both to the number of responsible ministries and authorities, and to their provision with finances and personnel. Alone two of the ministries are concerned with the economic development of the island. One is the *Ministry of Agriculture, Trade and the Environment* (MATE), being responsible for the primary sector and trade. The other is the *Ministry of Finance, Economic Development and Sports* under chairmanship of the Chief Minister. There are further a number of authorities which are also concerned with economic issues: (1) the *Tourist Board*, established 1994, being a mixture of public and private sector institution, responsible for the promotion and development of tourism, (2) the *Offshore Finance Centre Unit*, as explained above responsible for the control and promotion of the offshore sector, (3) the *Department of Agriculture* and the *Department of Fisheries*, both supporting the farmers and fishermen, (4) the *Land Development Authority* (LDA), leases arable land, and the *Development Unit*, has an advisory capacity and also implements laws, develops economic programmes and coordinates negotiations with potential investors and international organisations.

If one analyses the economic development of Montserrat, it soon becomes clear that the island's infrastructural facilities pose a significant constraint. Plymouth's new harbour is not sufficiently suitable for cargo vessels even though about half of the imported goods reach Montserrat by sea. The airport is also rather small. It provides facilities for planes not larger than a DHC-8 with 37 seats. International flights which reach Montserrat via Antigua are often delayed considerably, affecting the transport of passengers and luggage just as well as cargo. The small size of the home market and the limited working population also place serious constraints on the economic development of the island. The situation is worsened by the continuous emigration and the related brain drain effect. The most significant constraint to the system is its economic dependence on the exterior. Montserrat, for example, heavily depends on the remittances of its emigrants, financial aid from Britain and international imports. It is also sensitive to any changes in the global structures which, more than anything, affect tourism and offshore finances. Also, Montserrat's economic activities are in their majority centred around Plymouth, resulting in a strong spatial imbalance. In the early 1990s alone 40 % of the island's places of work were located in the southwest.

Summary (Provisional Appraisal)

Some of the major characteristics of the complex system Montserrat can at this stage already be pinpointed. The evolution of the system was and is decisively formed by specific agents which are characteristic of a micro state. These include the vulnerability of the ecological system, the limited number of natural, material and human resources, the small capacity of the home market, the overproportionate significance of foreign trade and the continuing emigration with its brain drain effect (Sandner 1982). The potentials of micro states, such as their increased flexibility and comparatively small and simple administrative structures, are hardly applicable to Montserrat. Its local administration is, as we have seen, extremely bureaucratised and inefficient. Many of the decisions, furthermore, have to be coordinated with the British mother country or the integration partners of the OECS and the CARICOM. Britain, on the other hand, offers backing, particularly regarding technical and financial support. And the memberships in the OECS and the CARICOM facilitate a constructive cooperation, especially regarding environmental and resource management. This provisional appraisal shall in the following be supplemented by the analysis of the meso and micro level.

4.1.2 The Meso Mevel – Perceptions of the Gatekeepers, Plans and Strategies

As outlined in Chapter 2.3, the exchange between the macro level (basic structures) and the micro level (individual) is influenced by so-called gatekeepers. Palm (1990) defines them as individuals with important resources and powers, as such being able to influence the action of any individual. They act at mid or at the intermediate level. Questionnaires were developed which were designed to transmit the perceptions of the different gatekeepers in the four sub-systems just outlined (environmental, social, political, economic). The first 60 non-standardised interviews were held with politicians, civil servants, planners, scientists, representatives of the private sector (e.g. hotel and restaurant owners, solicitors, businessmen) in the summer of 1994. However, the meso level is not solely dominated by the perceptions of individual gatekeepers, but also by planning concepts, development programmes and strategies. The analysis of the meso level therefore demands, on the one hand, that the perceptions of the individual gatekeepers are ascertained and, on the other hand, that current plans, programmes and strategies are analysed.

The Perception of the Gatekeepers

The interviews with the gatekeepers focused on the potentials and pitfalls of Montserrat's development. Surprisingly enough, there was a large degree of agreement, despite the different functions and interests of the individual gatekeepers. The interviewees at all decision levels determined a sustainable ecological and economic development of the island as the central objective. Special emphasis was placed on the creation of qualified job vacancies and a rising living standard. Most answers concerning the economic development potentials referred to tourism, agriculture and manufacturing (especially light industry). Apart from promoting economic development, the protection of the natural and cultural resources were of specific concern to most gatekeepers. The decrease in population was recognised as a serious constraint. The emigration of the younger and qualified Montserratians was often specifically mentioned. Internal political coordination was in many cases criticised. A number of interviewees felt that this situation was aggravated by the fact that political issues have to be coordinated with the British mother country. Others again assessed the cooperation with Britain as very positive. There was a general tendency towards economic independence which should later on be followed by political autonomy. These general perceptions about Montserrat's future development were supplemented by the analysis of numerous aspects concerning the sub-systems. These aspects shall in the following be briefly outlined.

Asked about the environmental system, most of the decision-makers commented to the effect that they have high hopes of the new *Forestry, Wildlife, National Parks and Protected Areas Act* (No. 3, 1996). They called for a quick establishment of a National Parks Administration providing the means to effectively implement the law. Such an administration should be designed to coordinate all measures concerning the protection and management of the resources. Asked about forestry, it became apparent that hardly any of the interviewees saw commercial interests. The indirect economic benefits obtained by protecting the watersheds, avoiding soil erosion and enhancing the attractiveness of the forests for recreation and tourism purposes were evaluated far higher. A pilot project for farmers was planned in order to relieve soil erosion. The establishment of a marine park in the north of the island was thought of in order to prevent conflicts between consumptive and sports fishery. Special attention was given to the necessary enhancement of sewage and waste disposal. A number of experts were working on relevant studies at the time of the field work.

Asked about the social system, many interviewees emphasised the importance of a considerable population growth. They expected a better rentability of social services and a bigger market for the local economy. A number of gatekeepers suggested to realise an immigration policy in order to attract qualified immigrants or those likely to invest (e.g. Chinese from Hong Kong). A higher demand for employees would, according to the gatekeepers, also reduce the brain drain effect. It was further suggested to employ head hunters who should assist in asking highly qualified emigrants to come back. Many decision-makers also expressed that their attention focused on the modernisation of the public health system and an education system which serves the needs of the labour market. The ideas put forward included new schools, a community college, a teachers' centre and the modernisation of Glendon Hospital. It was also suggested to generally enhance the curricula, offer further vocational training for teachers and realise a closer cooperation between public education and the private sector.

When asked about the political system, the interest focused on the infrastructural development. Businessmen and government representatives alike referred to the need to improve and develop the road network, to build a new administrative and parliamentary building, to modernise the harbour and to extend the airport. These measures were to a large extent financed by the British government. Improving the infrastructure was evaluated as one condition for the stimulation of investments. Extending the airport was, for example, supposed to promote tourism and facilitate exports. The British authorities had at the time of the field work also put a reform of the inefficient administrative body forward. The gatekeepers interviewed expressed that they paid special attention to this issue. It was the intention to reduce the number of public servants to 550 in the scope of the so-called *Civil Sector Reform.* Even though the necessity of such a reform was commonly recognised, many interviewees naturally worried about their post and expected a rising unemployment rate.

The strategic objective for Montserrat's economic system was economic autonomy for the island, this being declared by both the local and the British side. Many emphasised that they considered it necessary to enforce the functional relationships between the economic sectors. The economic-political measures suggested included the diversification of the tourism sector, developing a new industrial park, producing high-quality agricultural products and generally supporting enterprise on the island. Many investment projects had been granted since the early 1990s, but were not finalised at the time of the field work in 1994. These projects included the construction of 71 villas at Bransby Point, 40 flats in Providence Estate, 29 houses and 10 apartments in Isles Bay Plantation. The Government had for a number of years also been planning a large tourism project in Little Bay. It was supposed to comprise a yacht harbour, a hotel, a golf course and 800 villas and flats. So far all that has been realised is a declaration of intent, signed by the local government.

Plans and Strategies

A number of documents have been issued since the early 1990s, comprising development plans and strategies for Montserrat. The earlier mentioned *Philosophy for Development* which had been issued by the then governing party NPP continued to serve as the basic strategy. It referred to agriculture, tourism, manufacturing and construction as the central sectors of growth. The strategy also focused on the extension of the offshore finance sector. Intention of the government programme was not only the economic and infrastructural development, but also the protection of Montserrat's natural and cultural resources. The *National Environmental Action Plan* (NEAP) was to this end worked out in collaboration with the *World Bank* and the *United Nations Development Programme* (UNDP). The plan is designed to analyse the natural environment and to propose management measures, particularly for the forests, the water sheds, waste disposal, the coastal resources and the national parks and historical sites.

In the early 1990s a British consulting agency from Birmingham worked out the *Public Sector Investment Programme* (PSIP). It presented the proposed development projects for the period 1994 to 1996. A total of 120 projects had been planned, most of them depending on external financial assistance. The overall costs were calculated at US$ 75 million. Infrastructure as the biggest post accounted for about US$ 44 million (58 %). Alone US$ 30 million were supposed to be spent on the extension of the airport. However, the PSIP was not completely accepted by the local Government. A further strategic analysis was undertaken by the British *Overseas Development Administration* (ODA). In its *Strategic Review for the Government of Montserrat* the individual economic sectors were evaluated in a differentiated way. This concept generally reached a greater acceptance, but many government representatives still criticised that they had hardly any opportunity to put their own ideas and visions forward.

On the basis of the ODA report Montserrat's local government finally outlined the *Country Policy Plan* (CPP). It contained all development plans and projects to be realised until the end of 1997. The plan was settled amongst the British and the local government, especially with reference to Britain's financial support (Table 4.9). The CPP was a completely new instrument, initiated by the so-called Good Government Debate which early in the 1990s characterised the relationship between the governments of the dependencies and their mother country. Objective of the Good Government Debate was a closer cooperation and a generally enhanced mutual confidence. The development objectives for Montserrat were in the sense of this attitude worked out by mutual agreement (Government of Montserrat and Government of the United Kingdom 1994).

Table 4.9. Proposed expenditure by sectors, 1994–1996 (in million EC$). (After Government of Montserrat and Government of the United Kingdom 1994)

Sector	1994	1995	1996	1994-1996	%
Project preparations (resource and environment management; airport)	5.2	0.06	0.2	5.46	2.61
Productive sectors (agriculture, forestry, fishery, industry)	0.8	2.3	2.4	5.5	5.64
Economic infrastructure	10.6	46.5	63.8	120.9	57.86
Social infrastructure	18.9	18.2	12.2	49.3	23.6
Apparatus of state	10.9	12.9	4.0	27.8	3.3
Total	**46.4**	**75.96**	**82.6**	**208.9**	**100.0**

The CPP and all other plans, programmes and strategies were all integrated in the previously mentioned *National Physical Development Plan*, which became the comprehensive and integrative planning instrument for Montserrat. The plan was designed on a medium-term basis and as such supposed to regulate all development activities between 1995 and 2010. Central objectives of the plan were (1) the control of new built-up areas with a special emphasis on the enhancement of the functional relationships between rural and urban centres, (2) the integrative planning of all current and future land use. As will be illustrated in the following summary of the key objectives, the plan represented a continuation of other concepts.

The development objectives of the National Physical Development Plan:

1. Guaranteeing a sustainable population growth;
2. Developing a land utilisation strategy which guarantees a balanced development in all districts and, at the same time, the sustainable utilisation of all resources;
3. Promoting a stable, autonomous and spatially integrated national economy which does justice to social requirements;
4. Providing extensive and best possible infrastructural supply to all households;
5. Maintaining the ecological intactness and the unique landscape of the island (Physical Planning Unit 1995c).

Montserrat's *Physical Planning Unit* (PPU) developed three spatial scenarios which were evaluated and formed the basis of the general development strategy as outlined in the National Physical Development Plan. The first scenario accounted for a concentrated development around Plymouth. The second scenario depicted the spatial expansion of the development on the entire island's surface. The third scenario envisaged the development to concentrate on certain selected centres. The decision-makers finally opted for this third scenario. As suggested in Figure 4.9, a major strategic element was the development of three different categories of settlement with a declining degree of centrality. Plymouth's function as major centre was to be kept. The centrality of Harris, St. John's and Salem was supposed to be enhanced so that they could function as district centres. The other villages were meant to remain as sub-centres, with Cork Hill, St. Peter's and Lee's having a higher degree of centrality. This hierarchy in estates was complemented by the determination of agricultural land, national parks and protected areas. Figure 4.10 shows the respective land uses. The following development projects had also been planned:

1. Extension and modernisation of the fishery centres in Plymouth, Isle Bay and Carr's Bay;
2. Rationalisation of the industrial park in Plymouth and development of further industrial areas in Salem, Gingoes and at the airport;
3. Development of Little Bays as centre of tourism; further the promotion of eco-tourism in Tar River, Paradise Estate and the Silver Hills.

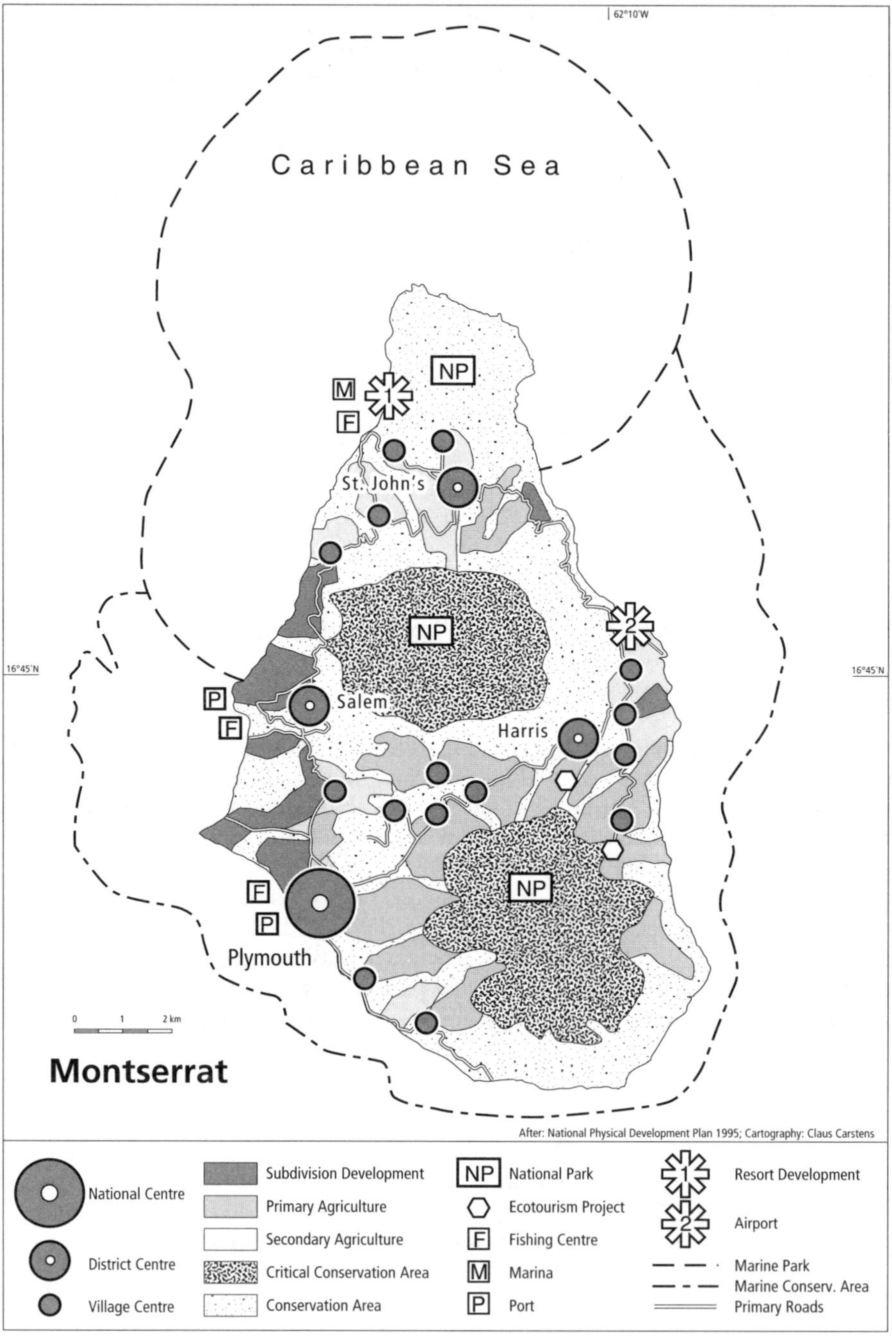

Figure 4.9. Strategic projects according to the Physical Development Plan of 1995

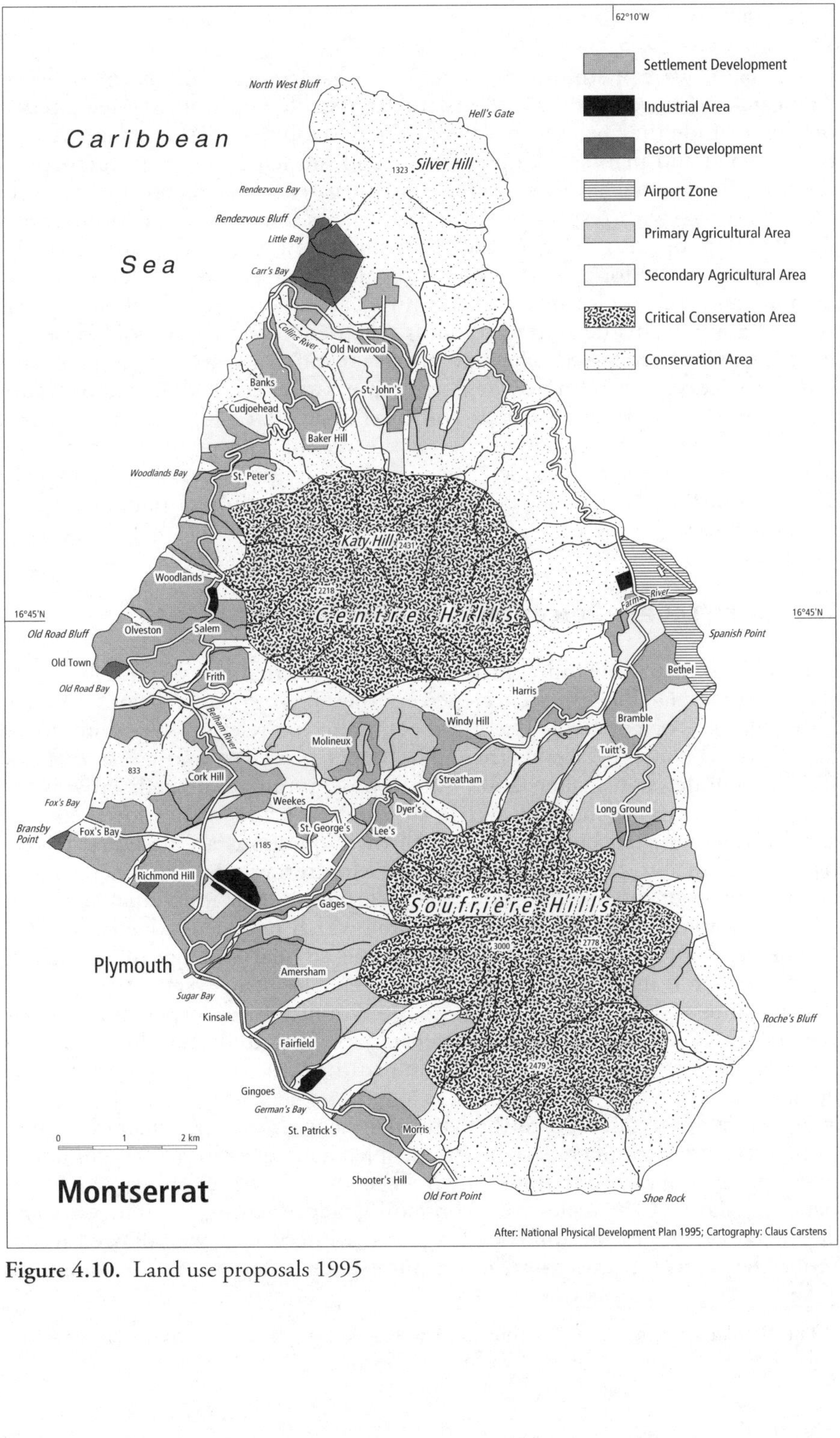

Figure 4.10. Land use proposals 1995

Summary

Once again, we can summarise significant characteristics of the meso level. The gatekeepers generally see the island as a stable and linear system. Based on this attitude they believe that the future development can to a large degree be predicted and planned. Unpredictable and sudden events and fluctuations are as a consequence hardly considered by the gatekeepers. There are relatively few differences with respect to the individual assessments about Montserrat's development options as there are surprisingly few conflicting interests. All gatekeepers strive for a sustainable ecological and economic development. Stability and security for present and future generations are a focal point of all considerations. This becomes very clear when analysing the various development plans and strategies. They all aim at creating or reaching an imaginary state of balance. For example, the latest conservation law is designed to secure a balance of the natural resources, and the Physical Development Plan strives for a balanced spatial development on the island. This pattern can also be observed in other specific cases, such as the immigration policy, being designed to further the balance between demand and supply of public and economic services.

4.1.3 The Micro Level – Perceptions of the Montserratians

Palm (1990) places the individual in the centre of attention when referring to the micro level. The experience, perceptions, interests and visions of each individual are of concern. There are a number of relevant questions to be asked in this context: How is the environment perceived and assessed? How are the local decision-makers and political programmes evaluated? How is the economic situation perceived? What are the hopes and wishes for their personal future and that of Montserrat? All these questions are designed to recognise and analyse the perceptions of the Montserratians. These perceptions are at the end of the day the factor that determines their individual activities. The linear approach of quantitative research which tests deductive hypotheses would thus not be suitable. This is why the circular strategy of qualitative social research is in the following applied. It enables us to systematically reveal new relations and structures[3]. Such an 'expedition of discovery' requires a certain receptiveness towards the research object. The understanding which is required implies a continuous adaptation to the structures which are discovered.

As a first approximation to the Montserratians' *lebenswelt* qualitative questionnaires were conducted with 70 people. The questionnaires contained open, pre-formulated questions which all focus on the four sub-systems (Appendix). Many of the issues were consciously addressed in different ways and at different stages of the interview so that the interviewee was allowed to describe the matter in question from a number of perspectives. The questions

[3] The circular strategy applied within the scope of this work is to a large extent based on the methodical considerations of Kleining (1994).

were designed as simply as possible in order to avoid predetermining the answer in any way. All of the interviewees were without exception given the opportunity to mention issues they themselves considered important. The questions were on the whole formulated variably, using descriptive expressions. The people interviewed could thus chose which part of each question they wanted to answer (Kleining 1994). The questionnaire comprised 50 qualitative and some quantitative questions. The data obtained from the answers to the quantitative questions served for the characterisation of the groups that were interviewed. We made sure that a wide and varied range of people were asked, their occupation, family size, sex and age being the decisive criteria (Table 4.10). Each interview took about 45 to 60 minutes. The answers were written down by altogether four interviewers[4].

Table 4.10. Statistic of the sample in summer 1994 (N = 70)

Sex	female	33
	male	37
Age (in years)	10 - 19	5
	20 - 29	15
	30 - 39	24
	40 - 49	7
	50 - 59	8
	60 - 69	4
	70 - 79	2
	80 - 89	2
	no statement given	3
Size of family (members)	1 - 2	18
	3 - 4	25
	5 - 6	15
	7 - 8	4
	9 - 11	3
	12 - 17	3
	no statement given	2
Occupation by sector	Tourism	20
	Public sector	9
	Trade	8
	Social services/public health	7
	Fishery/agriculture	6
	Transport sector	2
	Construction	4
	Financial sector	5
	Pensioner	6
	Pupil	3

[4] The interviews were conducted within the scope of the DFG-sponsored research project "Complex Research Management on Small Caribbean Islands!". The whole project team was involved, thus ensuring a good variance between the interviewers, an aspect which is of specific importance when conducting qualitative interviews. Some of the interviews were, furthermore, conducted by a Montserratian, this being of importance as the local is embedded in the *lebenswelt* and its context.

All interviews were used for the final evaluation. In contrast to the objectives of quantitative research methods we were concerned to track down the common factors rather than the differences in the answers given. To this end, all answers were read thoroughly and then purposefully analysed for common ground. The analysis was facilitated by the structure of the questionnaire. Characteristic answers which highlighted common views were noted separately, and also those which expressed extremely different views. There are four significant areas which were of specific interest within the scope of the research: (1) the perceptions of the natural environment and resources, (2) the perceptions of acute problems and future challenges, (3) the assessment of the economic situation and (4) the assessment of the political situation.

Perception of the Natural Environment and Resources

The answers to seven questions (Appendix) were used for the analysis of the Montserratians' perception of their natural environment and resources. The questions aimed to identify the personal evaluation of the state of the environment and also tried to determine which natural resources the interviewee was aware of. It was also the intention to ascertain the necessity and possibilities of environment and resource management.

About half of the people interviewed answered the question "What is your personal view about the environmental situation on your island?" as follows: "The environment in Montserrat is still pretty unspoilt". The environment was on the whole evaluated positively. The people's impression was that there had so far been hardly any overexploitation of the natural resources. Comparisons were made, pointing out that the environment on neighbouring islands was often in a critical state. Most of the interviewees saw the cause in the mass tourism on those islands. However, 23 of the people asked also recognised environmental problems on Montserrat. The following answer shall illustrate this in an exemplary manner: "You cannot say that it [the environment] is unspoilt. No, definitely you can't. Some of our natural resources have been taken away already. For example: We used and still cut some of the trees for the charcoal production. For farming and construction purposes land had to be cleared. Some of our land is really destroyed or not the way it used to be. The beaches are also a problem. People used to take sand for construction and this was certainly not managed properly. Our shoreline is very fragile now. This was very clear after Hugo. We had erosion and the marine environment was polluted. We had break offs and the waves could easily approach and erode all kind of things". Other aspects mentioned as a cause for concern were the inefficient public waste management and the careless behaviour of many people, simply disposing of waste where and whenever it suited them. Loose livestock was also perceived as a problem to the vegetation and private gardens.

This result has been verified by a number of specific questions asked about environmental problems that have a direct influence on the daily life. Again, about half of the people asked did not perceive any environmental problems. The other half, in contrast, saw problems which directly concerned them.

These included, again, the problem of waste disposal, loose livestock and careless behaviour. The illegal quarrying of sand was also mentioned by a few interviewees. Interestingly enough, most of the answers referred to the terrestrial environment. Only very few people considered the marine environment as endangered, the following answer being an example: "*The fishermen don't catch as many as they used to. Fish stocks, the marine environment as a whole suffered. I guess forestry and birds are not a problem here really. Littering is a problem, that's for sure. They throw all kinds of garbage in the ghauts, motor vehicles, even poisoned stuff. This is sooner or later all washed into the sea. We really should work on a proper solid waste disposal management concept. We also need a system to control and monitor things such as pollution, the impact of Hugo, the impact of development and other things that have an impact on the natural environment*". Only one of the people interviewed referred to the volcano as a potential threat: "*The volcano can erupt any time, you never know. That's a threat, but you cannot do anything against it*".

The answers to the question "*What do you consider to be the natural resources of Montserrat?*" illustrate that the volcano was before its eruption, if at all, assessed as a resource. The people interviewed referred to its qualities enhancing the unique and beautiful landscape of the island, enjoyed by both Montserratians and tourists. The hot springs were also named as a potential for tourism. The resources most commonly mentioned were the lush vegetation, the beauty of the landscape, the sea, the beaches and agricultural products (Figure 4.11). The soil's fertility and the good water supply on the island were also often referred to. Some of the people mentioned typical national dishes, such as the goat water stew, or the wildlife, particularly the Montserratian Oriole, as National Bird known to most of the interwievees. Characteristic of the answers was that altogether 16 people assessed the Montserratians themselves as a special resource: "*The people. We have many things here in Montserrat, nearly everything, beaches, forests, birds ... I mentioned the people, because that is really special on Montserrat. All the other things are also on other islands. By saying people I certainly have their culture, their history and everything else in mind*".

The next questions intended to ascertain whether the people interviewed had a concrete interest in the protection of the environment, respectively resource management. The great majority, i.e. 65 people rated the preservation of the environment as "*very important*" and only three answered "*not important*". Many instantly started explaining what they above all considered worth protecting. The protection of the forests was emphasised, as the following answer illustrates: "*We should protect our forests. They are important for the water. We have to create areas that are purely designed for watershed management. We need rules and regulations*". The majority (58) also answered the next question positively: "*Do you think that environmental policies or management are necessary for your island?*". Five people negated. The subject was further consolidated at other points of the interviews. The questions tried to

[5] If answers do not add up to 70 this is caused by either not mentioning or the interviewees responded with "I don´t know". These answers are not listed separately in the following.

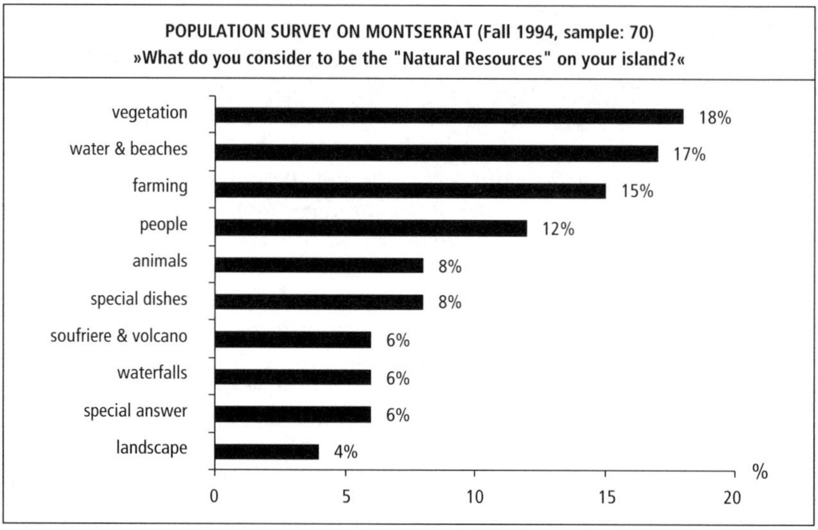

Figure 4.11. Resource perceptions by the Montserratians

ascertain the extent of personal responsibility, the willingness to act and individual ideas for the implementation of environment and resource management. It became apparent that the majority of interviewees had precise ideas.

The question "*Can you think of ways to protect the environment*" was in 21 cases answered with "*no littering*". The people who had given this answer at the same time pointed out that they were not only referring to themselves, but that they were also encouraging other people to follow their example. Some people referred to recycling and expressed their regret that these processes were hardly taken advantage of on Montserrat. Five interviewees saw a necessity to implement conservation laws and 13 referred to the importance of education to heighten especially the children's awareness for environmental issues. When evaluating the answers to the question "*Which possibilities do you see to use and protect resources at the same time?*", it became apparent that the majority allocated the responsibility to the public authorities. Many called for comprehensive environmental legislation and the establishment of protected areas which are to be regulated and managed by public authorities. Only four of the people that were asked expressed a personal sense of duty, respectively considered the issue (also) part of the locals' responsibilities. These people had concrete suggestions at hand, e.g. ecological agricultural production. Eight of the people interviewed did not see a chance at all for sustainable resource management. One of them emphasised the insular problematic: "*A small place such as Montserrat faces a lot of problems in doing so [to use and protect resources at the same time]. You cannot take it as a model. Although it seems it would be much easier in a small country to do so. I guess Australia, the US, Canada or the UK, in bigger countries you can do it. You have the space there, the expertise, the knowledge, the techniques and the money*".

Perceptions of Current Problems and Future Challenges

The natural environment and its resources were pushed into the background when the people were asked about the current problems encountered by Montserrat or its population. The answers to the following question clearly illustrate this: "*What are the common concerns and interests of the people of your island today?*". The answers focused on the labour market (32 answers) or the difficult national economic situation on Montserrat (14 answers). Six interviewees perceived the political administration as a major problem. They criticised that the administration was too bureaucratic and that this situation was aggravated by the (obligatory) coordination with the mother country. Some of the interviewees used the opportunity to argue about political autonomy for Montserrat. The answers given were, on the whole, well differentiated. The range of problems perceived reached from social cohesion via the situation of the younger population to the agricultural methods applied.

41 of the interviewees considered economic development as the central future challenge, especially in view of the prevailing problems. Above all, emphasis was placed on the availability of job vacancies for the younger population. Some people mentioned tourism as a potential motor for economic growth. Another aspect which was considered an important future challenge by altogether 12 of the interviewees were political reforms. Their main objective was political and economic autonomy: "*A lot of people are talking about not being on buy of the British. I think it is important to be independent and being able to deal with the investors directly. The people in the government are helpless. It is a problem that we get money from the British, but at the same time restrictions*". The remainder of the answers varied extremely and ranged from the implementation of a law and order policy, the promotion of education and training, the reduction of import quotes or reforms of the national tax system to the expansion of agricultural production or the promotion of protective measures for the environment.

The questionnaire tried to ascertain the exact priority. The following question was asked: "*What is your number one priority for Montserrat in the future?*". Altogether 46 of the people referred to economic development, most of them emphasising the expansion of agriculture and tourism. The interviewees furthermore called for a more effective protection of the environment (6) and basic improvements in the national education system, particularly with respect to the lack of institutes of further education on the island (6). The answers given to the question about the wishes for their personal future were similar. Most of the people asked desired a better personal economic situation and improved education and career opportunities for themselves and their children. Some also expressed the wish to have their own business, home or car. Apart from that the desire for a happy life was at the fore, in most cases defined as a harmonic family life and good friends (Figure 4.12).

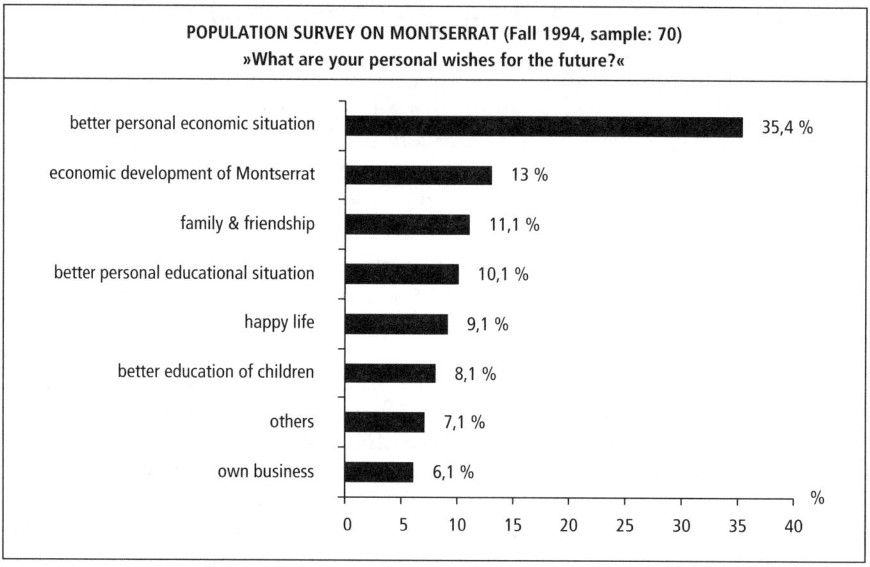

Figure 4.12. Personal wishes for the future

A further intention of the survey was to ascertain the personal ideas the Montserratians had about the desired economic development. We also wanted to identify whether they have a feeling for the sustainability or non-sustainability of development processes. Again, questions regarding this subject were asked in different phases of the interview. Sustainability always implies the aspect of necessities and interests of future generations. The interviewees were therefore asked: "*If you think about economic development on the one hand and the natural beauty of your island on the other, what, in your view, should be passed on to future generations?*". 15 of the people had difficulty answering this question. 18 people set priority on environmental protection whereas nine people gave priority to economic development. 16 interviewees, however, answered that both matters should go hand in hand: "*I think it is a combination of both. Development and protection of the country. My children need to get a job, but they should be able to enjoy nature also*". At times the questions were criticised. The interviewees remarked that one aspect does not function without the other. Six interviewees answered that education and knowledge were the most important things to be passed on to future generations. A further six people said that they wished to maintain Montserrat in its present state: "*Keeping the way it is*" was the answer.

The answering pattern changed only slightly for the next question: "*Considering your children, what would you like passing down on them?*". 16 answers referred to good education and formation, 12 to preserving Montserrat's nature and culture and nine desired a better national economy in order that their children might find qualified jobs when leaving school. Some of the interviewees wanted their children to inherit possessions, but only land property seemed important as this can be handed down from one generation to the next. On the whole the interviewees expressed great interest in preserving Montserrat's present condition for their children. When asked "*What can you personally do to guarantee that whatever you have chosen can in future also be enjoyed by your children?*", the majority answered that they could transmit knowledge and traditions, preserve nature and culture and pass on family possessions. The following quotations give an insight in the nature of the answers given.

- "I have to teach them how to cope with life generally. My grandma for example taught me everything about gardening. Sometimes I did not give attention to her. Young people don't like to listen. But today I remember what she said and I practise it. So I think it is important to transfer your knowledge from one generation to the other. I would also teach them not to use drugs."
- "No, money is not that important. A piece of land is good. You can pass it from generation to generation. You can't move it. If my daughter has a boy-friend later on and she has her own land, she is independent and has something if he leaves. Yes, my parents also had some land. Also, the parents of my wife transferred some land over to her."
- "I can be careful with my surrounding and I can educate my children about traditions and history."

Evaluation of the Economic Situation

The questions directed at Montserrat's economic system attracted great interest. A high consciousness about economic problems became apparent. The majority (65) gave a clear affirmative answer to the question "*Are you interested in the overall economic development of your island?*". Questions directed at the assessment of the benefits each of the most important economic sectors (tourism, agriculture, offshore finance etc.) produce were generally answered in a very differentiated way. The economic benefits of tourism were evaluated positively by 38 people. Most of them answered as follows: "*Tourism is good. It brings money in, foreign exchange and it creates jobs*". However, 21 interviewees restricted the positive economic benefits of tourism, pointing out that there were many problems linked with this sector. These people questioned the real benefit to the national economy. They wondered how much of the receipts actually remained on the island. Some diverted their criticism to the management structures and many feared negative effects on their culture and nature. These and other reasons prompted seven interviewees to even assess a further touristic development as very problematical and not desirable. The following exemplary answers illustrate that the Montserratians consciously evaluate a problematic such as tourism.

- "Too many outsiders are holding the higher positions in Montserrat";
- "Tourism is a problem. All the figures they are publishing about tourism, I don't believe them really. How much money is really contributed? How much money stays in the country? And if I think about all the money that

has to be spent to make people come to this place. I like having visitors, but it is not the solution for our economy as people like to believe. Yes, it provides jobs, some revenue, but you have to keep in mind that this is a small place. We are certainly not a place for any sort of heavy tourism development. They talk about ecotourism a lot, but even for that you have to invest a lot of money. We also have a transportation problem. And I have the fear that tourism can do harm to our culture, our people".

- "Tourism must be done on the island. The problem we have are the tourist packages they sell internationally. If everything is included in such a package, tourists won't spent much money on the island. The travel agent is the main winner in the game";
- "Tourism is okay, but it has to be monitored properly. The type of people coming to this place is very important. We don't like riffraff people".

The economic relevance of the agricultural sector was in contrast to tourism assessed very positively. Focus of the interest was guaranteeing local supply whilst reducing food imports. "*We have to feed ourselves*" was the most common answer. However, some of the people interviewed pointed out that amongst the young population there was a lack of interest to work in agriculture and they also pointed out that the local market for agricultural products was limited. A few suggestions were also made: "*Farming is great. I think the department in charge needs to do more. They should help the farmers in various aspects. Marketing is a problem, exporting and so on. We have wonderful volcanic soils. Farmers should get some education to use and manage it*". When it came to evaluating the offshore finance sector, a different picture was presented. Only 13 people assessed offshore finances as positively without any restriction. 36 interviewees saw it as very problematical, often mentioning the collapse of this industry in 1989. The necessity to control and monitor this sector was often emphasised.

The majority answered positively to the question "*What do you think about the economic contribution of expatriates?*". Altogether 47 of the people interviewed saw the expats as a benefit for the national economy. The fact that the foreigners who live on Montserrat contribute financially and also create jobs was often mentioned. Some people emphasised that many expats bring in their knowledge, this on the whole being advantageous. Other people, in contrast, said that the number of foreigners should be limited and that it was of importance to ensure that no social ghettos were formed. It was generally remarked that the majority of expats were well integrated into the Montserratian society, this being true for both the snowbirds and working immigrants of other Caribbean regions. On the whole, it became apparent during the interviews that the Montserratians have considerable interest in the economic development of the entire Caribbean region.

Evaluation of the Political Situation

The results so far obtained show that many interviewees had already uttered their opinion about Montserrat's political situation. Nevertheless, 40 people claimed that they were not interested in politics at all when the subject itself was mentioned. The following answers suggest that the reason for this is most probably a general discontent with the political programmes and the politicians themselves. When the question was asked whether the political parties responded to the needs of the local people, only five people affirmed. All other people either gave very different answers (26) or answered with a definite "no" (29). Altogether 36 interviewees evaluated the political strategies and programmes of the then governing party as negative and not adequate. Only 13 people said that the government was doing a good job. These people also commented that they were under the impression that most of the population was not aware of the governing party's endeavours and that the expectations were far too high.

- "I guess everybody expected more from the new government. They are the best educated group we ever had. Therefore, the expectations of people voting for them were very high. They started in a time of economic hardship, but still they should have been more successful by now."
- "I believe the people get the government they deserve. The problem is that they are honest. The people do not want that. The last government made a lot of promises and the people believed them, because they love to dream. This one is now telling the truth and they do not want to hear the truth."

On the whole there was hardly any confidence in politicians, no matter which party. Many replied to the question "*Do you have confidence in the capability of the politicians in the problem-solving-process?*" saying that one should never trust in politicians, an exemplary answer being "*No, I don't trust politicians anywhere in the world. Not in Montserrat, not in Germany, not in the UK, not in the US or wherever*". Quite a number of interviewees explicitly referred to the corrupt attitude of the previously governing party. Alone 31 of the people saw a general corruptive tendency on the island. Others, in contrast, did not consider corruption a significant problem: "*Not to a severe extent. There used to be some corruption in the Offshore Sector, the Medical School and in the Hotel Industry, but the island is still a little paradise. It is one of the best countries under god's heaven!*". 19 of the people interviewed did not see any problems at all .

The answers given with reference to the advantages and disadvantages of being a British dependent territory were very informative. 33 people preferred this political status. They were aware of the following advantages: financial and technical assistance, political and economic stability being guaranteed by the mother country. The latter being of significant importance, particularly for tourism and offshore finances. However, 13 of the 33 supporters of dependence expressed that they would prefer independence at a later stage, providing Montserrat had the necessary resources. Many people generally said that Montserrat was not in a position to become independent since there was no economic basis to support this idea. 18 of the people asked rejected the current political status and called for autonomy. Others gave very differentiated answers. Some of the replies shall at this point be quoted in order to illustrate the opinions.

- "Dependence is a waste of time. The British only use us. Think for example about the civil service! All higher positions are run by a British guy. The Police Commissioner for example is British. They bring us back to slavery in a way. We are ready to go on ourselves!"
- "Montserrat couldn't survive without the British."
- "Yes and no. Being a dependent territory you get aid and assistance from the UK and that's positive. But Montserratians are not treated as the British persons in the UK and this is negative. We should have the same rights, medical care and education."
- "The main problem we have in this context is the lack of communication really. Also, the government bureaucracy is too complicated and causes many problems. You know we have the British Westminster style, that includes special legal and political frameworks such as the general orders that are set up. But I think you cannot transfer this kind of thing to a small place like this, it is not applicable. I consider it is a problem that the political parties elect the person that is going in the country elections. I'd rather prefer a system where the local people in the various constituencies elect their favourite person to go for the elections. We should decide and not political parties that elect one member that is maybe not well known or accepted in the public."

Summary

When analysing the interviews one aspect, above all, becomes apparent: the Montserratians wish to preserve their nature and culture for themselves and future generations. The majority therefore emphasises environmental and resource protection. However, they feel that this is mainly the responsibility of the government. The majority of the population also calls for a significant improvement of the economic situation, focusing on the creation of qualified jobs. Important motors for such an economic development are believed to be agricultural production and tourism. A judicious and sustainable development of the tourism sector should, in the belief of the interviewees, be striven for. The general strategies, plans and programmes of the current government are by the majority evaluated negatively. Confidence in politicians is on the whole limited. It has also become apparent that unpredictable events are in their majority excluded from the reflections at the micro level. Very few people address uncertainty and think about the effects of a hurricane, the probability of a volcanic eruption or the instability in the offshore sector. The majority of the interviewees strives for security, stability and order in a harmonic life in accord with nature.

4.1.4
Interactions between the Levels

We have to relate the three levels of analysis to each other in order to be able to understand the interplay between natural environment, society and influencing agents and individuals. The system Montserrat shall in this way be characterised for the time of the field studies, i.e. the summer of 1994. The central question of the research shall serve as the main connecting thread: how do humans cope with complexity and uncertainty?

There are some features which have a significant influence on the system's components. One of the most characteristic features is certainly the small size of the island. It conditions the limited availability of space and resources, a general dependence on external factors, the limited size of the population, the small capacity of the local market and, on the whole, a considerable vulnerability. Vulnerability in this context not being a static variable but always the result of a dynamic interaction human-environment. Many of the ecological constraints outlined are not only the result of the utilisation of resources but also react upon it. Plantations offer a good example for this dynamic interplay. First, the deforestation of the original forests results in the erosion of the soils. This has, in effect, conditioned that the north of the island is today hot and dry, its landscape being dominated by dry woodland and cactus scrub. These climatic and vegetative conditions have in turn made the region unattractive for any economic use or settlement. Montserratians and tourists alike prefer the environment offered in the southwest of the island. The landscape is varied, the vegetation lush, the micro climate pleasant, and the fertile volcanic soils are suitable for agricultural production. In 1991 65 % of the population lived in this region, a fact that highlights the pressure exerted onto its natural resources. Tourism is to a large extent involved. New hotels, villas and a golf course have overformed valuable areas adjacent to the coast. The airport is located on flat land, rare to be found on the island. The plans for its extension aggravate the situation. Shifting land utilisation away from agriculture implies the loss of agricultural production for the supply of the local population. As a result, the majority of food stuffs has to be imported which again drastically raises import spending.

The analysis of these correlations shows that the interactions between humans and their environment are far more direct in a small island state like Montserrat than in larger area states. It would thus be wrong to believe that small islands such as Montserrat were less complex. The interplay between their agents is often even more intensive as the influence single agents have on the structure, function and evolution of the system can be rather great. The constraints outlined (small population, emigration, resource utilisation, small home market, dependence etc.) decisively determine the evolution of the system in the sense of its sensitivity to the initial conditions. For a long time there were relatively few and minor alterations in the system's structure and function. Up until the 1960s agriculture predominated on the island, changes being limited to the different products that were cultivated over the centuries. Parallel to these phases there were also various emigration waves, more or less intensive over the decades. However, it was not until the 1960s that Montserrat was subject to relatively fast shifts, e.g. from agriculture to a service oriented economy. This transformation, again, is decisively determined by the initial conditions.

However, not only the constraints or variables outlined above determine the island's development. The influence single individuals have is also quite considerable on such a small island. A typical example are Montserrat's Chief

Ministers over the past three decades. Each of them left his mark on the socio-economic development of the island. Let us recall: W.H. Bramble prepared the basis for residential tourism in the 1960s. His son Austin Bramble put the processing industry to the fore. It was his politics that drew the *International Air Studios* and the *American University of the Caribbean* (AUC) to Montserrat. Together with many local businesses that were established, they greatly contributed to the national economy. John Osborne was also quite successful with his economic policies during his 12-year-term. His main objective was the abolition of import restrictions. He was, on the other hand, as head of his party responsible for a number of corruption and drug scandals. These scandals not only resulted in the collapse of the offshore sector but also worsened the relationship with the mother country. John Osborne also dismissed environmental protection even though it had been established quite early and was pursued actively on the island. It was not until the NPP was enforced that environmental issues regained significance. The good collaboration between Chief Minister Reuben Meade and Governor Frank Savage also cumulated in normalised relations with Britain.

The comparison of the interests and objectives of the gatekeepers and the individuals show that in 1994 there was, interestingly enough, hardly any diversity within and between the two groups. Both the gatekeepers and the individuals desired a sustainable ecological and economic development. Both groups placed emphasis on employment, social and health services, a diversified economic structure and the preservation of the local nature and culture. Nevertheless, it was apparent that the population was extremely dissatisfied with the political plans and programmes. The reasons are manifold: (1) the population is often not informed early and well enough about current political projects, (2) the population is involved in planning at too late a stage, resulting in a low acceptance of the plans and strategies, (3) the local government usually has difficulty to quickly and effectively implicate its plans and strategies. Despite the population's discontent, few conflicts have been ascertained on the local level. This can be explained by the limited number of interest groups. The unions, the *National Trust*, the Chamber of Commerce, church circles etc. operate either to a large extent independent of each other or stimulate societal debates in a creative manner. The greatest potential for conflict has to be sought between the political parties, and between the local government and Britain.

Despite the large correspondence with regard to the development objectives there are a few characteristic disparities between the perceptions of the gatekeepers and those of the individuals. These disparities refer mainly to the environmental perceptions and the management styles. An explanatory pattern is offered by cultural theory (Chapter 2). The gatekeepers are, on the political and administrative level, certainly bureaucrats. They consider the environment as a matter which can be controlled by management. This includes the preservation of the system and the rational allocation of resources, all being regulated by the relevant institutions and plans. The individualists in the group of the gatekeepers are (amongst other) represented by the businessmen. Their management style flexibly adapts to changes in the environment, and they tend to make the best of any given opportunity. At the same time, they use technological and preventative measures and insurances to minimise risks. In contrast, at the micro level we tend to find fatalists and egalitarians. The former do not believe in man's capacity to control his environment. All positive and negative implications depend solely on fate. Egalitarians, on the other hand, consider their environment a fragile system and emphasise the necessity to carefully and judiciously manage the resources. As we have seen, the responsibility for resource management is primarily passed to the state. Their perception of the environment is egalitarian, but their attitude towards the actual application of management strategies is rather bureaucratic.

It can be concluded that there is no predominance of a risk culture when analysing the perceptions of an uncertain, dynamic environment. It has to be pointed out, however, that there is a predominance of a bureaucratic style on the level of planning and management. The island is on this level perceived as a stable system with largely predictable and planable development. The various political-administrative plans and programmes clearly illustrate that the main objective is a balancing stabilisation. The gatekeepers are, in fact, aware of some fluctuations and instabilities, such as the fragility of the ecosystem, emigration waves, boom-bust phases of certain economic sectors. However, there are hardly any flexible or integrative strategies at hand with respect to these issues. Fluctuations and instabilities are often not even perceived until they have caused serious consequences. The utilisation of resources is a prominent example. It is often adapted when internal fluctuations or disasters (e.g. exhausted soils, plagues of insects, floodings, hurricanes, drug and corruption scandals) or external fluctuations (e.g. collapse of an industry, international crises etc.) call for a reaction.

This attitude implies that fluctuations, instabilities and unexpected events are hardly ever considered in management and planning strategies. This includes the identification of fluctuations which can trigger off innovative self-organisational processes and also extends to the necessary reduction of vulnerability. Even the experiences made with hurricane Hugo has not brought about any constructive changes in this respect. To the contrary, the hurricane is perceived as a rare and extreme event that will not repeat itself for a long time. This is the reason why hazard management played a secondary role during the reconstruction phase. Reconstruction was characterised by linear approaches since all endeavours were directed at recovering a state of order, security and stability. The gatekeepers concentrated primarily on the economic and infrastructural development of the island whereas the individuals set priority on their jobs, their families and the preservation of the island's nature and their culture. An essential integrative approach was not considered. Planners and managers continue to view the various sub-systems or sectors as isolated entities. Or they concentrate on certain constraints, like erosion or the over-utilisation of resources.

On the whole, Hugo has as a case of extreme change triggered off numerous internal fluctuations. These have transformed the system's structure, i.e. the flora and fauna of the environmental system were destroyed to a large extent. However, they started recovering relatively quickly. The same applies to the economic system. Tourism collapsed for a few years. The *International*

Air Studios and the company *Sea Island Cotton*, which produced high-quality cotton, were severely affected and have not yet been rebuilt. The same happened to the international radio and television institution *Deutsche Welle* which was located on Montserrat before Hugo struck the island. Agricultural production had also not recovered by 1994. Apart from these negative effects there have surely been some positive outcomes, as for example the boom in construction following Hugo, or the improvements of the social and economic infrastructure. This had, five years after Hugo, quite a high standard in comparison to other Caribbean states, mainly owing to the extensive financial and technical assistance of the British mother country. However, in contrast to the hypotheses earlier outlined, hurricane Hugo did not cause a new order in the entire system's structure. There are obvious self-organisational processes (e.g. the regeneration of the environment or the political administration), but these have only resulted in transformations on the level of the subsystems. The identity and structure of the overall system Montserrat has to a large extent been maintained.

Let us briefly resume: Montserrat is a complex system, characterised by numerous interactions and fluctuations. Due to the small size of the island most of these interactions between humans and their environment and between society and individual are very direct. Hence, unexpected events and single individuals often have a far greater influence than in larger states. Complexity and uncertainty are, nevertheless, hardly considered by planners and managers. The prevailing plans and management strategies are sectorally oriented. Uncertainty is reduced and only dealt with in certain critical components of the system, the maintenance of the system always remaining the main objective. Also, the dynamics of the system is to a large extent cut out. Five years after Hugo, i.e. in the midst of reconstruction, all endeavours concentrate solely on the recovery of security, order and stability. The complete resources available are mobilised whereas potential disasters are hardly taken into account.

4.2 Montserrat During the Volcanic Crisis – Coping with Uncertainty

It was only six years after hurricane Hugo that the Montserratians were struck by another natural disaster. On 18 July 1995 the Soufrière Hills Volcano erupted. The population was caught unawares by the steam and ash venting from the northwest crater, even though seismic activity in the years 1989, 1992 and 1994 had indicated to the fact that volcanic activity had commenced. Scientists of the *Seismic Research Unit* (SRU) in Trinidad had installed two seismometers in 1989 in order to monitor these activities. However, Hugo destroyed the instruments, and another scientific study about the volcano was also abandoned, most probably because of the hurricane (Wadge and Isaacs 1989). In this study the geographer Wadge and the seismologist Isaacs simulated possible disasters of the Soufrière Hills Volcano in a computer model and drew up a risk map and a phase model of a potential volcanic crisis. They also explicitly pointed out that Soufrière was an active volcano, calling for its urgent consideration in Montserrat's national hazard management. They warned: "The Soufrière Hills Volcano is active and will erupt again. It is a potential threat to many of the people that live in southern Montserrat" (Wadge and Isaacs 1989, p. 1).

In addition, there were some regional hazard management attempts, but, again, without the desired effect. In 1986 a seminar about hazard management in situations of volcanic eruptions was conducted within the scope of the *Pan Caribbean Disaster Preparedness and Prevention Project* (PCDPPP). A scientific study was published evaluating the probabilities of volcanic eruptions on various Eastern Caribbean islands (Shepherd et al. 1988). In a comparison of, altogether, eleven volcanoes and the probability of their erupting, the Soufrière Hills Volcano was in ninth place. This is one of the reasons why it was not mentioned in the various national disaster plans, such as the 250-page document of the year 1995. Another reason is one of the most dangerous features of volcanoes in general – their long periods of inactivity. Radio carbon analyses suggest that the Soufrière Hills Volcano last erupted about 320 to 370 years ago. This last eruption seems to have been a minor one. The last major eruption will have taken place around 24,000 to 17,000 years ago (Wadge and Isaacs 1989). This means that not even Montserrat's first settlers had actually experienced a volcanic eruption. It is therefore not so astonishing that nobody was prepared for the slowly escalating volcanic crisis. Before we come to the chronology of the crisis, it will be necessary to briefly outline the type of volcano we are dealing with.

The Soufrière Hills Volcano

Montserrat's Soufrière Hills Volcano would be categorised as one of the earth's typical volcano. It is a located in a subduction zone of tectonic plates, i.e. the Atlantic and the Caribbean plate. The predominant form of these subduction volcanoes is the stratovolcano, composed of alternating layers of ash and lava. Chances Peak with 914 m (roughly 3,000 ft) was, until 1995, the Soufrière Hills' highest elevation. Highly explosive activity is typical for stratovolcanoes. The volatility of a volcanic eruption is generally determined by the chemical composition of its magma and its reactions with the ground or sea water. The gas content and the viscosity of the magma as well as the content of silicon dioxide (SiO_2) are decisive. The more silicon dioxide it contains, the more explosive the eruption. The magma from stratovolcanoes in subduction zones usually contains a high amount of gas and silicon dioxide. The latter provokes a high viscosity of the magma and low degassing properties. As the magma rises to the surface, it often forms a dome. The lava is relatively viscous and piles up to form a dome or cupola-like construction above the main vent. It closes the vent off like a champagne cork, the pressure thus not being able to escape for a long time. In the sense of self-organised criticality, the entire system then evolves towards a critical condition. The pressure gets out of control and the molten material is ejected out of the crater with explosive force. Glowing lava fragments are ejected to form a turbulent mixture of ashes, bombs, pumice and scoria[6].

Very characteristic are pyroclastic flows, which pose the most dangerous volcanic phenomena and can cover thousands of square kilometres of land. Humans have hardly any chance of survival. The gases which continuously escape from the hot mixture give it such high mobility that there is little warning (Schmincke et al. 1993, Wadge and Isaacs 1989). Pyroclastic flows are fragments of magma, such as ash and pumice that together with gases rush downhill at high temperatures and speed. The pyroclastic flows may originate in the collapse of eruption columns or the gravitational collapse of a dome. Such flows typically travel at velocities of 100 to 500 km/h (about 62 to 310 mi/h) and reach temperatures of 800° C (1,472° F). Often the flow separates downslope into a dense underflow which closely follows valleys and an upper, billowing cloud which spreads more widely. The eruption of the volcano Mont Pelée on Martinique in 1902 is one of the most famous events of this type. For this reason, potentially the most disastrous form of eruption is in technical terms called peléan type. In a single event the 1902 eruption of Mont Pelée wiped out the 29,000 inhabitants of the port of Saint Pierre. It was the most devastating volcanic disaster of the 20th century. Only two people are known to have survived this event, one of them being a prisoner who was protected by the rock walls of the local prison. He gained world fame with his accounts of the disaster. The crew of a ship which was anchored at a safe distance witnessed the eruption.

The Soufrière Hills Volcano may appear as Mont Pelée's little brother, but it is, nevertheless, an extremely dangerous volcano. Wadge and Isaacs (1989) established the connection: "Future eruptions will involve the creation of pyroclastic flows and surges and mudflows on the flanks associated with the growth of the dome of andesite lava within a specific region around the summit of the volcano. The deposits and effects of such eruptions are very dangerous. The 1902 eruptions of Mt. Pelée, Martinique that destroyed the town of St. Pierre provide a good analogy for this future activity" (Wadge and Isaacs 1989, p. 1). The potential danger has become apparent during three significant phases of seismic and fumarolic activity which have been recorded since the first colonisation of the island in the early 17th century: 1897–1898, 1933–1937 and 1966–1967.

In the following the complexity of the system Montserrat during the volcanic crisis[7] shall be analysed. It will, again, be a step-by-step approximation of the system, i.e. research of an ongoing process. We have to be aware that complexity is greatly increased in a dynamic and uncertain disaster situation, especially in the case of a volcanic crisis. A hurricane might pass in a day

[6] Volcanic ashes are fine particles of mineral matter from a volcanic vent with a max. diameter of two mm. Larger particles are called *lapilli* (4 - 32 mm). A volcanic bomb is a rounded mass of lava exploded from a volcano. Volcanic bombs can be catapulted up into the air for several kilometres and can have diameters of some metres. Pumice is a volcanic glass full of cavities and very light in weight.

[7] The term volcanic crisis is applied as the following outlines the entire phase of the previous volcanic activity, i.e. not only the onset of the disaster.

whereas a volcano can be active over months and years, thus posing considerable difficulty for any adjustment and adaptation processes. Uncertainty with regards to the volcano's next eruption, its effects and the long duration of the active phase are significant aspects. In the event of another eruption, the people would have to react instantly in order to protect their lives and belongings. Time pressure, fear and the lack of alternatives often result in conflicts. The natural disaster is accompanied by social and political catastrophes. The chronology of the events since the first eruption in 1995 will illustrate this.[8]

4.2.1 Chronology of the Volcanic Crisis

July to September 1995 – Immediate Reactions at the Beginning of the Crisis

In the afternoon of July the 18th, 1995 the population in the southern part of Montserrat heard a strange, rumbling sound, at first resembling that of a jet fighter passing by. But the noise continues and seems to come from the direction of the Soufrière Hills. The air is filled with an intensive stench of sulphur. Some of the residents are seized by fear. All of a sudden large amounts of steam and ash are vented in a sequence of phreatic eruptions[9]. The unthinkable has become reality – the Soufrière Hills Volcano has resumed activity. The immediate reactions are quite calm. Local decision-makers reactivate the *Emergency Operation Centre* (EOC) and contact the *Seismic Research Unit* (SRU) of the University of the West Indies (UWI) in Trinidad. Many decision-makers who want to prepare a potential emergency at this stage realise that the national disaster plan does not refer to the volcano. In the evening of the same day the Governor Frank Savage addresses the Montserratians on the radio and asks the residents living in the area adjacent to the volcano to prepare themselves for a possible evacuation that night. The international press meanwhile reports about the activities of the volcano. Montserrat's telephone network is soon overrun as hundreds of calls from friends and relatives try to reach the island. Some Montserratians are not even aware of the event until they are informed by these telephone calls.

The next day the first scientists from the SRU arrive on Montserrat for a closer examination of the volcano's activities. Soon they are supported by further scientists arriving from the UK (*British Geological Survey*), the United States and some Caribbean neighbouring countries. Since the activities continue, a scientific observatory station is established with British finances, the *Montserrat Volcano Observatory* (MVO). The volcanologists' four main activi-

[8] The chronology of the volcanic crisis refers to both the volcanic activity and the human reactions. The contents of the information outlined is based on interviews with decision-makers and the local population, discussions in the Montserrat Newsgroup on the Internet, the information of the Montserrat Volcano Observatory (MVO) and the summary of the events during the first weeks, published by Buffonge (1996).

[9] Phreatic eruptions cause the rising magma to come into contact with ground water and this results in steamy explosions.

ties are: (1) The analysis of seismographic data. Seismographs have been installed around the volcano. These record all seismic activity significant for the indication of potential eruptions. (2) The analysis of measures obtained by an Electronic Distance Meter. Any changes in the position or inclination of the volcano's slopes are recorded. Deformations of the volcano and possible magma movements below the surface can thus be monitored. (3) Measuring and monitoring the sulphur dioxide concentration at ground level and controlling the chemical composition of ash and rainwater samples. (4) The scientists also use GPS (Global Positioning System) to detect deformations over a wider area, particularly those which may be due to a deeper source.

The remaining residents are anxiously awaiting the first scientific results but otherwise carry on 'as usual'. Shops, offices and factories are open. There are only a few aspects reminding everybody of the special situation, e.g. ash falls, tents being erected in the north of the island and the daily arrival of scientists. Confidence in the scientists and their work appears to be very high. Many people even go to endangered regions, e.g. Tar River Valley, in order to be able to see the activities in the Soufrière Hills with their own eyes. On 26 July 1995 the necessary measures for the event of an evacuation and a system of alert stages are explained to the population. There are four stages of alert: (1) initial alert, i.e. the volcano shows unusual activity, (2) yellow alert, i.e. volcanic activity is expected, (3) orange alert, i.e. volcanic activity is scientifically confirmed, (4) red alert, i.e. there are sufficient indicators prompting an evacuation of the area. At this point in time orange alert is given. Two days later a new crater opens up in the Soufrière Hills. Long Ground, the settlement on the eastern slope of the volcano, is evacuated. Radio Montserrat the same evening informs that 1,200 people have left Montserrat since the beginning of the crisis, amongst them many tourists and students of the *American University of the Caribbean* (AUC), the holiday period just having commenced. Officials emphasise that there is no necessity to leave the island.

The phreatic eruptions continue during August 1995. The volcanologists recommend the evacuation of about 5,000 people from Plymouth and its neighbouring region. For more than three weeks the evacuated population live in churches, schools, tents or with friends or relatives. Many of them move to Antigua. The telephone network is once again completely overrun and many of the worried callers do not even know where their families are staying. Some volunteers organise an information service about the whereabouts of the evacuated. This private initiative evolves to become an important centre of information, later even affiliated with the EOC. On the 21 August 1995, the so-called 'Ash Monday', large ash eruptions blanket Plymouth in a dense ash cloud for about 20 minutes. Even public administration now has to leave Plymouth. Both, the government and the population are shocked and helpless. But volcanic activity then calms down again.

Twelve days later, in the midst of the ongoing evacuation, preparations have to be made for hurricane Luis. Meteorologists have forecasted that it will be even stronger than hurricane Hugo. But, after all, the situation could have been a great deal worse for Montserrat. Luis does not strike the island with its entire force. Only parts of its forests, beaches and some houses are damaged. Most of the tents erected for the evacuated population, however, do not resist and the emergency dock under construction at Little Bay is destroyed. During the recovering phase it is announced that the evacuated can return to their homes. Only the eastern settlements remain evacuated. Although the scientists expect further phreatic eruptions, they rule out the possibility of magmatic eruptions. On 15 September 1995 hurricane Marilyn reaches the island, but relatively little damage is incurred. The worst seems to be over.

October 1995 to February 1996 – A Dome builds up

Only few weeks later a new phase of volcanic activity commences. A dome starts growing in the crater. Day after day andesite magma is pressed out of the vent. It piles up higher and higher. This process is accompanied by strong ash falls. The volcanologists' concern about potential and gradually developing volcanic activity deepens. They refer to Wadge's and Isaacs' (1989) study which characterises the following phases for an eruption of the Soufrière Hills Volcano. (1) earthquake swarms and phreatic eruptions continuously increase in intensity and frequency; they can last months, (2) gradual dome building commences and is accompanied by larger explosive eruptions and pyroclastic flows, (3) phase of retarded non-explosive dome growth, possibly with occasional pyroclastic flows, the production gradually diminishing over months or years. In the light of the dome growth there is, at the point in time, reason for apprehension, but no further evacuation is ordered. However, some departments, instruments and patients of Glendon Hospital in Plymouth are precautionally transferred to St. John's as the north of the island is considered safe territory.

As a result of these events the insecurity of the population grows and an increasing number of complaints is recorded for the first time since July. Many people fear that the volcanologists are holding back important information. They are anxious to know more about the potential developments of the volcanic activities. Rumours start spreading over the island. The volcanologists, nevertheless, carry on with their previous practice to inform the local decision makers and leave it to them to inform the population. Much remains in the dark so that unofficial information channels establish themselves next to the official ones. Mobile radio operators and people with access to the internet contact friends and relatives abroad. The internet is also used to search for information about volcanic issues. Later on a news group is established on the internet. Its members live in all parts of the world and regularly discuss the developments on Montserrat. Finally, the official information policy is altered. Local decision makers and scientists now meet daily. Every morning and evening the MVO broadcasts a daily scientific report on the radio. Volcanologists also give public lectures and explain the scientific background on the radio or on television. The government organises the publication of information leaflets and also thinks about modernising the emergency shelters in the north of the island, respectively constructing new ones. For the first time since the emergence of the crisis clearer preparatory endeavours for a potential escalation of the situation become apparent. In December 1995 the second evacuation of the settlements in the southern and eastern part of the island, including Plymouth, is initiated. This second evacuation clearly benefits from the improvements in organised emergency planning. Again, a phase with less volcanic activity follows. Dome growth slows down considerably so that the evacuated can return home as of New Year's Day. The situation once again calms down.

April to August 1996 – Intensification of Preparatory Measures

In April 1996 the first major pyroclastic flows moved down the Tar Rivery Valley at average speeds of 240 km/h (almost 150 mi/h) (Photo 4.1). Moreover, a series of earthquakes suggest that the dome is growing faster now. The volcanologists recommend a third evacuation of the southern and eastern settlements. Residents and working population are restricted to stay in Plymouth during the day only and have to be on the alert. In the beginning there is hope that the evacuation will be for a limited period of a few weeks. But the dome continues to grow rapidly and parts of it keep collapsing. The volcanologists also note a fissure on the eastern face of the dome and fear the transition into an explosive phase. This time the population will not return home for a long time. About 5,000 people have to spend not only weeks, but months or even years in the formerly sparsely populated north. There is a lack of shelters, but the snowbirds' villas prove valuable. They are let to richer Montserratians, ministries and authorities. The financial burden on the population is heavy. The cost of living continues rising. Especially rents for

Photo 4.1. Tar River Valley in August 1996. (Possekel 1996)

houses in the safe zones are exploding. The problem for a considerable number of Montserratians is that they have to carry on paying the mortgages for their houses in the Exclusion Zone. 1,500 Montserratians are allocated to 25 emergency shelters and live for months or even years in schools, churches and tents. Dozens share a bathroom and a kitchen and there is no privacy.

The situation on Montserrat worsens and is aggravated by the economic crisis. The best arable land is located in the Exclusion Zone. Some farmers still cultivate their land, but agricultural production is declining. Even subsistence fishery is on the decline since the ash has damaged the coral reefs and fish population. Much of the food is already supplied by the neighbouring countries. Tourism is the sector which is most adversely affected by the volcanic crisis. Apart from some German cruise ship tourists who stay for a few hours, there is hardly any other tourist or snowbird coming to the island. The number of visitors declines by 49 % between August 1995 and February 1996. Many employers shut down their business and move to neighbouring islands. Despite the volcanic crisis, there are about 30 of them who seek a new location in the north of Montserrat. The transfer of the American University of the Caribbean to St. Maarten has serious economic consequences. As a result of all these developments Montserrat experiences steadily rising unemployment (Development Unit 1996).

The extent of the crisis becomes more and more obvious. A transformation from short to medium-term adjustments begins. Chief Minister Reuben Meade comments on the situation in an interview: "There are farmers who can no longer farm, so therefore they have no income. There are business people who are not making sales and there are people who are unemployed. So clearly we need to provide a lot more in terms of sustenance for the majority of the people on the island even those who have themselves relocated. With a volcanic situation such as this, we now have to look at life and livelihood and creating a balance between the two. And currently we are looking at life (...). And there will be significant rebuilding required in the future whether we go back into town or we are going to be doing whatever we are doing in the north of the island, that rebuilding process must take place" (Government Information Unit 1996, p. 1). The necessary infrastructural development of the north takes precedence over anything else. The needs of the affected population have to be satisfied as best as possible. Britain funds 20 emergency shelters and the local government initiates a programme for the construction of small family homes, financed by the *Royal Bank of Canada*, *Barclays Bank* and the Montserrat Organisation on the British Virgin Islands. 100 hurricane-proof prefabricated wooden houses are to be built, each comprising two bedrooms, a bathroom, a lounge and a kitchen. The costs for each house amount to EC$ 28,000 (Photo 4.2).

The Governor informs the population on 24 April 1996 that Britain has offered a two-year residence permit to any Montserratian willing to live in the mother country. The permit guarantees social security, assistance when looking for a job or housing, and the children are entitled to attend British schools. However, the Montserratians are required to pay the travelling costs, a prerequisite which is harshly criticised, above all by those who cannot afford to buy a ticket. Only 100 Montserratians take advantage of the offer when it

Photo 4.2. A family unit in the north. (Possekel 1996)

is first announced. Apart from Britain other member states of the *Organization of Eastern Caribbean States* (OECS) also offer work permits. But the Montserratians generally fear higher costs of living abroad. The uncertainty of what is to be expected drives many to stay, believing life with the volcano to be the more secure and maybe even only alternative. Also, local identity has grown with the crisis. A growing number of poems, songs and films are produced, all praising the island's beauty. They help to cope with the crisis and motivate people to carry on (see for example Textbox 4.1, Soca King Arrow, Ah just can't run away).

Textbox 4.1. Ah just can't run away

> On the map you can hardly find it. But it's always my paradise. And everyone who have seen it has always been hypnotised. But beneath the green and the mountain, a volcano has been sleeping. And now he is awakening. I don't know what to do, what can I do? Ah just can't run away. No, ah just can't run away. As long as breadfruit and mango down there, I'll be holding on. As long as I can leave my door wide open, I'll be holding on.
>
> I know people are uneasy, some are depressed some frustrated. But only the Lord Almighty controls our destiny. What is to be will be. I'll never forsake my country. You might run away from volcano and end up in hurricane and tornado. Ah just can't run away. Arrow ain't going no way. As long as hot hot Soca down there, I'll be holding on. As long as I can drink a cup of goatwater, I'll be holding on. Montserrat, sweet Montserrat, hold on!

> Every time I fell the earth shaking like it's doing its final dance. The smoke and ash keep coming, but still I take my chance. Oh Montserrat, oh Aliouagana, is this your farewell song? But I'm still planting my banana, I'm still holding on. You know why? Ah just can't run away. Ah just can't run away. As long as St.John's have good shelter, I'll be holding on. As long as Foxes Bay still down there, I'll be holding on. Hold on Montserrat! Hold on to what you've got. What a lovely country, I'll be holding on. As long as Runaway Ghaut still have water, I'll be holding on. How could I run away when I've never seen the National bird, the Oriole? I ain't going nowhere
>
> (Arrow Music Ltd., Montserrat 1995).

In August 1996 a commission is set up on the requests of the British government. The public is asked about its opinion regarding the elections originally due shortly. A census is also prepared and the population asked about their wishes and needs. A few weeks later the results of both questionings are announced. The majority of the population approves of holding elections and wants to carry on living on the island. As a direct consequence, Baroness Chalker, British Director of the *Overseas Development Administration* (ODA), announces that £ 25 million have been granted for a development programme in favour of northern Montserrat. A number of projects are initiated, e.g. the construction of new factories for small and medium sized companies, extension of the temporary hospital in St. John's, improvement of electricity and water supply, construction of schools, provision of arable land and the construction of an emergency jetty at Little Bay, serving as alternative to Plymouth in case the crisis escalates.

The projects are all slow to start. The reasons are mainly the bureaucratic structures, interest conflicts and a general policy of wait-and-see which is adopted by many decision makers (Chapter 4.2.2). The local businessmen who have decided to stay on Montserrat appear far more flexible in their decisions and activities. They recognise the need for new locations at an early stage. The telecommunication company *Cable & Wireless* for example was the first to transfer its head office to St. John's. Other companies follow and their investments affect the region. Initially, they build provisional huts which are used as shops, restaurants or bars. Salem (meaning peace), formerly a small quiet village, soon looses its peaceful character. The complaints of its residents grow. The market has become noisy, traffic has increased and Soca songs can be heard out of any rum shop. Salem grows quickly and is now the island's new centre. Life starts returning to run its usual course. The people help each other and they keep with traditional events such as the yearly beach cleaning programme, the 'Best Village' competition or numerous barbecues, the only difference being the smoking Soufrière Hills Volcano in the background. There is only one wish which remains despite all adjustments: soon being able to return to Plymouth, to their own home or their own company.

September 1996 to May 1997 – The Situation Worsens

In the night of 17th to 18th September 1996 a lateral explosion is caused by a series of dome collapses. The pressure upon the underlying magma is released and ash is catapulted out of the crater in a cloud reaching heights of up to 14 km. Pumice and volcanic bombs are catapulted through the air and destroy many houses at Long Ground (Photos 4.3 and 4.4). Cork Hill and Richmond are affected for the first time. There has been no advance warning, but, nevertheless, both settlements can be evacuated within three hours. The Governor immediately addresses the residents on the radio, urging them to stay in their houses. Despite this warning many still leave their houses, jump into their cars and drive northwards. Ash and pumice hit the cars and there are numerous traffic jams, particularly on the curvy road to Little Bay. Hundreds of people simply sit in their cars, waiting for daybreak.

Textbox 4.2. Ah had to run

Ah see myself getting covered in ash
And totally confused when the lightning flash
Red, as blood, with a hissing sound
And the thunder roll
Put a shiver in me soul
Listen to the mountain roar
That's not rock breaking? You sure?
Look up in the sky
Ah know it's dark, but that is ash, not clouds over me
Run for your life, everybody going to die
Grab your bag with your passport, some food and clothes
Throw the children in the car, we have plenty gas
Let's go fast! Forget about the door, hurry, let's go.

Stop! Said a masterful voice.
You rushing like crazy.
You forget we in the safe zone, we don't have to go nowhere
Let's be calm, go back inside the house, wait it out
We safe here as long as we lock up from the ash
Fearful eyes and uncertain feet plod up the steps
Master starts to whistle to calm them down
An unholy sound, extraneous to the mood
Because the mountain up there behaving rude
The lightening crack louder as the ash get thicker
Right over the house, see everybody scatter
That is stones falling on the roof?
Yes, one just hit me on my head, here's the proof
Rush to the car in nightgown, barefeet and empty hand
Safe zone? Maybe for you, not for me

That's why I run
Ah had to run

Look, other people running too
The Allens down the street, their pick-up full
See a long line of cars heading out of town

The ash will kill us in this traffic jam
Watch the people from the shelter, they too running
A man in his drawers and a poor baby crying
Lean on your horn, drive faster, hurry man
Keep calm, plenty cars infront, and you should see behind
This is the big one for sure this time
Thank God we had the car full with gas
Ah sorry for those people who have to walk
How come we didn't get no warning, somebody fall asleep?
Hear the stones on the car roof
People gone this time
If the ash don't stifle them, stones go kill them dead
Put on the radio. Oh, They off the air?
Had, good thing we have a car, sorry for those who don't

Lord, watch them run
Allyuh, run, run, run.....

(Rachel Collins, Montserrat 1996)

The event causes growing uncertainty in the population and confidence in the volcanologists dwindles. The majority believes that the scientists should have been able to predict the eruption and issue suitable warnings. After this frightening night the population wants to know what '*the big one*' will look like if this last eruption was not a major one. Many also question their safety in the so-called secure zone. The volcanologists try to mediate. They explain that 'secure' only implies that there is no danger to life. A comprehensive risk map is designed and distributed to the islanders in October (Figure 4.13). It is based on the results of Wadge's and Isaacs' (1989) computer simulation and shows a combination of seven zones (A to G) and different alert stages (white to lilac). The alert zones reflect the actual volcanic activity and determine access to and utilisation of the different zones. Risk increases from zone G to A, i.e. zone A being most endangered.

Although there has been an enormous reduction of pressure after the dome collapse and subsequent magmatic explosion, the volcano remains very active. Dome growth continues at a rate of about $6m^3/s$. Further pyroclastic flows reach Tar River Valley. All of the residents living south of Belham River are now requested to leave their houses, Cork Hill being the only exception. The situation aggravates visibly. The north is not sufficiently prepared to receive further evacuees, and more and more people leave the island. The population has declined to about 7,000. Those staying on Montserrat are mainly civil servants who do not want to lose their pension, also owners of houses, land or companies and many elder people who are not willing to leave their home country. Those leaving Montserrat do this less because they fear the volcano but because of the worsening living conditions and their resultant personal difficulties. They hope to find work on St. Kitts, Nevis, Anguilla or Antigua. In the meantime a few hundred have taken advantage of Britain's offer for a two-year residence permit.

Photos 4.3 and 4.4. A house in Long Ground before and after the lateral eruption. (Possekel 1994 and 1996)

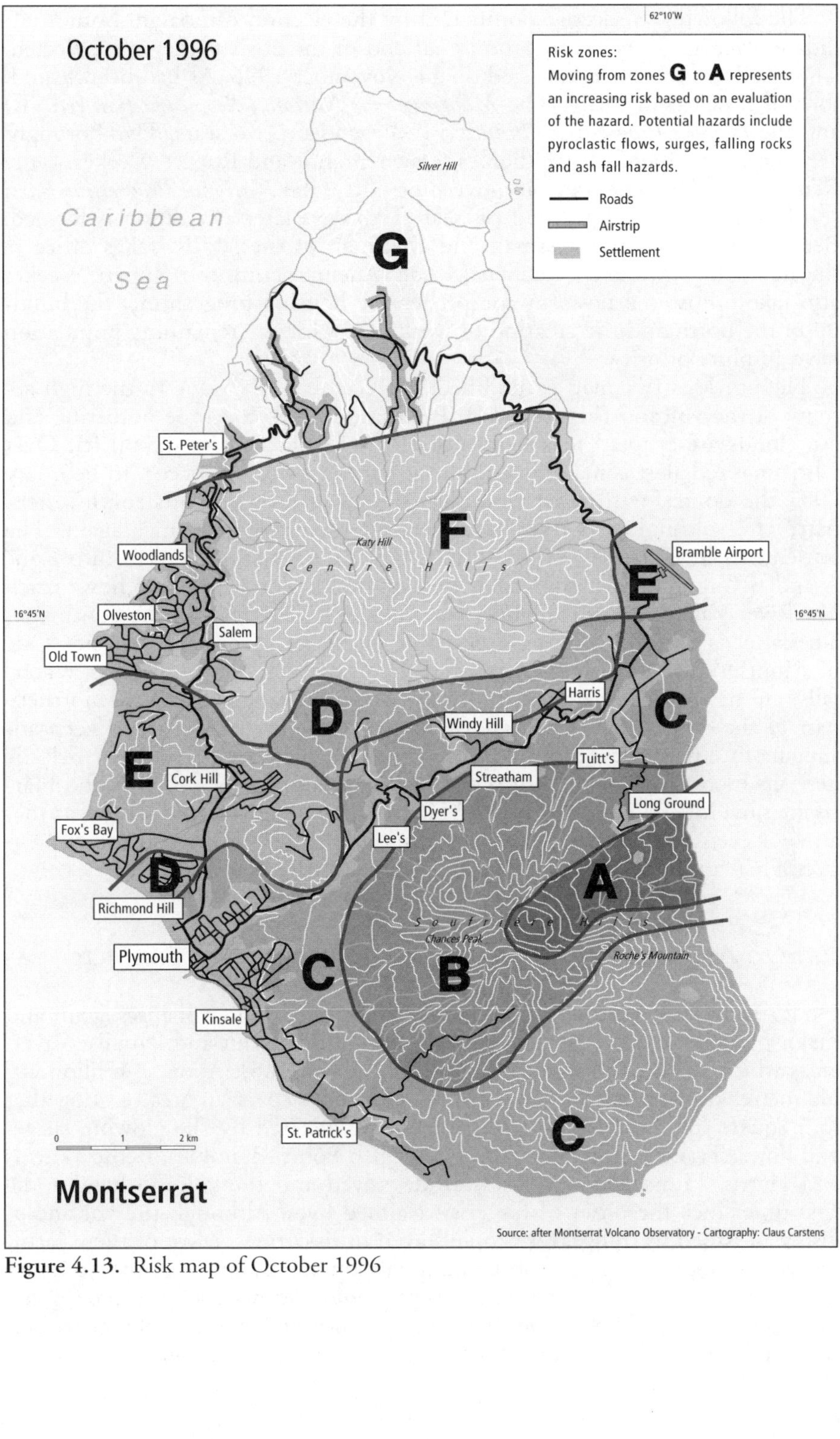

Figure 4.13. Risk map of October 1996

The following weeks are dominated by the election campaign. Some candidates even go so far as to promise an end of the crisis if they were elected. The election is finally conducted on 11 November 1996. As had been feared, there is no clear majority. The *Movement for National Reconstruction* (MNR) and the *People's Progressive Alliance* (PPA) each gain two seats. Two boroughs are won by independent candidates, David Brandt and Rupert Weekes. Only Reuben Meade of the previous governing party, the *National Progressive Party* (NPP) can successfully defend his seat. Two days later a coalition is formed. Bertrand Osborne, businessman and chairman of the MNR, takes office as the new Chief Minister. Reuben Meade, Adelina Tuitt and Rupert Weekes also take ministerial posts. A comprehensive housing programme, the build-up of the north and the creation of work places for the remaining population have absolute priority.

The previous risk map is modified in December 1996 due to the high activity of the volcano (Figure 4.14). Pyroclastic flows continue occurring and ash clouds are erupted to a height of 10 to 12 km (up to 40,000 ft). Over Christmas red alert is maintained, but no greater eruptions occur. In February 1997 the dome for the first time shows signs of growth in its southwestern part. The volcanologists thus have to adapt their risk map once again. The settlements St. Patrick's, Gingoes and Aymer's are now integrated into Zone A and B (Figure 4.15). By Easter the expected major pyroclastic flows reach Tar River Valley and, for the first time, White River Valley in the Southwest. The latter is buried by one million cubic metres of pyroclastic deposits. One of Montserrat's most significant touristic attractions, the Great Alp Waterfalls, are destroyed. Again, the volcano's activity changes. Now the northern part of the dome grows at a rate of about $4m^3/s$. Pyroclastic flows keep advancing into the valleys around the volcano. Each time the level of pyroclastic deposits increases. The risk zones are adapted again. Farrel's, Tuitt's and Harris are now integrated into Zone A (Figure 4.16). The volcanologists fear that a major event is imminent. They continuously issue warnings to the population. First rumours about an evacuation of the entire island spread.

June to August 1997 – The Disaster and Escalating Conflicts

On 25 June 1997 the disaster becomes reality. The dome collapses again and major pyroclastic flows advance towards Mosquito Ghaut and Paradise River, areas which have, before, not been affected (Photo 4.5). About 8 million cubic metres of material are deposited. The deposits cover an area of altogether four square kilometres. Eight settlements are affected by the glowing surges and flows: Trant's, Streatham, Dyers, Spanish Point, Brambles, Bethel, Farms and Harris. Hundreds of houses are destroyed and the volcano has for the first time since the onset of the crisis claimed lives. Although the volcanologists had issued warnings, 64 people stayed in risk zones. Most of them farmers who wanted to take care of their animals and fields. Those in risk zones on their way by car hear the warnings on the radio and manage to escape, some of them, however, being seriously burned. 40 people are rescued by helicopter task forces. 19 people cannot be rescued. Some are still missing.

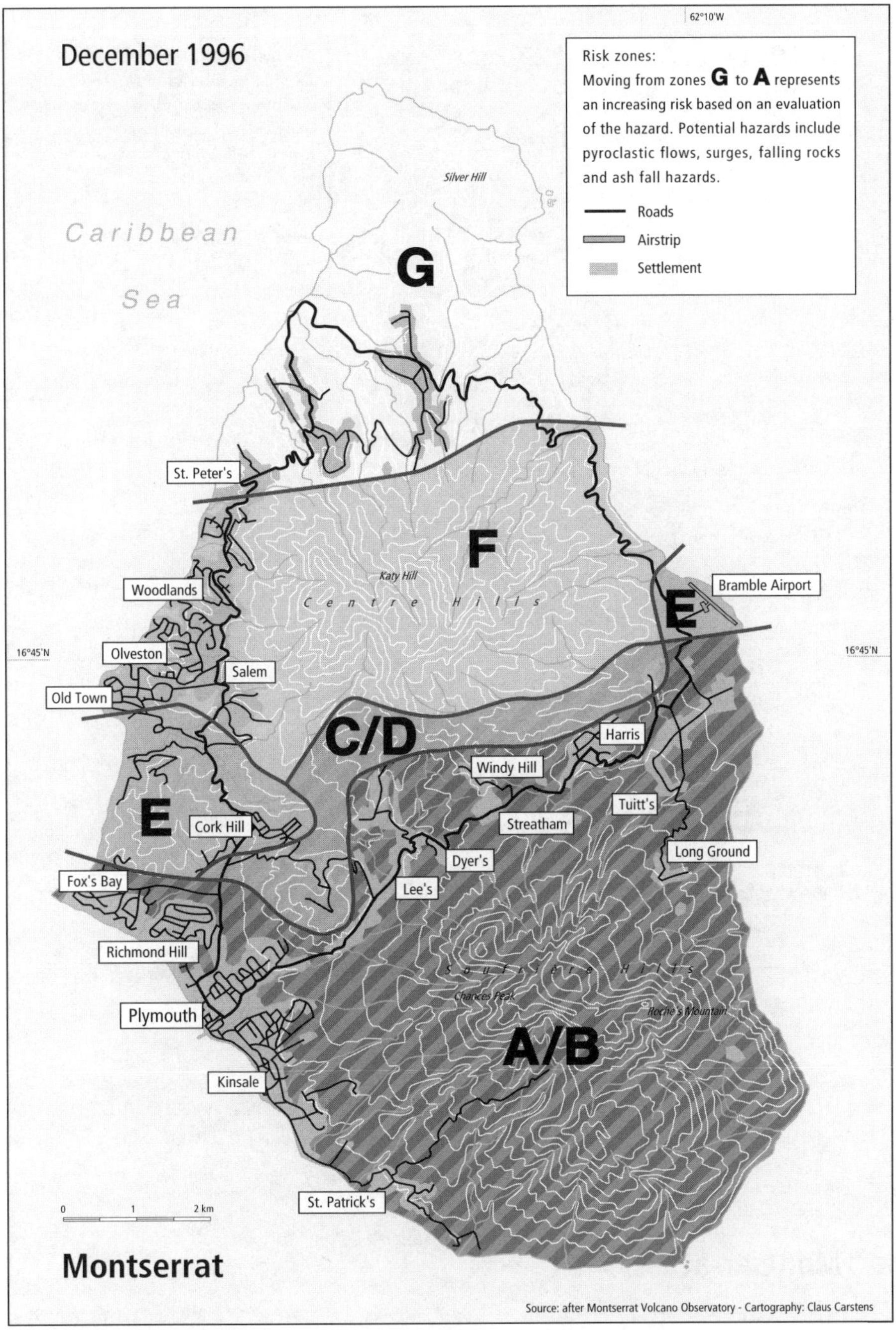

Figure 4.14. Risk map of December 1996

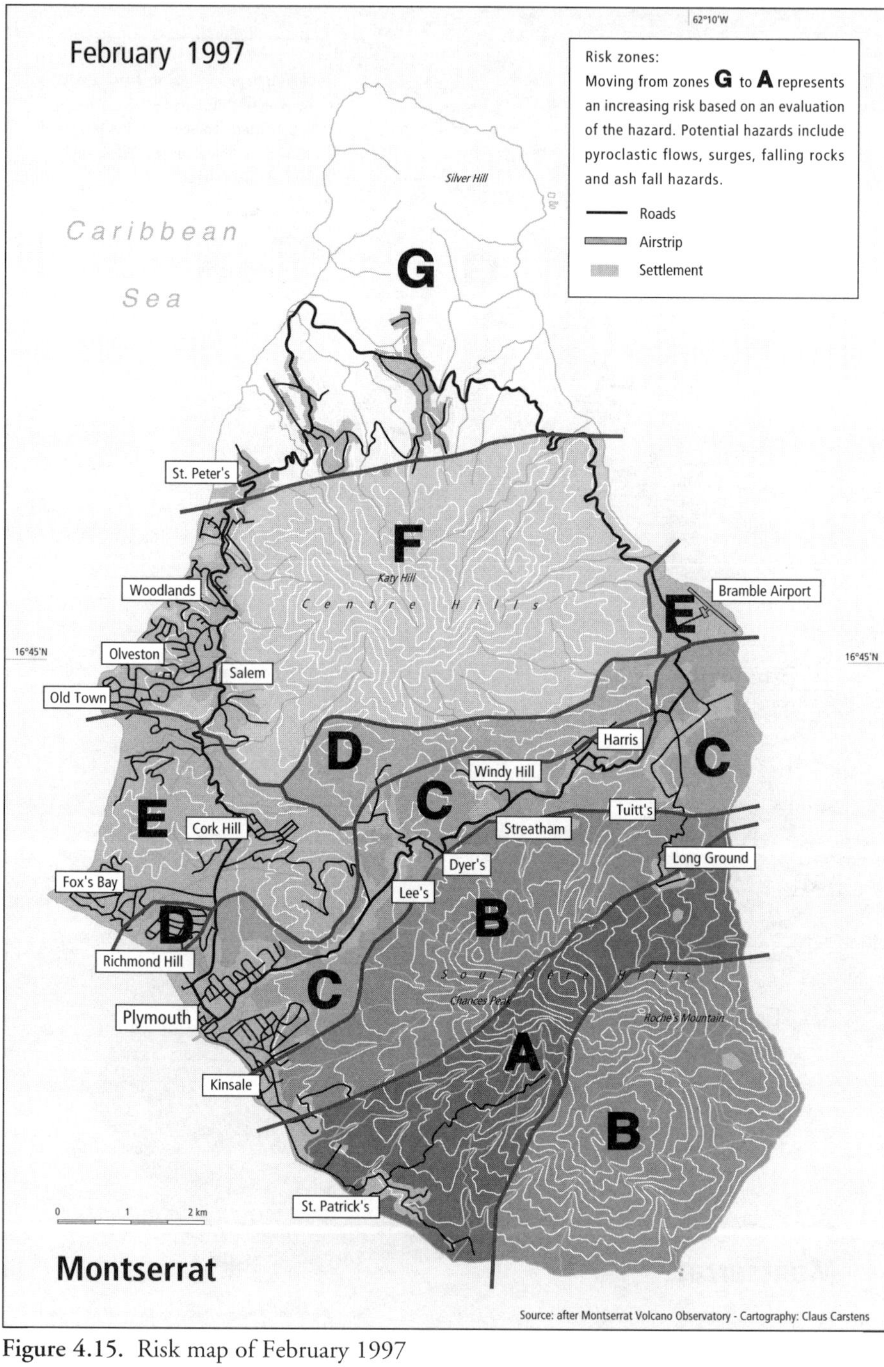

Figure 4.15. Risk map of February 1997

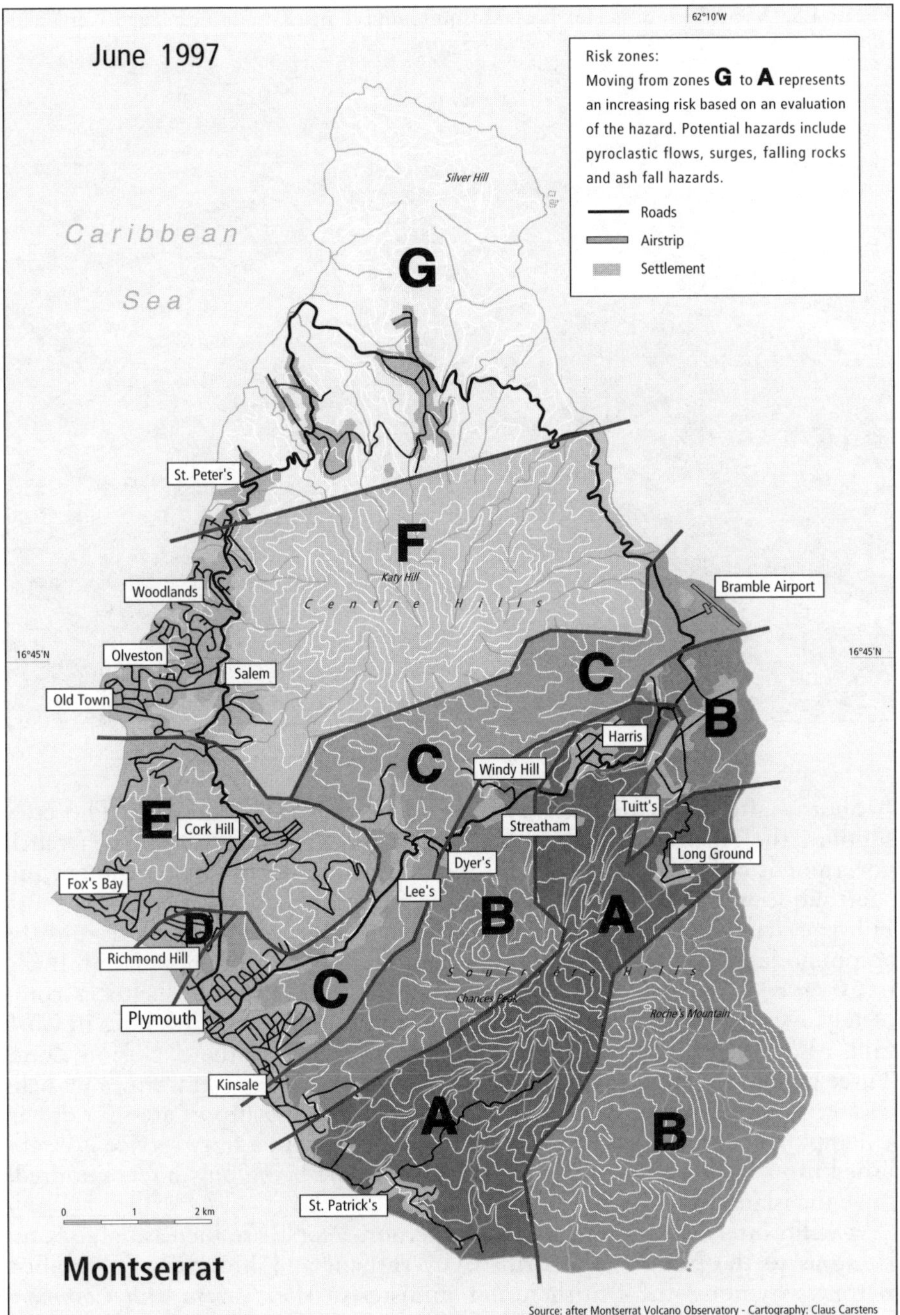

Figure 4.16. Risk map of June 1997

Photo 4.5. View from over Harris of Mosquito and Tuitts Ghaut with Tar River Valley in the distance. (Lea 1997)

Shortly after this event a Dutch vessel berthes in Montserrat. The crew supports the relief activities and helps to search for the missing. The French government offers its assistance and the injured are flown to Guadeloupe and Martinique where particularly the burnt are treated in special clinics. Cork Hill is evacuated a day after the disaster. A further 1,300 people have to be accommodated in the north. By now there are 50 % of the population living in 20 % of Montserrat's former total housing. The available shelter is completely overcrowded so that hundreds now refuse to leave their homes in Cork Hill. Also, many evacuated from other regions return to the Exclusion Zone. Three people are even flown out of the Exclusion Zone several times by helicopter. The rice mill in Plymouth, the harbour and the airport are shut down. Unemployment has risen to 60 %. The same weekend a ferry service is established from Little Bay to Antigua, but for the time being only a few hundreds leave the island.

A radio interview is given by the Governor. People are increasingly speculating as to the plans of the British government and Montserrat's viability. Baroness Symons, the British junior minister of the *Foreign and Commonwealth Office* (FCO), arrives on Montserrat on 30 June 1997 in order to gain an impression of the situation. During her negotiations with the local government she emphasises that the new British Labour government is prepared to offer comprehensive assistance, providing a detailed strategy can be presented. It is agreed that a plan is made to awake hope in the Montserratians and to avoid a mass exodus. But the local population is increasingly displeased. The local government as well as the British government are criticised. It is said that they do not represent the Montserratians' interests well. The critical situation in the shelters is also subject to harsh criticism. Suitable measures have been neglected although they were called for even before Cork Hill had to be evacuated. Some now request that Plymouth finally has to be given up. The British Parliament also discusses the ineffective resettlement policies on Montserrat. One Member of Parliament claims that the inefficiency was responsible for the many dead: "Had there been proper housing, had there been proper arrangements made for their resettlement, they would not have lost their lives. In the past there has been a lack of concern for the long-term interests of the people of Montserrat" (Reuter, 30 June 1997). Great Britain finally grants a further £ 6,8 million aid.

Uncertainty lasts for the whole of July. The dome grows fast and the residents of Isles Bay are requested to leave the area. Once again, there is resistance and some residents even write letters of protest to the Governor. Rumours increase that the British government strives for an evacuation of the entire island and is not willing to fund acceptable living conditions in the north. Speculations about Montserrat's future also increase on the internet Newsgroup. The *Emergency Operation Center* (EOC) finally decides to publish a daily report about the events and activities on Montserrat on the internet. By now 1,244 people have registered to leave the country and go to Britain. Half of all the nurses leaves the island because of the critical working conditions. They are substituted by nurses from Jamaica and Martinique. The risk map is, this time, completely modified. The previous system with seven zones and 47 different options is too complicated. The new map (Figure 4.17) shows only three zones: (1) the exclusion zone to which access is strictly prohibited, (2) the central zone that is subject to permanent alert and (3) the secure zones in the north. This is the only zone in which development activities are allowed, but its total surface extends to no more than 36 square kilometres.

The volcano is again very active at the beginning of August. It resumes its cyclical behaviour with hybrid swarms followed by pyroclastic flows. Strong magmatic explosions shatter pumice all over the island (Photo 4.6). On 3 August 1997 parts of the dome again collapse. Three massive pyroclastic flows for the first time advance towards Plymouth and destroy most of the town (Photos 4.7 to 4.10). At night the evacuated watch the reflection of the burning houses and have to abandon their hope for a quick return to Plymouth. On 7 August a number of pyroclastic flows head in different directions. It becomes increasingly difficult for the volcanologists to predict the volcano's activities and to issue timely warnings. Evacuation of the entire central zone which had before been assessed safe is now initiated. Old Town, Frith and Salem have to be left.

The events escalate in the following days and weeks. The natural disaster step by step turns into a political and social disaster. On 12 August 1997 civil servants demonstrate in support of better working conditions and higher salaries. On 16 August a further evacuation is initiated. The only area now assessed as safe is that north of the River Nantes (Figure 4.18) At the same

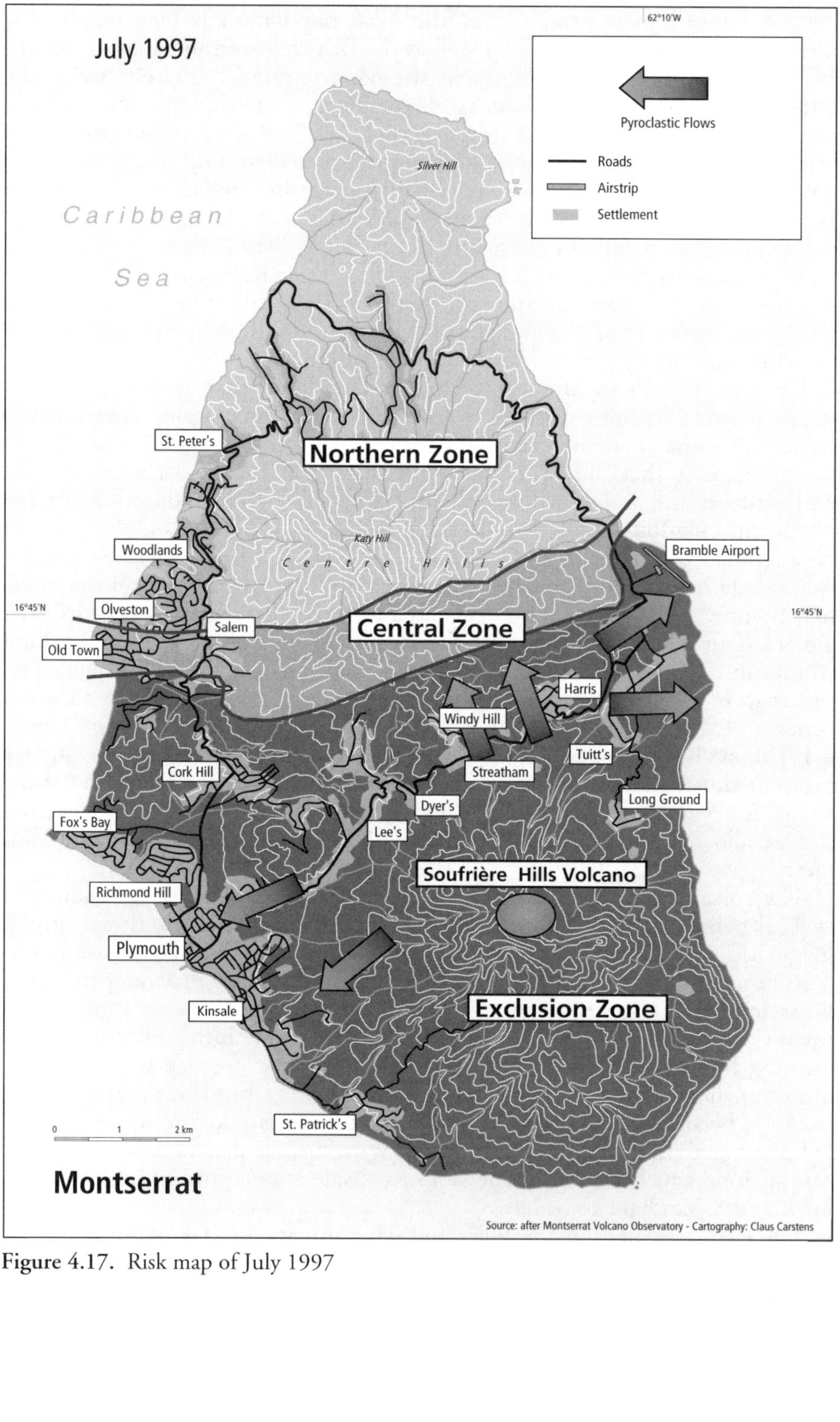

Figure 4.17. Risk map of July 1997

Photo 4.6. Eruption of August 6th. Pyroclastic flows can be seen going down Mosquito and Gages Ghauts. The Hollender Estate is in the foreground. (Lea 1997)

time the British government announces that it will financially support all Montserratians willing to resettle to neighbouring islands. This information is misinterpreted by the international press as an evacuation of the entire island. Rumours to this end are supported by a statement made by Georges Foulkes of the *Department for International Development* (DFID): "... the possibility of a massive, cataclysmic-intensive eruption of the volcano can no longer be ruled out. The consequences of such an event, if it did occur, would be extreme and anywhere on the whole island could be significantly threatened ..." (Georges Foulkes on 18 August 1997, Reuter). This statement actually was a misinterpretation of an MVO-study, but the reassurances of the volcanologists that the north of the island was safe could in the light of the general escalation do little to calm the situation.

Montserrat's population is, in view of the ongoing speculations, extremely uncertain. There is also still no precise information about the extent of financial aid granted to Montserratians willing to leave the island. First protests are organised in Salem the night of 19 August 1997. The protests increase over the next few days. A financial programme for resettlement to other Caribbean islands is finally announced. Adults will receive £ 2,500 each (unless they have more than £ 10,000 savings) and children £ 600 each. Additionally, flight tickets will be paid by the government. This offer is far below the sums the local government had requested and does not meet the expectations of the

Photos 4.7 and 4.8. The War Memorial in summer 1994 and August 1997. (Possekel 1994 and Darby 1997)

Photo 4.9. Aerial view showing the damage along Ft. Ghaut on August 6th. (Lea 1997)

Photo 4.10. Destroyed houses in Plymouth. St. George's Hill and the Soufrière Hills are in the background. (Darby 1997)

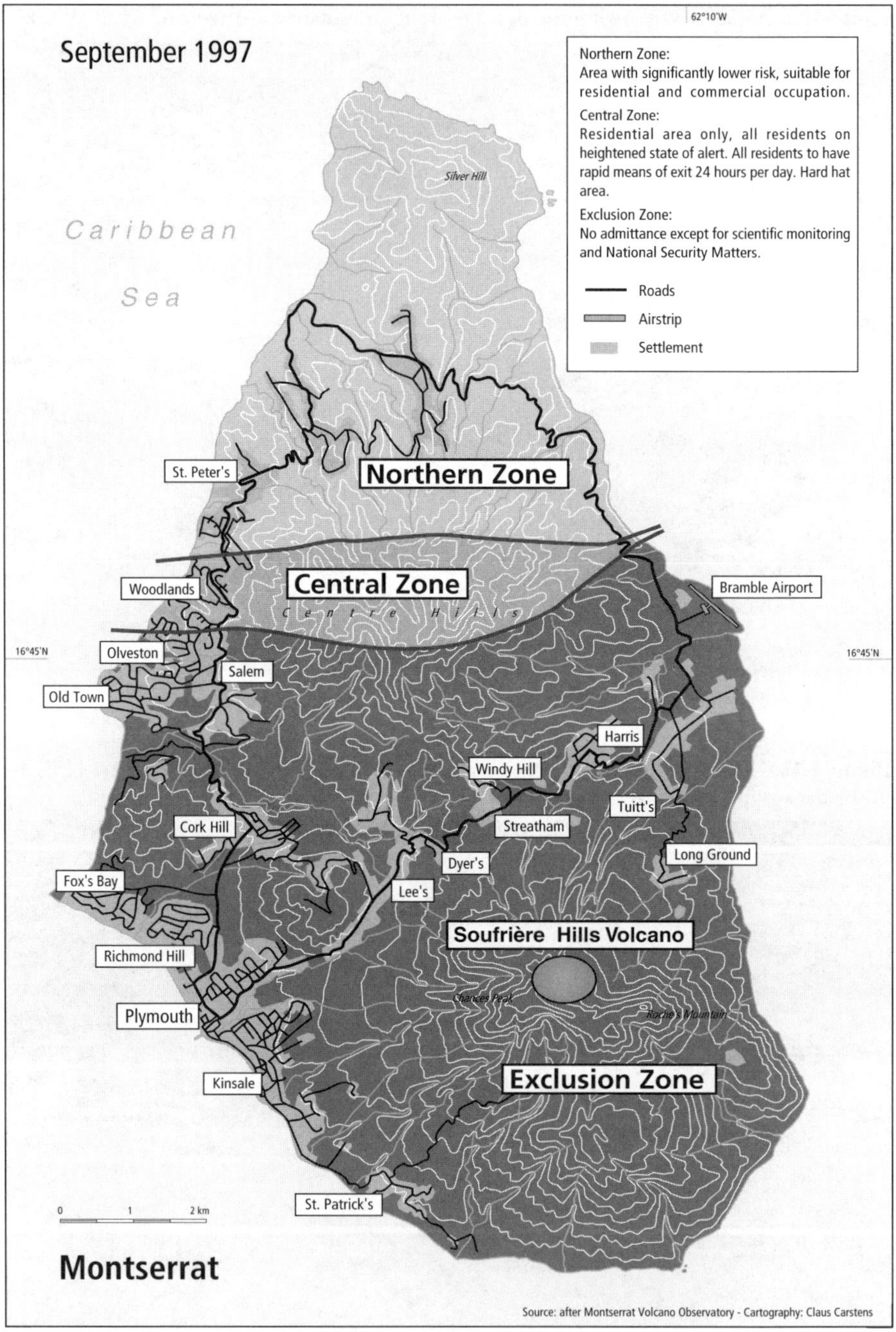

Figure 4.18. Risk map of September 1997

affected, many of which have lost all their possessions. After further demonstrations Bertrand Osborne finally tenders his resignation. David Brandt assumes the post of Chief Minister. He will soon travel to London in order to take up negotiations with Premier Tony Blair.

At the same time it is reported that a contract with a US-American company for the construction of 250 prefabricated houses has been cancelled. The dramatic reports in the international press also prompt all insurance companies to terminate their policies for Montserrat. The population is upset and questions about the whereabouts of the British funds amounting to £ 48 million come up (Chapter 4.2.2). The people want to know-how they are expected to build up a 'new Montserrat' without any insurance cover. The British Minister of Foreign Aid, Clare Short, comes under fire. She gives an interview and remarks that it is to be expected that the Montserratians will be demanding 'golden elephants' next. Again protests are made on the island and reports about the mismanagement of the volcanic crisis are published by the international press. When the Foreign Minister Robin Cook is on a state visit to Asia he is repeatedly asked about the situation on Montserrat. Put under pressure, he promises a detailed inspection and improvement of the present management. An investigating committee is set up, involving high-ranking decision makers of six ministries. At the same time British funds for the construction of houses in the north of the island are increased by a further £ 10.4 million.

September 1997 to April 1998 – Present Peak of Eruption and Intensified Efforts to Develop the North

By September 1997 the dome has expanded to a height of more than 1,000 m and a volume of about 80 million cubic metres. It continues to grow at a rate of $7 m^3/s$. Explosions are a recurrent phenomenon. Volcanologists have alone counted 76 explosions in the period between 22 September and 21 October 1997. Pyroclastic flows have during this time destroyed the airport building, parts of the runway and the remaining houses in Bethel, Tuitt's and Spanish Point. The new Governor Tony Abbot orders the water and electricity supply to be cut off to the houses in the Exclusion Zone. He wants to ensure that no further lives are endangered. A comprehensive plan for a possible evacuation of the entire island is worked out.

By now 4,138 Montserratians have registered for the resettlement project. 1,824 of them have already left Montserrat to move to other Caribbean islands. The population shrinks to 4,000. In November Britain requests the MVO to prepare a detailed analysis of all potential volcanic activity and its risks for the remaining population. The volcanologists believe that there is only a slight, insignificant risk of an eruption affecting the region north of the River Nantes. The British government releases some of the frozen funds, amongst others those for the financing of 50 houses, low interest mortgages and loans for businesses. Construction activities in the north grow from day to day (Photo 4.11).

Photo 4.11. A new settlement with prefabricated houses. (Darby 1998)

On 26 December 1997 the most violent eruption to date occurs – it goes down in the annals of history as *the Boxing Day Event.* 20 million cubic metres of material are blown up after a huge dome collapse. Pyroclastic flows inflict devastations on the volcano's entire western side. A number of evacuated settlements are destroyed, amongst them Morris, Reids, St. Patrick's, Trials and Fairfields. The economic damage incurred so far amounts to about US$ 120 million. As before, a new dome begins to grow directly after the eruption. In March 1998 dome growth finally comes to a halt. Volcanic activities in the following months are limited to smaller earthquake swarms, rockfalls and landslides. Another major pyroclastic flow advances towards Tar River Valley at the beginning of July 1998. On this occasion about 20 % of the dome collapses. Since then there have been numerous further collapses. The volcanologists put this down to the fact that the dome is unstable. There is no evidence of renewed magma movement.

All activities now concentrate on the development of the northern region. More and more houses are constructed so that the number of those accommodated in provisional shelters is reduced to 512 people. The British Foreign Minister Robin Cook visits Montserrat in March 1998. He confirms that his government takes its responsibility for the development of the north seriously. Great Britain has since the onset of the crisis provided funds of more than £ 51 million (Table 4.11). In view of the reduced volcanic activity the first Montserratians have started to return to the island. Plans for the construction of a new hotel in Woodlands are announced in April 1998 and speculations about a possible return into the Exclusion Zone and about the reforestation of the destroyed areas are coming up.

Table 4.11. British foreign aid and disaster funds for Montserrat between June 1995 and March 1998 (in million Pounds Sterling). (After International Development Committee 1998, p. 31)

	Funds granted	Sums spent up to September 1997	Sums spent up to March 1997
Finalised projects	1.8	1.8	1.8
Pending projects	9.4	2.9	7.5
State budget	12.1	12.1	12.1
Disaster funds*	22.5	12.8	22.5
Resettlement funds	7.6	—	7.6
Total	**53.4**	**29.6**	**51.5**

* including £6.5 million for houses and £0.5 million for the hospital

4.2.2 The Meso Level – Perceptions of Gatekeepers, Plans and Strategies

As the chronology of the crisis shows, complexity in a dynamic and uncertain disaster situation is extremely high. The interaction between humans and the volcano is very direct, being intensified by the small size of the island. This interaction is characterised by unpredictable, unknown and not immediately comprehensible processes. In the following these processes will be analysed on the meso and micro level. When approximating the meso level different temporal phases will be distinguished, corresponding to the gradual escalation of volcanic activity. The perceptions of the gatekeepers are derived from interviews with about 30 politicians, civil servants, planners, scientists and representatives of the private sector. These interviews were conducted on Montserrat in autumn 1996 and January 1997. Many of the gatekeepers' perceptions were followed up by personal contact or discussions with the Montserrat Newsgroup on the internet. Furthermore, discussions were held in London in spring 1998 with representatives of the *Foreign and Commonwealth Office* (FCO) and the *Department for International Development* (DFID). As in Chapter 4.1.2, next to the perceptions of the gatekeepers, the relevant plans and strategies are taken into consideration.

First Phase: July 1995 to April 1996

The majority of gatekeepers, particularly the politicians, had one central objective during this phase. They intended to keep the population at a level that guaranteed the viability of the societal functions. An early mass exodus was to be avoided since volcanic activity was perceived as a temporary phenomenon. Amongst other arguments, the volcanic crisis on Guadeloupe in 1976 was referred to. At the time an early evacuation of about 72,000 people caused enormous expenses and the volcano never erupted. Such a case was to be avoided for Montserrat. This is the reason why the first evacuations were ordered at short notice. The then Governor Frank Savage was the only one to express at this early point in time that there was a necessity to provide sufficient emergency shelters and to build houses in the northern part of the island in order to be prepared for increased volcanic activity. The local decision makers did not share his view, and British aid was limited to emergency aid in the case of acute evacuation and the funding of the MVO. The volcano had in this first phase of the crisis been completely underestimated.

As a result of this general perception, life more or less carried on 'as usual'. The programmes, plans and strategies (outlined in Chapter 4.1.2) were strictly adhered to. Amongst others, the *Forestry, Wildlife, National Parks and Protected Areas Act* (No. 3, 1996) had been passed but its implementation was postponed. After the two hurricanes had passed in autumn 1995 the beaches, mangroves and parts of the forests had been cleaned up and reforested. Around US$ 10 million had been granted by Britain. Designated measures comprised the extension and modernisation of the Forestry Unit, reforestation of ecologically sensitive regions, the Agro Forestry Programme and the management of some plantations. The endeavours in this phase concentrated on the reforestation of plots in the northern region and managing the watersheds since the water supply was prone to becoming a problem due to intensified volcanic activity. Merely the redevelopment of the waste disposal site at White's was in the light of volcanic activity postponed.

Tourism had in the first half of 1995 increased considerably, but collapsed at the onset of the volcanic crisis. The gatekeepers therefore concentrated on marketing this sector. It was their aim to motivate the snowbirds to continue visiting the island and to maintain cruise ship tourism. Day tours to the northern region were organised, new walking trails prepared in the Centre Hills and picnic areas established on some of the beaches. The plans to equip the harbour with a number of tourist facilities were also adhered to. The appearance of the harbour road was improved, shops and toilets were installed. On the whole, the volcano was at a very early stage used as a marketing argument for eco and adventure tourism. Adverts were, for example, published in magazines like *Outdoor* and *Outback*. The *Tourist Board* was soon requested to stop its active marketing of the volcano since visitors were, for safety reasons, not to be admitted to the Exclusion Zone.

In summer 1995 the land utilisation plan (*Physical Development Plan*) was finalised. The goals set had, despite the occurrences, hardly been changed. The gatekeepers even viewed the volcano as a catalyst for the planned decentralisation. The first evacuation, however, showed that suddenly decisions had to be made which severely affected the northern region. As a result a *Reappraisal of Development Strategy in Response to Volcanic Situation* was carried out from October to December 1995 in order to determine the necessary adaptations (Physical Planning Unit 1995a). To this end first of all three scenarios were designed for the potential course of volcanic activity.

1. *Status Quo Scenario*: volcanic activity remains but no part of the island has to be evacuated;
2. *Emergency Scenario*: volcanic activity calls for a temporary evacuation of parts of the island;
3. *Relocation Scenario*: volcanic activity requires a medium-term evacuation of the entire risk area.

Representatives from ministries, authorities and the private sector were confronted with the scenarios and asked for their plans. Finally, some measures for a safety zone in the north of the island were determined: (1) extension of the school in Salem and the temporary hospital in St. John's, (2) improvement of electricity supply and roads, (3) search for alternative locations for the police station, radio transmitter, prison, MVO and a jetty and, finally, (4) land acquisition for short-term emergency projects like the construction of shelters, warehouses etc. The extension of the airport was also postponed for the time being. These measures appeared sufficient to cope with a temporary phase of volcanic activity. However, after the second evacuation in December 1995 it was taken into consideration to resettle the population to the north for a longer period.

Second Phase: April 1996 to May 1997

It was with the third evacuation in April 1996 that the extent of the crisis became completely apparent and a transformation began from reactions at short notice to medium and long-term adaptations. The gatekeepers realised that they had to balance the newly evolved needs of the evacuated population and the sustainable development of the northern part of the island. The *Physical Planning Unit* (PPU) reacted instantly and created a strategic development plan for the north in collaboration with other ministries, authorities and the private sector. The plan aimed at resettling the evacuees and relocating shops, the administration and various infrastructural facilities to Salem, St. John's, Carrs Bay and Little Bay. Permanent housing, schools and factories were to be constructed, emergency shelters provided and an agricultural development programme issued. Despite all these activities the gatekeepers still expressed their desire for a quick return to the south, especially to Plymouth. The volcano was still assessed as a catalyst for the spatial decentralisation strategy: "The present volcanic crisis has presented the northern area of Montserrat as a centre of opportunity and the setting for the dynamic expression of human potential and human development. The crisis has let to a rapid concentration of people in the northern area, and has placed an extraordinary strain on the ability of the national government to meet the needs of residents. It is important to note that the situation in the north of the island has presented planners with an opportunity to plan the development with innovation" (Physical Planning Unit 1996, p. 7).

The dilemma incurred in this phase with the management of the volcanic crisis was that the resources for developing the north were to be used as effectively as possible whilst at the same time the central facilities in Plymouth had to be maintained. It has to be kept in mind that some of these facilities had been installed or thoroughly modernised just after hurricane Hugo. It was also decided to stick to the development goals determined before the onset of the volcanic crisis. On the other hand preparatory measures were called for in the light of the increasing activities of the volcano (Physical Planning Unit 1996). As a result, the interest conflicts between the decision makers grew. Some of them intended to declare Salem as new primary centre of the island and organise the previous uncontrolled construction of emergency huts, shops etc. in the north. Large investments were thus directed to Salem for the construction of new schools, shelters, shops, offices and houses. Other decision makers assessed the probability of larger eruptions as far greater and preferred resettlement to the far northern settlements, accompanied by the construction of an emergency jetty at Little Bay. A third group feared the destruction of the resources and the unspoilt landscape by such developments. They preferred an early return to Plymouth and the utilisation of all existing facilities, like the jetty at Belham Valley.

Particular difficulties arose when it came to the purchase of land and the realisation of the housing programme in the north. The strategic plan for this region provided for 94 hectares of land, necessary to meet the demand for housing in the safe zone. The realisation of this plan would have required an enormous financial effort. Some of the land owners also demanded high prices to such an extent that the acquisition of land and as a consequence the construction of houses was blocked. Nevertheless, the Governor for a long time waived the government's rights to insist on enforced sales of private land at a fixed price. Until the end of September 1996 only 20 further emergency shelters and 32 prefabricated houses were built. The common design of the houses was subject to much criticism and it was often emphasised that private initiatives were a better alternative. Some key persons even questioned the need for the construction measures. Britain at this point offered to also cover the flight costs for Montserratians willing to leave the country. The local government rejected this so-called Assisted Package Scheme since it tried to prevent a mass exodus prior to the forthcoming election.

Britain accepted this and supported the local government with £ 25 million for the development of the northern region. It placed absolute priority on the construction of an emergency jetty at Little Bay in order to ensure the supply of relief goods and allow for an evacuation in the case of an escalation. Once a US-American institute had confirmed that the project was ecologically tolerable, it was soon put into practice. The general supply of food stuff was a further priority set by the British government. Food vouchers were distributed in this phase of the crisis. Many decision makers criticised this concept and it was finally replaced with individual income assistance. A large amount of the funds went into St. John's hospital, schools, the national budget and emergency facilities in general. The construction of permanent housing was excluded since the local government believed it could source the necessary funds elsewhere. The Governor, at this point in time, once again emphasised that it was essential to build enough houses in the north. However, he was not given a hearing neither by the local government nor the *Foreign and Commonwealth Office* (FCO) or the *Department for International*

Development (DFID). It was not until the new government came into office that housing became a focal point of activity. But, again, conflicts arose. The British government requested to build the houses on state-property at Little Bay whereas the local government wanted to avoid construction activities at this early stage since this region was to be developed touristically. Little Bay was also considered the only location for a new town, should Plymouth against all expectations be destroyed.

Many of the people in charge of the emergency and development projects in summary adopted a policy of wait-and-see. This was mainly caused by interest conflicts between the gatekeepers and the prevailing uncertainty as to the volcanic activity. The following statement from a civil servant illustrates this: "We have developed and continue to perpetuate a wait-and-see attitude. Wait until the volcano quiets down and see if things get back to normal, then we will make decisions, then we will act. We cannot continue to pretend that the volcano will go away and put things off until such time" (personal questioning 1996). Furthermore, many short-term decisions were made which in the long run had negative consequences. A concrete example is the relocation of the public waste disposal site and the collecting tank for sewage from the emergency shelters to Little Bay. This location had severe ecological consequences as toxic substances washed out into the ground and the sea. The local decision makers decided to request experts of the *Caribbean Environmental Health Institute* (CEHI) and the German *Gesellschaft für technische Zusammenarbeit* (GTZ) to determine an alternative waste disposal site in the north-east. Nevertheless, waste was for many more months dumped at Little Bay and has not been cleared away up to now.

Due to the extreme pressure exerted on the resources in the north and the destruction of resources in the south, the gatekeepers in this phase paid more attention to ecological matters. The *National Trust* and the *Ministry of Agriculture, Trade and the Environment* (MATE) concentrated on the protection of ecologically particularly sensitive regions. A biodiversity study was supposed to be carried out in collaboration with the *World Wide Fund for Nature* (WWF). However, the funds of £ 100,000 were because of the ongoing volcanic activity finally transferred to the Cayman Islands. Some studies about the effects of the volcanic activity on the island's flora and fauna were, nevertheless, realised. Environmental-political programmes were furthermore initiated and broadcasted on the radio or presented in exhibitions or school projects. Information was given as to the effects volcanic ash has on human beings, flora and fauna and which species are particularly endangered. These include the Oriole, the Iguana and the Mountain Chicken Toad. All of these endangered species were to be protected in the remaining forest areas around the Centre Hills. The population's attention was also drawn to the fact that loose livestock endangers the natural environment.

Communication was assessed as successful, despite of the prevailing interest conflicts between the decision makers. The volcanologists, the Governor and ministers met twice weekly to discuss the actual volcanic activity and its implications. Once a week all public representatives (ministries and authorities) met and a further meeting in a smaller circle was also held to agree on their decisions and activities. Contrary to their impression, the gatekeepers of the private sector put many of the management problems during the crisis down to inefficient communication structures. For example, they mentioned the absolute necessity to base the management of the volcanic crisis on different scenarios. This shows that the gatekeepers were not aware of the *Physical Planning Unit's* (PPU) activities. The collaboration of British, Caribbean and in the early stages US-American volcanologists was criticised because of a number of contradictory opinions.

Third Phase: June 1997 to April 1998

The perceptions and activities of the gatekeepers underwent a visible transformation after the first deaths, the destruction of Plymouth and the steady growth of the Exclusion Zone. The local government now accepted the Assisted Package Scheme proposed by Britain. A system of financial aid for the people willing to resettle to neighbouring islands was at this stage established. The new Chief Minister David Brandt uttered harsh criticism since the various options were not of the same value. Montserratians who had gone to Britain for example received welfare whereas those who opted to live on neighbouring islands were excluded from such benefits. Brandt requested the British government to considerably improve the living conditions in the north and he also held it responsible for many wrong decisions. The questions he addressed to the British government illustrate this: "Why after two and a half years people are still in inhumane shelters with no permanent housing? Why there are still no permanent government offices built? Why no land has been purchased in the north for agriculture? Why after two years the hospital is in the condition that it is in? Why, two years after the crisis started, it took the death of 19 Montserratians to prompt the announcement of the first permanent housing project?" (International Development Committee 1998, p. xvi).

When the volcanic activity escalated it prompted severe communication problems between the decision makers involved. The volcanologists, for example, had to inform the Governor, the local government, the FCO and the DFID about the volcanic activity, a procedure they considered highly ineffective. Furthermore, the gatekeepers in charge at the FCO and DFID often changed and it was difficult for any of the newcomers to understand the situation on Montserrat. Many volcanologists also expressed their belief that the British government was, in effect, not really that interested in Montserrat, particularly at the time of the British election. Even after the new Labour government had been elected in spring 1997 there was no apparent improvement. The volcanologists continued their proactive work, constantly contacting the various departments, but without much success. Many misunderstandings ensued, often aggravated by reports by the international press. In contrast, the information flow between the volcanologists, EOC, local ministries and the Governor's office was further evaluated as very efficient. This was to a large extent put down to the work of the then Governor Frank Savage. The situation changed with the local election as the new Chief Minister and Governor were at first not able to keep the efficient structures upright.

As a result of all these events, endeavours towards a coordinating management and long-term planning strategy were finally intensified. A land utilisation plan was worked out for the safe zone, i.e. the *Physical Development Plan for North Montserrat.* It constituted the framework for all public and private activities. According to the plan 416 prefabricated family homes and 43 emergency shelters had been built since the onset of the crisis. The construction of a further 135 prefabricated houses was initiated, 35 of them funded by the Caricom States. The plan also comprised activities in education and health and the promotion of fishery at Carr's Bay, agriculture and manufacturing in Brades and tourism in the Silver Hills. A location for a smaller runway was determined. It was to replace the airport in the Exclusion Zone. The main objectives of the plan are the economic development and improvement of the living conditions whilst at the same time protecting the natural and cultural resources. Most of the people interviewed believe that it can contribute to maintain at least the conditions prevailing before the volcanic crisis (Physical Planning Unit 1997).

The ongoing intensive volcanic activity led to a review of the plans for an evacuation of the entire island. Although such a plan existed, the majority of the gatekeepers believed that the people were not sufficiently prepared for an emergency of that scale. Some claimed that neither they themselves nor the population had been aware of such plans. There were also severe communicative problems with other Caribbean islands, for example St. Kitts or Nevis, which were both preparing to receive Montserratians in an emergency case, but their governments had apparently not been informed that there also existed appropriate agreements with Antigua and Guadeloupe. Antigua's government was in this phase financially supported by Britain in order to be able to cope with the rising national budget caused by about 3,000 Montserratians who had chosen it as place of refuge. The funds were provided for health and education projects and the promotion of small and medium-sized companies the Montserratians were establishing on Antigua (International Development Committee 1998).

The reproaches of the international press induced the British Foreign Minister Robin Cook to generally revise the relations between the mother country and its dependent territories. He initiated an in-house examination of the FCO and DFID and an investigation into the previous management of the volcanic crisis. The latter resulted in an extensive report that illustrated the difficult coordination processes between the ministries and authorities involved (International Development Committee 1998), namely: (1) the FCO, responsible for the general political coordination with Montserrat, (2) the DFID, coordinating foreign aid, (3) the *Emergency Aid Department* (EMAD), being a subdivision of the DFID and responsible for the direct emergency aid, (4) the *Dependent Territories Regional Secretariat* (DTRS), a department of the *British Development Division* (BDDC) in Barbados, as a local institution in charge of expediting the coordination of development aid and projects, (5) the *Aid Management Office* (AMO) in Montserrat, established by the EMAD with similar responsibilities to those of the BDDC. Montserrat's Governor and local government were also included (Figure 4.19).

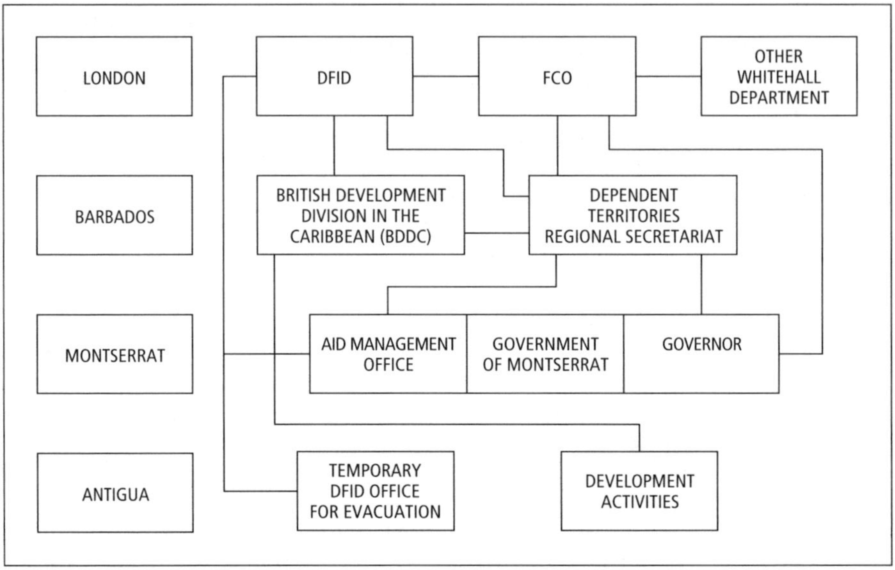

Figure 4.19. Institutions responsible for the management of the volcanic crisis. (After International Development Committee 1998, p. 37)

It became apparent that too large a number of institutions and decision makers were involved, a situation which recurrently posed severe coordination problems and delays when it came to the implementation of essential measures: "... we had to work within the existing machinery. You cannot change the machinery of government or its budget in the middle of an emergency (...) there are so many players in this thing that it is very difficult to have authority over people who make the decisions or know the answers ..." (extract from Clare Short's statement before the International Development Committee 1998, p. xvii). Personal rivalries between the decision makers were also revealed, both on Montserrat and in Britain. Many gatekeepers furthermore criticised the fact that there should have been a shift from pure disaster to development aid at an earlier stage. It was finally agreed that there should be closer coordination between the AMO on Montserrat, a newly established *Montserrat Unit* at the DFID, a *Montserrat Action Group* at the FCO, the Governor and the local government. The DTRS and BDDC were thus excluded from any further involvement.

The British government also initiated a further MVO risk study in order to secure the lives of the Montserratians and to back its own decisions and activities. The study addressed the main question "How safe is safe enough?". The results suggested that the potential risk for the regions north of Lawyers River was assessed minimal (in the ratio 1:100,000 and 1:1,000,000). The risk for the region north of Nantes River was evaluated low (in the ratio 1:1,000 and 1:10,000) whereas the people in and around Salem faced a moderate risk (between 1:100 and 1:1,000). Additional studies were carried out about the long-term consequences on health caused by the continuous ash falls.

As volcanic ash contains large proportions of silicon dioxide it can cause severe pulmonary diseases when inhaled over a long period of time (silicosis). Usually impairment of the circulatory or cardiorespiratory system are the consequences.

The risk of incurring any of these impairments is quite high for people living south of Nantes River whilst it is relatively low in the northern region. Britain arranged for the re-checking of these results by further experts. They confirmed the results obtained and recommended to evacuate about 200 people still living in and around Salem and a further 600 people living in Woodlands. They also pointed out that probabilities were not appropriate to reduce the uncertainty of decisions: "... we should be wary of placing too much faith in the predictions of future activity as a foundation of making decisions. The additional work has emphasised the unavoidable uncertainties. The report itself points out that such a complex system remains at some levels of its behaviour, inherently unpredictable. Ultimately, all that can be offered are distribution of probability, whose shapes and magnitudes are themselves uncertain" (personal communication).

Whilst the British government's priority was safety, the local government put Montserrat's and the Montserratian's future to the fore. It argued that there was no security anywhere in the world: "The question is how safe is Montserrat. We cannot guarantee the safety of anyone in Montserrat. And I do not think anyone can guarantee the safety of persons in New York, Miami, Boston, London or elsewhere. The prognosis of the scientists is that the risks are not considerable in the north of the islands" (personal questioning). The local government interpreted Britain's activities as another attempt to evacuate the entire island, this being the cheaper alternative. British decision makers in response emphasised that this had never been their intention and that a complete evacuation was costly since the existing infrastructure had to be maintained for a later return and military forces also had to be sent to the island to protect it against looters, drug dealers, gunrunning etc.

Summary

When analysing the various phases, it becomes apparent that the gatekeepers constantly select between different behavioural alternatives in their attempt to adopt to the escalating volcanic activity. Time has become a critical factor in the process. Sudden events call for instant reactions and are then followed by weeks and months of apparent peace and quiet. Time is a factor that facilitates or limits the options. Long-term planning is constantly interrupted by wait-and-see phases or chaotic reorganisation. Coping with the socio-economic and political problems that aggravate in the course of the ongoing crisis on the one hand calls for flexibility but on the other hand demands continuity in order to satisfy the needs of the remaining population. The spatial aspect has also become a critical factor since the events gradually destroy the *lebenswelt* of the population. There is pressure to decide, to take action and there are conflicts of interest between the gatekeepers, all aspects which intensify in the course of time and are generally aggravated by the numerous different perceptions.

4.2.3 The Micro Level – Perceptions of the Montserratians

The second approximation to the *lebenswelt* of the Montserratians at a later point in time, i.e. in autumn 1996, also based on qualitative interviews with 70 people. It aimed at determining the individual risk perceptions and adjustments. It was also ascertained how the decisions and activities of the gatekeepers were evaluated and which anticipations and hopes the interviewees had for their future (see annex for questionnaire). Furthermore, some subject matters of the previous questionnaires were again treated in order to assess alterations in the individual perceptions between 1994 and 1996. The previous results were categorised to also include quantitative analytical methods. The focal point of this comparative analysis was the perception of the environment and resources, the assessment of the various economic sectors and the evaluation of the political situation.

The interviews were conducted between the end of September and the middle of October 1996, i.e. during the third evacuation. Only a few days earlier the lateral explosion of 17 September occurred, for the first time affecting Long Ground and its surrounding area. Plymouth had by that time been buried beneath a 40 cm (1.3 ft) deep layer of ash. It was basically deserted with only a few exceptions. The harbour, the rice mill and the electric power station were still in operation. Most of the island's economic and social activities concentrated on Salem and Old Town. The administration was either accommodated in rented holiday villas in Old Town and Woodlands or found room in wooden barracks in the north. This is where the first prefabricated houses, schools and emergency shelters were build. 29 of the people interviewed owned a house in the safe zone. 41 people had to resettle to the north, 18 of which lived with friends or relatives, 14 in shelters and nine had rented or built their new accommodation on their own initiative (Table 4.12).

Perceptions of the Actual Situation at the Time and with Regard to other Disasters

Despite ongoing evacuations, the majority of the interviewees evaluated their living conditions as good to average. Only three people evaluated them as bad. The individual answers obviously depended on the type of accommodation and its conveniences. Those who lived in emergency accommodation or had rented new lodging, generally assessed their situation as worse. Occupants of emergency accommodation usually complained about a severe lack of privacy. Especially women criticised the inadequate sanitary facilities. They reported that they had to queue up as early as five o'clock in the morning in order to ensure that their children could get ready on time for school. Many also complained about the permanent quarrels between the occupants: "*The shelter life is horrible. There is a lot of fighting and arguing. I live in a shelter*

Table 4.12. Statistic of the sample in autumn 1996 (N = 70)

Sex	female	37
	male	33
Age (in years)	10 - 19	6
	20 - 29	20
	30 - 39	14
	40 - 49	14
	50 - 59	8
	60 - 69	4
	70 - 79	4
	80 - 89	0
	no statement given	0
Size of family (members)	1 - 2	13
	3 - 4	30
	5 - 6	13
	7 - 8	4
	9 - 11	3
	12 - 17	3
	no statement given	0
Occupation by sector	Tourism	11
	Public sector	22
	Trade	6
	Social services/public health	3
	Fishery/agriculture	1
	Transport sector	1
	Construction	3
	Financial sector	3
	Other services trades	6
	Pensioner	3
	Pupil	5
	Housewives	4
	Unempolyed	2

with 82 people and we share two toilets and two bathrooms. I'm sure you can imagine how this affects my daily life. I live out of my bag and have nothing except a plank bed. We are waiting for a family unit. If we don't get one, we will leave for Nevis."

When asked "*What kind of destruction happened to your property due to the volcanic activity?*" the following picture was represented. 38 of the interviewees said that they had been lucky and not claimed any losses yet. Some remarked that there was ash in the interior of their homes but they did not consider this significant. Many others, in contrast, said that the ash posed a great problem. They reported that they desperately tried keeping their houses clean but there was no way of avoiding the ash to penetrate electric appliances and furniture, gradually destroying it all. It had often also been experienced that the roofs of houses collapsed under the weight of the ash. Many therefore kept returning to the Exclusion Zone in order to shovel the ash off their roofs. The flowers and vegetables in the private gardens and the crop on the arable land had also been devastated by the ash. Some people from Long Ground had lost their houses and animals as a result of the lateral eruption. Most of them were, nevertheless, optimistic that they could soon return home (see poem). There was only one woman from Long Ground who was rather depressed and tried to repress all memory of her loss: "*I never went back home.*

I just cannot see the place. But friends and neighbours told me that some rocks destroyed parts of my house. I wonder if I will ever go back there? It will be totally destroyed in case of a major eruption. I therefore try to forget home. It is too hard to think about it."

Textbox 4.3. Montserrat - A dot in the sea.

> Montserrat was a dot in the sea
> Just as calm and as peaceful can be,
> When, one day the great mountain
> spewed forth a big fountain
> of ash, which rolled down the sea.
>
> Frightened people called "Goodness! Oh my!"
> and gazed up at the threatening sky.
> "We'll run, helter skelter
> and stay in the shelter,
> Oh Lord! Give us courage tonight."
>
> "The volcano", you ask, "when will it cease
> to rumble and threaten and tease?"
> There'll be leaping and prancing
> and singing and dancing,
> that fine day, when it goes back to sleep!
>
> Then the bells ring out joyous-lee
> The Queen will ask Bertrand to tea,
> all the world will applaud
> at home and abroad
> for that brave little dot in the sea!
>
> Written by a woman from Montserrat (name is unfortunately not known)

When the question was asked "*What are the causes for the actual volcanic activity?*" a common attitude the Caribbeans have towards natural disasters became apparent. 15 of the people interviewed responded that the volcanic activity was an act of God, six named it an act of fate and 26 stated that it was an extreme natural event, against which mankind could not do much. Many of the interviewees were not in a position to answer this question. A few referred to the islanders themselves or the local government as cause for the volcanic activity: "*Bad people, bad government and the way they treat foreigners here.*" Apart from that there were only two other speculations: (1) "*The human beings are the reason, their behaviour. I think the volcano started because of the greenhouse effect*" and (2) "*We had a long drought and then suddenly rain. This must be the reason.*" The number of people effectively knowing the cause was extremely small. They tended to explain the tectonic plate movements in a detailed way. Most of these 'experts' had gained their knowledge after the onset of the crisis. To them the awakening of the volcano had also been a great shock. Almost all Montserratians knew the Galways Soufrière and had learned at school that they lived on a volcanic island, but the volcano itself was commonly considered extinct. "*Well, we believed that the Soufrière is dormant. We were taught in school that it is dormant. There were signs of activity at the Galways Soufrière, but this was no indication for a larger eruption."* Only 20 of the interviewees stated that they had been aware of the potential threat. They had, however, hoped that they would not witness an eruption: "*We didn't expected it really. I didn't expect it during my lifetime. We knew that Montserrat is a volcanic island and that some day the volcano might erupt, but we believed that it won't happen to us.*"

The expectations as to the further course of volcanic activity were multifold. 22 of the interviewees believed that the active phase would not terminate until after a larger eruption. They often described the recent experience with the first lateral eruption which had been frightening to most of them. For this reason they were anxious about the implications of a large eruption. 15 of the people questioned believed that volcanic activity would cease soon since they assessed the recent lateral eruption as the climax of the activities. Another 15 people stated that the volcano would be active at the same intensity for another few months. Many answered that nobody except God knew what was going to happen: "*I'm sure that God will help us through this crisis. He must have heard us during the national day of prayer. Montserratians hope, pray and stay until the time the Government says to leave.*" Only one person considered it probable that the volcano could be active for another three or four years. The different expectations as to the further course of volcanic activity did not, however, significantly influence the desire to leave or to stay on the island. The reflections about this question were quite balanced. 32 people had already thought about resettlement, whereas 38 intended to stay as long as possible. The latter group in most cases stated that their decision to resettle depended on official instructions. Preferred destinations were the United States, Britain and other Caribbean islands.

The awareness of other disasters was determined to be relatively high. The frequency of such events proved to be of specific influence in this context (Table 4.13). The awareness of hurricanes is extremely clear and can surely be put down to personal experience and a quite active national disaster preparedness strategy. Most of the people interviewed did have difficulty, however, in evaluating their individual assessment with regards to the recurrence of various disasters. The disasters expected most likely were hurricanes, eruptions and earthquakes. A general consensus was revealed in view of the consequences of the various disasters. 49 of the interviewees stated that they considered the volcano the worst disaster. The reasons given were the long temporal period and the great uncertainty as to the future course of volcanic activity: "*The volcano has a very strong impact. Families are separated, we experience a severe economic loss, the social problems are growing, and I'm sure this will be even more difficult in the future.*" To many the volcano represented a far greater threat than a hurricane. 18 of the people, in contrast, responded that they feared a hurricane more and that Hugo had had far more devastating effects than the volcanic crisis since it had, after all, destroyed the whole island.

Table 4.13. Individual perceptions of potential disasters on Montserrat (N = 70)

Type of disaster	Named by persons
Hurricane	54
Volcano	29
Earthquake	26
Flooding	14
Landslide	5
Tidal wave	4
Oil pollution	2
Air crash	2
Drought	2

When the question of possible adjustments was raised, the responses were clearly motivated by the current volcanic crisis (Table 4.14). Most of the interviewees stated that the best alternative was resettling to less vulnerable areas of the island (in this case the north) or leaving it altogether. One person expressed his concern, arguing that the island was too small in size: "*I did some research on this issue via the internet. I believe that Montserrat is not safe. It is far too small. Compared to other volcanic eruptions we should have an area of at least 50 miles around the volcano which is not populated. But even knowing this, I'm not leaving.*" When asked about potential hurricanes and earthquakes, the responses generally emphasised the importance of earthquake-proof buildings. Insurance policies ranked low in the interviewees' considerations, once again a consequence of the recent experience: "*Insurance is of no use. The insurance companies are reluctant already. Therefore, leaving would be the safest alternative. But it is very difficult to leave home.*" Most also did not rely on the effectiveness of public measures. Some even stated that nothing could be done: "*You cannot do anything. And even leaving wouldn't help because something other might happen to you. Everywhere are hazards.*"

Table 4.14. Individual perceptions of possible adjustments (N = 70)

Adjustments	Named by persons
Resettling to the north	40
Insurance policies	20
Earthquake-proof construction	15
Leaving Montserrat	13
Recommendations of volcanologists	2
Nothing can be done	1
God will help me	3
Decent emergency shelter	1
Wearing masks (over the mouth)	1
Securing windows (hurricane)	1
Seek shelter (hurricane)	2
No idea	7

Evaluation of the Public Measures and Reconstruction Prospects

The next bundle of questions was designed to determine the interviewees' evaluation of the authorities' endeavours to restore social relations (cf. Figure 4.20). The situation was judged worst with reference to the schools and the provision of shelters. Many homeless were, in this phase, accommodated in school buildings, therefore these two answers were closely related. Water and electricity supply were, on the whole, evaluated positively, indicating that the necessary infrastructural facilities had been installed promptly in the north. Great uncertainty generally characterised the responses with reference to the distribution of development and disaster aid. The answers regarding this matter were multifold. 28 interviewees declined to respond to this question. Many criticised the public (and also private) transport situation. Mini busses had irregular schedules and shops and administrative departments were scattered over a large area. To many interviewees this implied inconvenience in their daily lives. Plymouth's central character was generally found lacking. The British government's measures were on the whole evaluated as very good to good. There were only 17 people who judged British aid to have been modest and complained about the slow and restricted granting of funds. The endeavours of the Red Cross were in most cases also evaluated as very good to good. Emphasis was in general placed on the reciprocal help of the Montserratians. Conflicts were in this context hardly ever mentioned at all.

The interviewees' personal attitude with regard to the reconstruction prospects was also determined (cf. Figure 4.21). At first it was asked how they assessed the attitudes of their neighbours, friends and relatives. 35 people responded that the Montserratians intended to reconstruct the island as early as possible. They also expressed their belief that those who had left the island wanted to return as soon as the situation had improved: "*Many have left. But if they see that the place recovers, they will come back very quickly.*" 16 of the

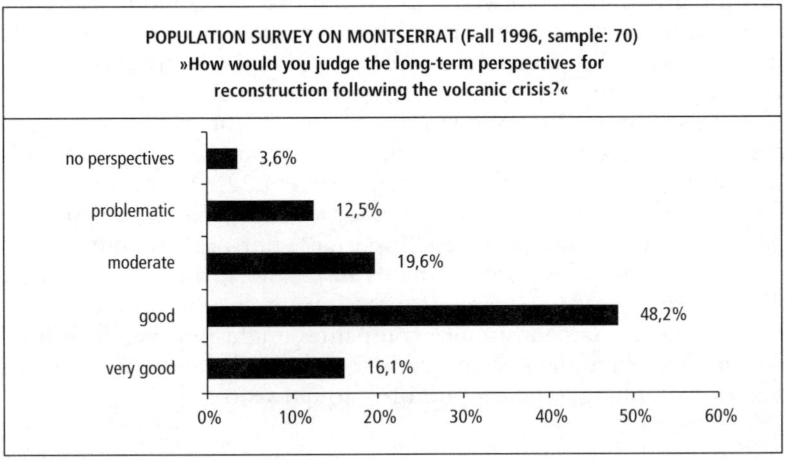

Figure 4.20. Evaluation of the public endeavours to restore the social relations

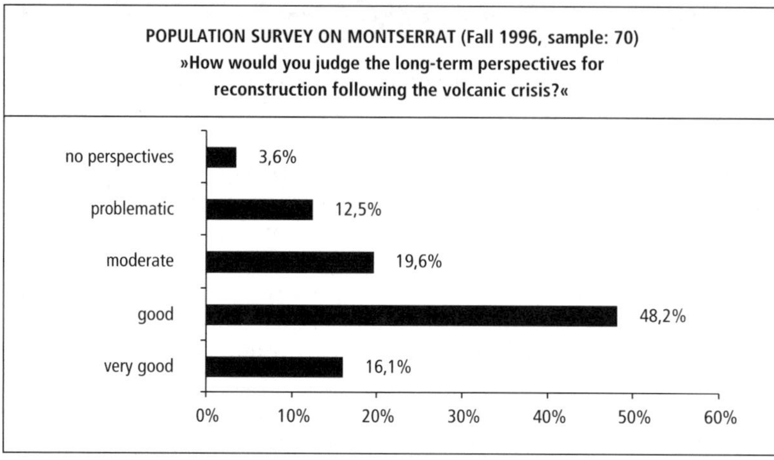

Figure 4.21. Evaluation of the long-term reconstruction prospects

interviewees stated that most people were frustrated and many were not in a position to cope with the difficult conditions for much longer. Six people were of the opinion that both the will to reconstruct the island and the frustration were the two main characteristics of the general impressions: "*Most of them want to stay and rebuild the place, but they are still very frustrated about the situation.*" The long-term perspectives for a full resumption of the predisaster state were evaluated as very good to good by the majority: "*It will be a great deal of work, but I'm optimistic. We will learn a lot.*" There was hardly anyone who did not see prospects. Some people mentioned problems, such as resettling to the north, employment and the ongoing uncertainty as to the volcanic activity. On the whole it was ascertained that the attitude towards the future prospects was not related to the individual's lodging situation, age or occupation.

The following question put forward "Do you think one could perceive the disaster as an opportunity for structural changes or do you prefer to restore what had been" intended to ascertain whether the people perceived the volcanic crisis as an opportunity for positive transformations. 34 people advocated structural changes, whilst 28 preferred to return to the predisaster state. When asked which were the most important structural changes, economic and social measures were put to the fore (cf. Figure 4.22). Vulnerability was to be reduced by risk zoning, whilst alterations to the spatial structure or building guidelines were considered less important, but not insignificant. In contrast to the earlier responses (adjustments to disasters), insurance policies were in this context assessed as being of great importance. Many people expressed the wish that the large insurance companies might improve their policies towards the Montserratians. The majority estimated the duration of the entire reconstruction phase to take around five to ten years.

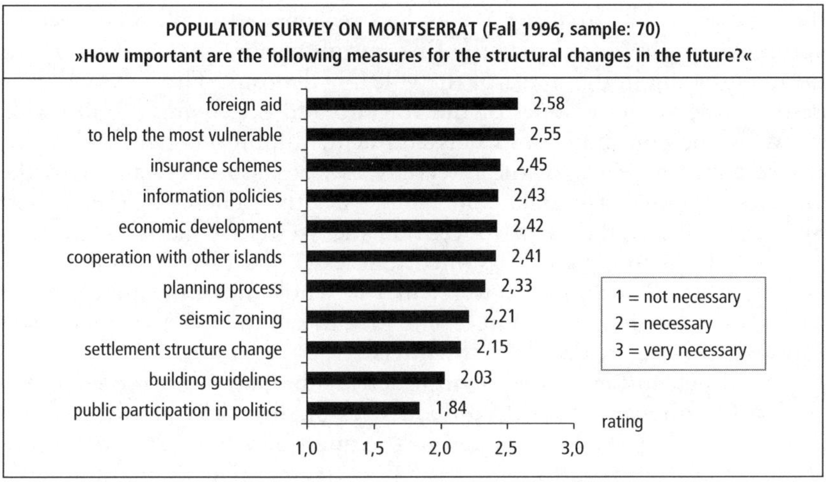

Figure 4.22. Evaluation of the measures deemed necessary for structural transformations

The volcanic crisis clearly determined the responses given to the question of the personal wishes. The majority wished that the volcanic activity ceased, that they could return home and that Montserrat returned to a normal and safe state.

- "I want to see Montserrat coming back to the situation before it all started. It was so peaceful and quiet."
- "Return to normal, safety for everybody here on the island."
- "That the volcano does what it has to do and goes back to sleep. Then everything can go back to normal."
- "That we can go back to normal, to the situation before it started."

Apart from the central need of security and normality, the other personal wishes remained to a large extent identical to those stated during the 1994 summer interviews. Most people desired a better income, a better job, a good education and an improved economic development for the island. Many mentioned that the actual economic crisis had aggravated their personal situation.

Perceptions of the Natural Environment and Resources

The first questions referred to the situation of crisis. Apart from this, other questions were again asked in view of the perception of the natural environment and resources. The natural environment in and around the Soufrière Hills had, at the time of the interviews, obviously changed quite considerably. The Tar River Valley, formerly a fertile and afforested valley, had been devastated by a series of pyroclastic flows (Photo 1). Many Montserratians were anxious about this development and asked themselves how long it would take until the vegetation in Tar River Valley had regenerated itself. Some people were optimistic and reported that the first coconut palms and banana plants were already growing in the midst of the volcanic deposits. The destruction of the forests on the western slopes of the volcano and St. George's Hill which was caused by the continuous ash falls and acid, sulphurous rain, did, however, arouse concern. Many of the interviewees were also uncertain as to the consequences of the ash falls on their health. There again, others remarked that the island was in its central and western regions as green as never before. They consequently assumed that lesser amounts of ash increased the land's fertility. The ash falls were on the whole not only perceived as threat but also as a potential resource. The idea to sell the ash as souvenir was also proposed and later initiated via the internet.

The general perception of the environment had, apart from these specific responses to the direct ecological consequences of the crisis, hardly changed (cf. Figure 4.23). 55 people responded to the question "*What is your personal view about the environmental situation on your island?*" that the natural environment was in a good to normal condition: "*It is still in good condition. Some areas have a lot of ash, but we have the feeling that the vegetation comes back very quickly.*" Only nine interviewees believed the environment to be in a bad condition: "*The environment was in good condition before the volcano started. Now it is a mess. And above all the hurricanes of last year ruined the bird sanctuary at Foxe's Bay.*" Asked about environmental problems which affected their daily life, the majority answered that it was not aware of any such problems. Only 27 people mentioned problems which affected them directly, such as waste, marine pollution, animals running loose and water supply. Most answers were, in summary, identical to those of the previous questioning. Amongst the new problems were toxic gases and the construction of the emergency jetty at Little Bay: "*Before the volcano it was getting better, especially the littering. They also want to built the alternative jetty in Little Bay, just where the reef is. In my opinion the reef should be protected for the tourists.*"

The perception of the natural resources had also hardly changed (Figure 4.23). As the comparison of the responses between 1994 and 1996 suggests, there were only slight shifts. Agriculture had lost significance as a resource whilst the sea and fishery were considered resources by a slightly larger number of people. This shift can be put down to the collapse of agricultural production and the development activities at Carr's Bay and Little Bay, both being significant fishery sites. The perceptions of the vegetation and landscape also changed moderately, surely, again, because of the effects of the crisis. Astonishing enough, eight people still evaluated the volcano as a resource. They tended to argue that it was an exciting experience to watch an active volcano grow day by day. It was also suggested to promote the volcano as a tourist attraction. Returning to the general questions, it became apparent that the majority considered it important to protect the environment and the natural resources. The results corresponded to those determined by the previous interviews in 1994. When asked "*What can you personally do to protect the environment?*", again, the majority answered "*no littering*", "*advise other people*" and "*educate my children*". 59 people considered it feasible to utilise resources whilst at the same time protecting them. The majority expressed that specific protective laws were needed to realise this goal.

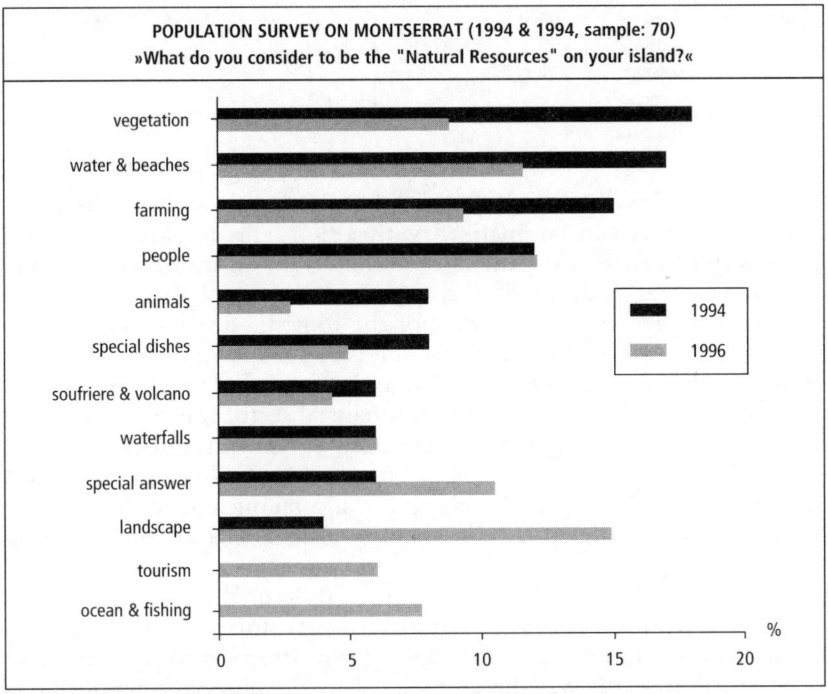

Figure 4.23. Perceptions of resources, 1994 and 1996 in comparison

Evaluation of the Economic and Political Situation

The economic situation was on the whole judged as unfavourable. Many feared loosing their jobs and to consequently have to leave the island. The economic development of the various sectors was thus considered even more important than in 1994 (Table 4.15). Marketing strategies for tourism and developing the agricultural sector in the northern region received high ratings: "*Tourism is important. We need tourists to come in during the crisis. The taxi drivers for example have nothing to do. Agriculture is also a problem right now. We have to import a lot. There is no land in the north to cultivate. The main areas for farming are destroyed or in the unsafe zone.*" In the light of the crisis the supply of food stuff received considerable interest, often with the effect that fishery on the whole gained importance, even though some people criticised the fishermen's practices: "*The fishermen set their nets too close to the natural harbours and coral reefs. This can cause ecological problems.*"

Table 4.15. Evaluation of the economic sectors, 1994 and 1996 in comparison

	1994			1996		
Sector/evaluation	**satisfactory**	**average**	**unsatisfactory**	**satisfactory**	**average**	**unsatisfactory**
Tourism	38	21	7	57	1	1
Finances	13	25	12	29	5	1
Agriculture	55	8	2	57	1	1
Fishery	–	–	–	54	1	2

When it came to evaluating the various economic sectors, less differentiated answers were given in comparison with 1994. This might, on the one hand, have been caused by the general economic crisis or, on the other hand, by the previous categorisation of the questionnaire which had obstructed extensive answers. Nevertheless, in view of the touristic development the interviewees generally still emphasised that the island should not be overdeveloped. Mass tourism was generally rejected and some people mentioned ecotourism as a significant niche. The offshore financial sector was often seen as a problem, most responses emphasising the need of control and monitoring. Compared with 1994, less interviewees knew about this economic sector, probably because the 1989 scandal was gradually falling into oblivion. The expats were, again, considered by the majority a beneficial factor to the national economy.

Corresponding to these evaluations, almost 70 % of the interviewees saw the economic development as the central challenge for the island's future. Tourism, agriculture and construction were given utmost priority. Emphasis on the construction sector was directly related to the necessary development activities in the north. A further matter of concern to the interviewees was the education of their children. Political matters were, at a mere 2,7 %, of considerably less significance than 1994, this especially in view of the forthcoming election being rather astonishing. Many of the people asked explained that each of the political parties encountered severe difficulty when attempting to manage the crisis. It, therefore, was not considered important which party was in power. The two questions "*Do you think the political parties respond to the needs of the local people*" and "*Do you think that politicians are capable of solving problems*" were both consequently answered to the following, quite balanced pattern. A third affirmed, a third negated and the last third stated that some parties or politicians did, some not. The vast majority of the interviewees now advocated that the island remained dependent territory of Great Britain, mainly because of the severe social and economic consequences of the crisis.

Summary

The following general results have on the whole been ascertained in view of the individual perceptions and evaluations of the volcanic crisis at the time of the second interviews in autumn 1996. The Montserratians in general show a deep affection to their native country. Many of them are thus inclined to stay on Montserrat, despite the risks they face. Those who have thought about leaving the island express that they would do this because of the difficult conditions and not because they fear the volcano. They encounter visible difficulty in leaving their country, families and personal belongings. Furthermore, future in another country appears to be rather uncertain to them. Against the background of these reflections, a resettlement abroad often seems to be a worse and uncertain alternative. It has also been revealed that the public adjustments are at this point in time still evaluated relatively positively. The exceptions are extremely contentious matters, such as the provision of shelters, schools and housing, and these are clearly enunciated. Financial support is rated highly important, by this time complemented by the majority's positive attitude towards Montserrat's status as dependent territory. On the whole, a decisive willingness for reconstruction has been ascertained, the future of their island being a major concern to most of the interviewees. Accordingly, the majority evaluates the island's prospects positively. Central wishes and desires are the return to normality, security, economic prosperity and that the island's nature and culture may be maintained.

The individual perceptions and awareness have in the course of the volcanic crisis, obviously, changed considerably. Until as late as spring 1997 the Montserratians declared that they did not consider the term 'volcanic crisis' appropriate, requesting to correct it to 'volcanic situation'. The latent threat had gradually become a 'normal' situation. But the events in summer 1997, especially the first fatalities and severe destruction of Plymouth and the eastern settlements had their consequences. Fear of the volcano's destructive powers and frustration surged visibly. Many reported that the island seemed to be getting smaller whilst the volcano was growing more and more. Herman Francis' song (Textbox 4.4) illustrates these impressions. The emotional aspects were accompanied by a worsening of the socio-economic situation. Unemployment had by that time risen to 60 % and the government was not in a position to provide adequate shelter. Many saw no other alternative but to leave the island. But the goals of both the refugees and those remaining on the island remained the same, with the modification that their 'new Montserrat' is to be rebuilt in the north of the island.

Textbox 4.4. One more river to cross

First it was the White's River
So they evacuate Long Ground
The weakening of the Gages Wall
Ft. Ghaut now threatens town
So next the whole population
Who were south of the Long Valley

No matter how many rivers we cross
This thing still follows me

One more river to cross
One more river to cross
Every time we start to count we loss
One more river to cross
One more river to cross
Every time we think we pay de cost
One more river to cross

First, it was White's
Next, it was Ft. Ghaut
Then it was Belham
Now they say Nantes'
Hear the announcement made
By the chief and the governor
In the next 72 hours, Nantes' to cross over
But the Risk Map say
Don't feel too much at home
From Nantes to Lawyer's
Is the Buffer Zone.

(Herman Cupid Francis, Montserrat 1997)

4.2.4 Interactions Between the Levels

The volcanic crisis illustrates the extreme vulnerability of a small island state. Two thirds of the island were at the end of 1998 uninhabitable, most of the agriculturally productive land has been devastated and subsistence fishery is on the decline. Tourists have not been received on the island for quite some time and most of the industrial centres have closed down or were destroyed. The remaining population inhabits no more than a mere 36 square kilometres of land. The pressure exerted on the space and the natural resources in the northern region continuously grows. The constraints on all four sub-systems clearly increase and aggravate the situation. This applies to soil erosion, water supply, waste disposal and the general vulnerability to other disasters. It also concerns the political coordination in general and the decrease in population with its brain drain effect. All these factors, especially the limited resources and space, severely obstruct the flexibility the people have when trying to adjust to the volcano. Their financial and personal dependence on Britain increases, in turn intensifying the apprehension to lose control over internal matters.

This situation is aggravated by the untypical eruptive behaviour of the Soufrière Hills Volcano. Other volcanoes usually erupt suddenly and then gradually decline. In comparison, the Soufrière Hills Volcano's activity has slowly increased with time and has come to its hitherto climax after two and a half years. Thus, the emergency situation cannot be temporally restricted as it is, for example, in the case of a hurricane. To the contrary, the disaster unfolds over a period of years and the typical phases of hazard management – mitigation, preparedness, restoration and reconstruction – are compressed temporarily. All phases are activated almost simultaneously. On the one hand, the volcano's activities have to be monitored, the population warned and constant adjustments found to cope with new and extreme events. On the other hand, a continuous development of the island's northern region is called for in order to satisfy the needs of the remaining population. Planning an appropriate reconstruction process during an ongoing crisis is an enormous challenge to all parties involved.

The prevailing uncertainty about the intensity, frequency and consequences of future events impede the entire decision making and implementation processes. The adjustments are usually related to the volcanic activity on a step-to-step basis. The situation constantly alternates between phases of wait-and-see and those of chaotic reorganisation. The initial phase at the beginning of the crisis was dominated by a waiting attitude, based on the assumption that the phenomenon was a temporary one. These hopes were intensified by the fact that reconstruction following hurricane Hugo had just been finished and that the majority of funds amounting to about £ 20 million had been invested into the zone which was now classified as endangered by the volcano. The Governor is the only one requesting extensive preparations for an emergency case, but most take this for exaggerated caution. This initial phase is therefore hardly used as an opportunity for necessary adjustments.

In the further course of the crisis it soon becomes clear that the people on all levels are not facing a short-term event but a rather permanent phenomenon. The local government now assesses the viability of the island as endangered. It strives to provide a secure zone for the population whereas the British government prefers a strategy of risk-minimisation and cautious decision processes with regard to the allocation of its resources. The diversity of interests grows with the crisis and when the disaster finally occurs in the summer of 1997, the interest conflicts also escalate. This escalation is to a large extent due to the former consultation and decision-making mechanisms between the local and British government. They are routinely adhered to during the crisis even though they are not flexible enough to meet the specific requirements of the situation. The realisation that these mechanisms do not work in an extremely complex emergency situation comes relatively late. Far too many gatekeepers are involved in the decision-making processes. They, furthermore, always tend to secure their decisions at the highest level, thus often causing temporal delays when it comes to the actual disaster and development aid. Many gatekeepers also do not have the essential know-how to interpret the volcanologists' scientific information and risk analyses. This situation is aggravated by sensational press reports during the summer recess 1997. Also, the governments of Britain and Montserrat, here including the Governor, change during the volcanic crisis. Some valuable and quite successful management adjustments which had been established in the learning process come to a standstill. An effective and integrative management is in general constrained by the complexity and dynamics of the disaster situation.

The gatekeepers also adhere to their established bureaucratic management style. They continue striving for control. After each relatively significant event the frame conditions are reexamined, revisions initiated, information gathered, and all plans adjusted to suit the prevailing situation. This is, on the one hand, sensible since it enables the gatekeepers to learn from their mistakes, understand the volcano better and develop and test new methods. The plans worked out before the onset of the volcanic crisis are also of valuable assistance since the former strategies for a spatial decentralisation can now be applied to the resettlement endeavours. On the other hand, many measures and activities are delayed where rapid reactions would be appropriate. Adhering to previous plans often obstructs flexible adjustment to changing conditions, as Clare Short's comment illustrates: "We have been dealing with a cumulating emergency rather than a stable situation where we can plan for the future" (International Development Committee 1998, p. xviii).

Coping with such an open situation with unknown outcome overburdens the gatekeepers since it implies discontinuity of the normal, the routine and the order. It is particularly the British government that strives to safeguard its decisions and requests precise forecasts about the further course of volcanic activity and a calculation of the risks at stake. But this approach will not reduce the uncertainty of decisions. Another factor is that the gatekeepers tend to individually interpret the risks. The different perceptions, mentalities and social values are decisive factors for the individual adjustments. The main concern of the Montserratians, for example, is to preserve their country and culture and protect their personal possessions. The majority is therefore for a long time prepared to accept the risks they face. There are, nevertheless, clearly perceptible differences between the individual strategies to cope with the situation, depending on individual personalities, circumstances and role in society. There is, for one, the rather fatalistic group that transfers all responsibility to the government or even God. These people accept their fate. They neither negate the risk nor do they passively accept it. For most of them it is not possible to actively participate or act. They leave the island when they cannot see further prospects and receive more or less acceptable offers for resettlement. Another group is, in contrast, activistic and problem-oriented. It tries to reduce the risks and the socio-economic consequences of the volcanic crisis as far as possible. Businessmen, for example, belong to this group. They have at an early stage either left the island or resettled to the north in order to seek their opportunities there. Many others also form part of this group of people, be it that they developed their individual strategies in the scope of their possibilities on the island itself or abroad. Many creative self-organisational processes evolve in this way, good examples being the media centre on Montserrat, the Montserrat Newsgroup's discussions or the reciprocal help. Mental and emotional digestion of the crisis in the form of poems, songs, film etc. is also such a self-organisational process, just as the self-help groups in Antigua, London, Montreal, the British Virgin Islands, Toronto and elsewhere. They support the Montserratians on the island and abroad with donations and advice. The expats who have lived on the island for a few years or still live there also show a strong sense of responsibility, actively supporting the Montserratians to cope with the crisis.

Sir George Martin, for instance, set an example. The former owner of the *International Air Studios* organised a benefit concert in September 1997 in the Royal Albert Hall with Elton John, Sting, Arrow's, Paul McCartney and Eric Clapton to raise funds for Montserrat. Sir Martin himself travelled to Montserrat to hand out the proceeds of £ 1 million. Other musicians who had produced albums on Montserrat also organised benefit concerts and, furthermore, many expats play an important role as informants. They have addressed the international press, local politicians or even the Queen in order to attract attention to Montserrat's problematic situation. They also keep contact to the Montserratians and can be considered an important channel of communication. These creative processes of self-organisation are a significant aid for the individual to cope with the crisis. It has to be said, however, that they are carried out in complete isolation of the administrative activities. The plans, concepts and risk studies have without exception been worked out by foreign experts and the local population is hardly involved in the management strategies. This leads to the effect that the basis develops its own strategies, as illustrated by the following statement of two expats living on Montserrat: "Es ist schon eine interessante Situation hier, fast wie in einem surrealistischen Theaterstück. Regierung und Volk sind komplett entkoppelt. Unsere Verwaltung verwaltet, regiert und plant. Und wir kümmern uns nicht im geringsten darum. Die Leute sind munter wieder nach Salem und andere Gegenden der 'Exclusion Zone' gezogen" (*We have encountered a very interesting situation here, almost like in a surrealistic theatre performance. Government and people are completely decoupled. Our administration administers, governs and plans. And we do not care in the least. The people have cheerfully returned to Salem and other areas of the Exclusion Zone.*) (Wolf and Inge Krebs, March 1998). Because of this separation numerous significant ideas, strategies and information are lost. The potentials of innovative self-organisational processes are hardly taken advantage of. The present hazard management rather concentrates on the possible control of the environment than the necessary encouragement and mobilisation of the individual skills and abilities of the affected people.

If one takes the interests and objectives of the meso and micro level into account, the implications of this decoupling are perceived even more clearly. The variations in the interests are, astonishingly enough, still low. Both groups desire to preserve the island's nature and culture, whilst at the same time striving for economic prosperity or, at least, resumption of the predisaster state. Both groups also evaluate the reconstruction prospects as positive and are prepared to accept and initiate comprehensive structural transformations. This preparedness was visible at an early stage of the crisis and turned to become a necessity after the severe destruction in 1997. And both groups also express a desire to maintain certain aspects of 'the good old times'. The objectives of both groups are on the whole identical, but what is lacking is the mutual implication of the goals. This situation can cause further problems and delays when it comes to the reconstruction process.

The success of reconstruction will also depend on the refugees' willingness to return to the island. From today's point of view most of the Montserratians abroad wish to return since most of them have not got accustomed to life in London, the United States or the neighbouring islands. They feel dislocated from their socio-cultural surroundings and wish to return home as early as possible. It is, however, common knowledge that the time factor is of significance in this context. The longer the crisis goes on, the more likely will the refugees adapt to their new countries. This is the case especially for the younger generation who is looking for a job or seeks further qualification, be it vocational training or university courses. Their willingness to return will decisively be determined by their opportunities abroad, but also by the success of the reconstruction on Montserrat. Should they not return, exactly those who are highly qualified and desperately needed for a successful reconstruction of the island would be missing.

Summarising the results and reflections, it becomes apparent that the complexity of the system Montserrat during the volcanic crisis is characterised by a number of components: a constantly changing number of involved agents, multilayered interactions and sudden fluctuations. The volcano is the cause of extreme change which, other than hurricane Hugo, leads to a complete restructuring of the entire system. The earlier spatial structure is completely reversed. The north, previously sparsely populated, has now become the only inhabitable area on the whole island. The previously densely populated southwestern region has, in contrast, to a large degree been destroyed. A return to this region might not be possible for years or even decades and a new town will, according to the gatekeepers, not be built in that region since there is too great a risk of further eruptions. But not only the spatial structures have changed. The social structures have also undergone severe transformations. The majority of the younger and qualified population has left the island whilst many members of marginal groups have remained. The latter are accompanied by politicians, civil servants, businessmen and construction workers. The volcano can thus be considered an impulse for the social and regional transformation. At the same time it also represents an opportunity for the conscious encouragement of structural changes, combined with a reduction of the society's vulnerability. For this reason it is recommended to not only consider the short and medium-term interests when planning the reconstruction process, but to also contemplate the long-term interests and objectives of the island.

Recovery is a link in a chain of events, which can be helped by carefully planned policy at each stage. At different points throughout the emergency management cycle, critical decisions are made, whether by design or by default, that have important consequences for the future of communities and the citizens who populate them. Interaction between citizens and government officials can foster or hinder the course of recovery for communities and individuals (Comfort).

5 Structuring the Future and Hazard Management – Scenarios as a Participative Planning Method for Reconstruction

Reconstruction after a disaster is the least examined and understood phase in hazard management. (Berke 1993; Rubin 1991; Drabek 1986). Only few scientists have so far addressed either individual or household adjustments (Drabek and Key 1984; Bates 1982; Bolin 1982) or those undertaken by governments (Kartez 1991; May and Williams 1986). Some scientists have examined the economic relevance of public financial support for reconstruction (Burby et al. 1991; Friesema *et al.* 1979; Kunreuther 1973). The question whether reconstruction can be carried out to reduce the vulnerability of a society to other disasters has also only been addressed by very few scientists. In practice, most reconstruction phases tend to satisfy short-term requirements and needs. The endeavours are in the majority of cases centralistic and hardly relate to the capacities of the affected population or the socio-cultural circumstances of the region concerned (Anderson and Woodrow 1989). However, balancing the short and long-term objectives and an active participation of the local population are decisive factors which finally determine the success of the reconstruction processes.

5.1 Reconstruction Determined by Long-Term Objectives

Haas, Kates and Bowden (1977) were amongst the first to theoretically approach the reconstruction activities of communities and societies. Their notable study titled *Reconstruction following Disaster* compares reconstruction processes in four different towns (San Francisco, Anchorage, Managua and Rapid City). The resultant data served to design a phase model. As Figure 5.1 suggests, there is a linear, orderly sequence of different reconstruction phases. Each phase takes about ten times as long as the preceding one. The authors show that ten years are needed for an entire reconstruction process. According to their opinion the return to a regular state and the reduction of vulnerability and uncertainty are the main objectives: "Disaster recovery is ordered, knowable and predictable. The central issues and decisions are value choices that give varying emphasis to the early return to normalcy, the reduction of future vulnerability, or opportunities for improved efficiency, equity, and amenity. Over-ambitious plans to accomplish these goals tend to be counterproductive. Major opportunities to improve the reconstruction process lie in early recognition of overlooked problems, people, functions, and areas; the reduction of uncertainty about the future for those who live and work in the city; and the preparation for reconstruction before the disaster comes" (Haas et al. 1977, p. xxvi).

This theoretical model has for a long time exerted influence on hazard management, but it has so far been verified by no more than one single empirical case (Geipel *et al.* 1988). Numerous other scientists who have tried to demonstrate the relevance of this model in practice could not confirm the linear sequence of the individual phases (Berke et al. 1992; Quarantelli 1989; Stuphen 1983). Rubin et al. (1985) carried out an extensive study, comparing the reconstruction processes of 14 US-American towns one year after they had been struck by totally different disasters. The authors' evaluation of the reconstruction processes based on qualitative interviews that had been con-

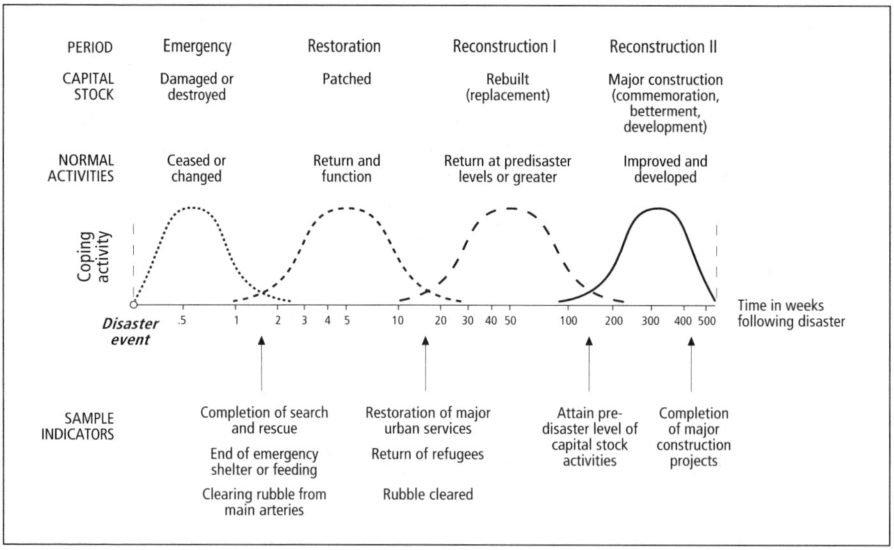

Figure 5.1. Model of the reconstruction phases. (After Haas et al. 1977)

ducted with important decision-makers. It became apparent that the behaviour of the decision-makers in the uncertain and dynamic disaster situation was influenced by a vast range of factors, including time pressure, uncertainty, complexity and different interests. The individual phases thus overlapped in practice. It also became clear that the success or failure of the reconstruction processes was decisively determined by the ideas, capabilities and activities of the individual decision-makers. Rubin et al. determined the following three major characteristics: *personal leadership*, *ability to act* and *knowledge of what to do* (Rubin et al. 1985).

It is certainly undisputed that the capabilities and activities of each individual decision-maker have a considerable influence on the reconstruction process. This knowledge has been affirmed by the events on Montserrat. However, there are a number of other factors which have to be taken into account. These include the decisions and activities of the affected population, communication and cooperation between the different interest groups and the extent and type of disaster aid. For cases of severe disasters it furthermore has to be considered that it is hardly possible to evaluate the success or failure of reconstruction as early as one year after the event. Research was to this end carried out at the Technological University Munich under the leadership of the geographer Robert Geipel. Reconstruction after the strong earthquake in 1976 in Friaul (Italy) has been monitored by Geipel and his staff within the scope of various projects over a period of ten years. The resultant studies comprehensively examined the effectivity of reconstruction planning in neighbouring communities (Geipel et al. 1988; Loda 1990; Geipel 1980; Steuer 1979; Geipel 1977).

A study published by Dobler (1980) put up the hypothesis that earthquake disasters not only restrict the development of the affected area but also have to be seen as an impulse to development, especially in respect of processes on the regional economic level. Dobler verifies his hypothesis with numerous examples where disasters have initiated an economic upsurge. These facts are, however, not all that new. Hazard research has since the early 1960s sought to understand disasters as an opportunity for the economic improvement of the affected regions. Innovative, nevertheless, is Dobler's request to consciously encourage regional structural transformations after a disaster by integrating long-term goals into reconstruction planning.

This approach is followed up by Geipel, Pohl and Stagl (1988) in their publication "Chancen, Probleme und Konsequenzen des Wiederaufbaus nach einer Katastrophe" (*Opportunities, problems and consequences of reconstruction after a disaster*). Every reconstruction is, according to the authors, public activity affecting the spatial dimension and, therefore, regional development policy: *"Reconstruction in any case, whether entirely at the cost of the affected or with the society's support, implies the compulsion and the opportunity to change structures. The obligation to act finally results in altered spatial structures. The type and extent of these changes depend on the goals determined and the selected and existent resources and means"* (Geipel et al. 1988, p. 145). Naturally, there are always a number of coexistent goals, as the volcanic crisis on Montserrat illustrates. Based on the long-term examinations at Friaul it has been possible, however, to determine different categories of objectives for reconstruction, all of them decisively influencing the future spatial structures: "*Either the aim is the complete and prompt restoration of the status quo ante, or reconstruction is seen as an opportunity to realise structural improvements, or there is no aim at all for the time after the disaster*" (Geipel et al. 1988, p. 145). As illustrated in Table 5.1, these aims correspond to three different reconstruction alternatives.

Table 5.1. Reconstruction alternatives according to Geipel et al. (1988). (After Geipel et al. 1988, pp. 146–152)

Criteria/alternative	1. Status quo ante	2. Structural reform	3. No objectives
Main objective	Return to normalcy	Structural improvements	n.a.
Subordinated objectives	Reconstruction of villages in the previous locations; compensation for individual losses	Decentralised settlement structure; equal assistance to all people concerned	Provisional accommodation of the evacuated
Organisational form	Decentralised planning on the local level; autonomous reconstruction by the owners	Central planning on the highest administrative level, co-operative and public reconstruction	Some influential individuals purposefully take advantage of their close contact to superior decision-makers
Instruments	Financial and other support for repairs or reconstruction at the previous location; funds for private people	Regional planning, development scheme, new constructions, public funding, guiding the population	No purposeful application of instruments

The attempt to attain significant structural improvements which also reduce vulnerability is particularly desirable. Geipel et al. (1988) nonetheless consider this still utopian. In practice, long planning phases and persistent perceptions and structures obstruct any such attempt. Experience with reconstruction in Friaul has shown that there was a clear preference towards resumption and preservation of the former structures, as opposed to the opportunities new planning strategies and restructuring would have offered. At the end of the day it was exactly this unreflective reconstruction that transformed the lebensraum: "*The construction of uniform buildings on a massive scale, the greater distance to the neighbour, the expansion of privacy (...) are processes which here in Friaul have taken place in less than a decade, instead of usually over generations. Adults still have the socio-psychological pre-earthquake situation in their minds whilst the actual environmental conditions have in the meantime altered completely. The environment today shows rather hostile tendencies than it used to; reference not being made to the threat of an earthquake but the social environment*" (Geipel et al. 1988, p. 156). Reconstruction in the Friaul region was economically quite successful, yet, had a number of undesired side effects since the regional structural and social effects of the reconstruction measures were neglected.

Based on these findings, Geipel et al. (1988) worked out 14 so-called "Do's and Don'ts" for hazard management. Those which are considered of specific importance to this work are listed below.

- “Conceptions and plans should be at hand which after a disaster facilitate a new version fit for the future;
- Minimum requirement of any foresighted planning are seismic zoning and building regulations;
- Reconstruction planning must not be simply a reflex to the emergency situation, but has next to (...) the spatial structures to pay attention to the demographic and social development. New problems will be encountered should this not be the case” (Geipel et al. 1988, pp. 152–155).

Geipel’s and also Dobler’s (1980) principles illustrate that any reconstruction should be oriented at long-term goals. Structural improvements certainly have to be realised in order to reduce vulnerability. The other two major tasks have to be the restoration of the people’s confidence in their future and ensuring a certain sensitivity towards the needs of future generations. In short, sustainability has to be a significant aspect of any reconstruction activity (Geipel et al. 1988).

Rubin (1995) has followed a similar approach in a recent study of hers. She distinguishes three reconstruction categories, the degree of measures, obligations and temporal extension being the criteria for her categorisation. Reconstruction is, according to Rubin’s theory, either (1) short-sighted and rather unpretentious, i.e. purely aimed at resuming the pre-disaster state, or (2) capable of integrating foresighted planning in order to reduce the vulnerability to other disasters, or (3) visionary and strives to principally improve the living conditions. The ‘Rocky Mountain Model of Recovery following a Disaster’ (Figure 5.2) illustrates these different categories. Rubin explains that the visionary type of reconstruction is the most difficult to realise since societies which have been struck by a disaster do not normally have the energy and foresightedness to realise such ideals. This is where hazard research and management are called upon. The opportunity to sustainably influence the future should not only be recognised but effectively realised. This is the way to reduce vulnerability and to ensure the resilience of social systems.

5.2 The Relevance of Participation

There is agreement on the point that the participants in planning, decision-making and management processes should come from of all sorts of interest groups. It is relevant to all sectors of social development, be it education, health, the economy or the environment, that the number of participants is kept as large as possible. Only a widespread participation can produce efficient development programmes and encourage the participants’ confidence in their own capacities (Oakley 1991). Such large-scale participation can even evoke modifications to the original programme or its complete transformation. Participation has over the past decade gained specific relevance with respect to the global endeavours for sustainable development. The *World Commission on Environment and Development* (WCED) in its 1987 report “Our Common Future” pointed to the significance of participation: “Making

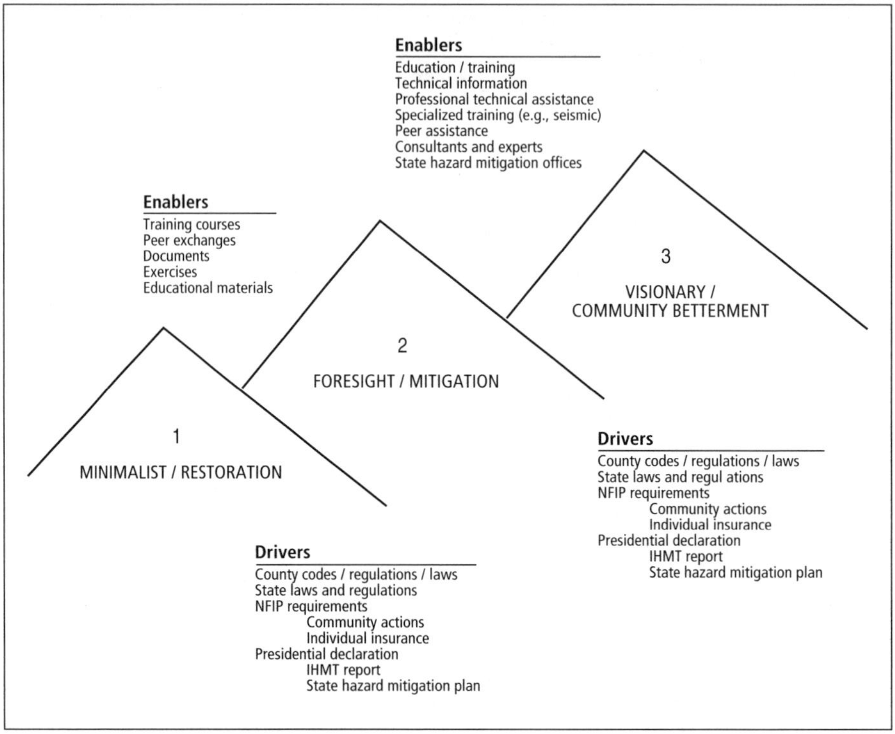

Figure 5.2. Rocky Mountain Model of recovery following a disaster. (After Rubin 1995)

the difficult choices involved in sustained development will depend on the widespread support and involvement of an informed public and of non-governmental organizations, the scientific community and industry. Their rights, roles and participation in development planning, decision-making and project implementation should be expanded" (WCED 1987, p. 21). The programme "Agenda 21", passed in June 1992 at the *United Nations Conference on Environment and Development* (UNCED) in Rio de Janeiro, represents the attempt to actually practise widespread participation. This programme focuses on the community's integration, i.e. it aims at involving the citizens in the determination of goals and the implementation of sustainable development at community level.

But what exactly does participation mean? Participation generally is an active process in which the people concerned cooperate in determining the direction of development programmes and later on assist with their implementation. They usually or ideally profit directly from these programmes and also determine their success: "... participation includes people's involvement in decision-making-processes, in implementing programmes, their sharing in the benefits of development programmes and their involvement in efforts to evaluate such programmes" (Cohen and Uphoff 1979, p. 35). Generally, two central functions of any participation can be identified. Oakley (1991) refers to them as *participation as a means or an end.* Participation as a means is a rather short-term process oriented towards a defined goal. Its main objective is to find a broad consensus for a development programme. The people concerned are mobilised and involved up to the point when the goals are clearly defined. The implementation of the goals is then again the decision-makers' responsibility. Participation as an end, in contrast, refers to a slowly unfolding process during which each participant's capabilities to structure and to manage are continuously encouraged. This form of participation is designed for long-term projects. It places the organisation and capabilities of the people concerned in the centre of attention and thereby goes far beyond the pure definition of goals.

Participation is by no means a new phenomenon, even if the current discussion suggests so. To the contrary, participation has a long tradition, reaching back to the 1960s when various methods were proposed and examined. The intention was to enforce the participation of social groups in the goal-defining, structuring and management processes of development programmes. One of the outstanding and successful methods developed during that time is Robert Jungk's so-called *Zukunftswerkstatt* (*workshop for the future*). A group of people meets over a certain period of time in order to find creative solutions to problems in a step-by-step dialogue and learning process. This method has been quite successful, not last because it motivates the participants to develop self-initiative and to actively organise their world (Jungk and Müllert 1990).

It is lamentable that such methods are rather the exception than the rule in current management strategies. Most of the so-called participative strategies refer to plans and ideas which are brought before the forum of public opinion with the expectation that the citizens may simply give their blessing to the planner's proposals: "*Élites define the citizens' desires and in a belated attempt try to obtain the people's approval for their ideas, suggesting that the people themselves want or have to want their realisation*" (Jungk 1990, p. 114). The citizens' capacities to develop own ideas and find solutions is generally neglected. Their selective involvement in individual projects does in itself not constitute proper participation. Uphoff (1986) has referred to this practice as pseudo participation, an opinion Oakley agrees with: "Current practise suggests that undoubtedly in many (...) development projects participation is stronger in rhetoric than in practical reality; that there is a lot of lip service to the notion of participation but less commitment to the changes in direction and style that would be required to implement it" (Oakley 1991, p. 118).

The criticism uttered has in the meantime in practice led to a reorientation. One can currently perceive clear endeavours to fundamentally change the direction and style of the previous strategies in order to enforce real participation. The *Agenda 21* is a decisive motor in this context. It has clearly encouraged the active participation of citizens in discussions about the objectives and forms of sustainable development. Above all, some common principles of participation have finally been agreed upon, the most important of them listed below (Birzer et al. 1997):

1. The people concerned have to be integrated into the planning and decision process at the earliest stage and released from the subsequent implementation process at the latest stage possible.
2. The decision-makers and the affected population are equivalent partners. Their experience, priorities and ideas shall influence these processes in the same way, ensuring that a mutual consensus for the objectives and strategies is found and acceptance thus improved.
3. The interaction between the decision-makers and the people concerned requires discursive and creative processes which are certainly prone to mistakes but this is deliberately accepted since it encourages continuous learning of and with each other.

These criteria have to be fulfilled for the realisation of a proper participative planning and implementation process. The evolution of new methods and the revival of traditional methods have therefore recently received considerable attention (Feindt 1997; Zimmermann 1997; Gessenharter 1996). But which relevance does participative planning have in hazard management, specifically in the reconstruction phase following a disaster, i.e. at a stage when the *lebenswelt* of the people concerned is characterised by complete disorder? This is a central question that has finally attracted more attention in hazard management. It has become apparent that the integration of ideas, information and priorities expressed by the affected population is not only feasible but an absolutely vital prerequisite for the following reconstruction process (Anderson and Woodrow 1998). At the end of the day, the measures designed to reduce vulnerability can only work when they are understood, designed in collaboration with and accepted by the local population. Furthermore, particularly the erosion of apparent securities after a disaster requires extensive discussions about basic values and objectives, enabling the people concerned to develop a new orientation about the constitution of their *lebenswelt.*

The individual and collective adjustments carried out on Montserrat are a clear practical example of these theoretical considerations. Being condemned to wait and to a passiveness has aggravated the worries and anxieties of many Montserratians. It has to be kept in mind that they are on the whole neither helpless nor did they suffer a severe shock, which would limit their capacity for decisions or activities. Many do show stress reactions, but the majority of Montserratians is clearly in the position to take the initiative, as the self-organisational processes have confirmed. Most of the Montserratians want to be actively involved in the reconstruction process. This would encourage their confidence in the future and help them to cope with the disaster. Nonetheless, most of the hazard management's responsibility is taken on by the public sector. The Montserratians themselves are generally perceived as victims who are hardly in a position to take their own initiative. An active participation in the planning and management processes has therefore more or less been ruled out. The ability and powers to make decisions and to act is confined to the local and British governments and numerous foreign experts – a situation the Montserratians mistrust.

Montserrat is no exception since this type of 'top-down' approach has dominated the world-wide hazard management for decades. Maskrey (1994) has summarised the pitfalls of such a purely centralist programme:

1. Vulnerability is not addressed. Most programmes concentrate on one hazard in one limited period of time. Vulnerability as a complex interplay between humans and their environment is completely neglected. Many determinant factors of the human decision-making processes initiated when dealing with a hazard are not taken into consideration.
2. Human participation is not adequately considered. Most programmes are technical and extremely costly. They depend to a large extent on the knowledge of foreign experts who often enough do not know the specific local conditions. These are best known by the affected themselves. Nonetheless, their activities and organisations for individual adjustments are excluded. Participation is reduced to self-help programmes, such as the private building of houses.
3. There is a danger of manipulation. The concentration of power, finances and knowledge in a few central organisations bears the risk that the programmes are manipulated by central political groups. To many governments hazard management is rather a measure to satisfy the economic and political interests than an end to respond to the people's needs.

The deficiencies of centralistic programmes often even cause a greater vulnerability of many affected people, especially of marginal groups (Blaikie et al. 1994; Susman et al. 1983). The population also experiences a growing feeling of impotence and helplessness. This is exceptionally true for extremely ambitious projects which are carried out entirely without participation of the population. The real needs and requirements of the affected are not only neglected but completely demoralised and demobilised. Furthermore, expectations as to the 'wonders of reconstruction' are often provoked which later can by no means be met. Typical consequences are for one a lack of acceptance of the new structures and further a waste of resources. Planning and constructing houses, towns and industries seems completely irrelevant since the priorities and requirements or the affected are not taken into consideration. Above all, an opportunity is missed since the most important resource of all remains unused – the local people with their specific knowledge and their invaluable capacities (Maskrey 1994).

Maskrey (1994) and other scientists have recognised the immense significance of participation in hazard management, specifically when it comes to reconstruction. Berke et al. (1993), for example, requests to place the affected population in the centre of all reconstruction endeavours. They are, after all, the experts of their life *lebenswelt*, and have the relevant competence to guarantee a successful reconstruction. Berke emphasises that the local population must not be perceived as victim but active and competent participant in the reconstruction process: "The community can assume the role of active participants rather than helpless victims. Local people can define goals, control resources, and direct redevelopment initiatives with long term economic and social benefits" (Berke et al. 1993, p. 93). Blaikie et al. (1994) also highlight the necessity to consider the experience and abilities of the affected citizens, particularly in the context of reduced vulnerability: "Reduction of disaster vulnerability requires full, day-to-day participation of ordinary people and their own popular organizations in the struggle to enlarge choice and reduce vulnerability. This participation must be asserted and protected as a right" (Blaikie et al. 1994, p. 238). Despite these realisations, there have so far been only a few practical attempts to effectively encourage the participation of the local population in the reconstruction process.

One of the few, but succinct examples is the participative elaboration of a reconstruction and development plan after the strong earthquake in Alto Mayo (Peru) in May 1990. Thanks to the close cooperation between the affected population and the various governmental organisations, the plan was finalised no more than two months after the disaster. The priorities and needs of the local population had clearly been put to the fore. New building technologies were applied which not only ensured a sustainable use of the local resources, such as wood and bamboo, but at the same time also facilitated the construction of earthquake-proof houses. The plan also integrated conservation guidelines and provided for a sustainable development of the whole region. The locals' positive attitude towards the plan resulted in numerous communal activities which were carried out in close cooperation with the local NGOs, governmental or international organisations. As Maskrey points out, "the experience of the Reconstruction plan (...) shows how working through and from the time/space specificities of people and their organizations in their localities and regions can help to reassemble the fragmented, disarticulated and dispersed elements of their own past experiences, present possibilities and future dreams into a new paradigm to mitigate disaster" (Maskrey 1994, pp. 110–111). The local population in Alto Mayo is despite the severe destructions in a position to define long-term goals for the sustainable development of their lebensraum and to actively implement them. Participation enhances the opportunities for structural improvements, resulting in a reduced vulnerability to other potential disasters. The decisive processes which are initiated by participation, are in the following summarised (Maskrey 1994):

1. Problems, needs and priorities of the local population are articulated and structured.
2. The opportunities for reflection, communication and learning are enhanced with the result that flexible and creative solutions to problems are found.
3. Visions about the future of the society are created.
4. The existing resources are mobilised.
5. Strategies involving participation meet the requirements of the local situation better.
6. Bureaucratic constraints and corruption are reduced.
7. Resilience of the society is improved.

Based on the practical experience, more and more scientists call for intensified endeavours to develop and test methods which facilitate (1) the mobilisation of the people affected by a disaster, (2) self-reflection of a society about its basic values, objectives and interests, (3) autonomous identification of problems and (4) the development and discussion of alternative solutions and strategies (Anderson and Woodrow 1998; Rubin 1995; Geipel 1993; Comfort 1988; Geipel et al. 1988). Berke et al. (1993) summarise: "Research is needed on the types of planning tools currently available for recovery in communities and on the existing roles of governmental and nongovernmental organizations. The feasibility of reorienting existing tools to incorporate recovery demands, as well as the adoption of new tools, needs to be investigated in terms of fiscal and political dimensions" (Berke et al. 1993, p. 107). The scenario method seems to be most suitable for the organisation of group-oriented investigation and adaptation processes. It can serve as a tool to unite the people in order that they may identify their problems and needs, design alternative futures, determine goals and priorities and develop flexible strategies.

5.3 Scenarios in Hazard Management

It has been known for quite some time that scenarios have specific potentials for hazard management. The first scientists who addressed scenarios in the context of hazard management were White and Haas (1975). They developed a four-stage method, designed to reduce the destructive potentials of different natural events for the affected population. Focal point of their method was the analysis of historic disasters in a certain region over a certain period of time, including the respective human adjustments, whether realised or not. White and Haas used historic data to determine the components and interactions of the hazard system, these again serving as basis for different scenarios. It has to be mentioned at this point that the authors did not design classic scenarios in the sense of future plots, but they created scenarios of the past. Each scenario described the consequences of the historic disaster as it would have been if this or the other alternative adjustment had been realised.

Haas and White (1975) dedicated part of their study to the devastating floodings in Rapid City (USA) in 1972. They put up the hypothetical assumption that extensive warning systems had been installed a few years prior to the event and came to the conclusion that this would have reduced the damage by at least ten percent. This and other hypothetical assumptions were then used for future implications. The authors recommended that a warning system be established and informative campaigns about the risks of floodings be carried out. Scenarios of the past can thus supply instructions for the future, as Foster also illustrates: "Used in this manner scenarios help to make explicit the reduction in damage and casualties that could have been, or will be, achieved if specific adjustments, such as better building codes or stricter air pollution standards, had been or are adopted (...) one of the chief objectives in writing scenarios is to illustrate the probable impact of certain adjustments or strategies on other components of the hazard system" (Foster 1980, p. 148).

Haas' and White's (1975) method attempts to attain long-term improvements of hazard management by analysing historical patterns of decision-making and activities, i.e. they learn from past mistakes. There is also another group of scenarios that address the direct consequences of disasters, i.e. the so-called disaster scenarios. They illustrate the alternatives for the destructivity of potential disasters so that scientists, emergency staff, planners, managers and politicians get an impression about potential events and have a clearer guideline as to their general scope of decisions and activities. Disaster scenarios are particularly helpful when it comes to assess the aptitude of disaster plans, building guidelines, land utilisation plans, warning systems, in short all preventative measures. Their purpose is thus similar to that of Haas and White, but they suggest a larger number of alternative consequences. Disaster scenarios can furthermore also arouse awareness in the local population and enhance its acceptance of necessary preventative or preparatory measures. These scenarios bear many potentials, but in practice often do not measure up to the demands placed on them. Only a few scenarios about potential earthquakes or volcanic eruptions have been designed so far (Merriman 1995; Schmincke 1993). One of them is Wadge's and Isaacs' (1989) study outlined in Chapter 4.2. The authors combined quantitative simulation models with scenarios in order to answer cause-and-effect questions, e.g. "what would happen with Montserrat if the Soufrière Hills Volcano erupted towards the north?" However, the willingness to envisage potential disasters in advance and to work out appropriate strategies is on the whole rather low. Disaster scenarios are usually applied by scientists, e.g. volcanologists, meteorologists, seismologists, whilst politicians, planners, managers and the population tend to dispel them (Merriman and Browitt 1993; Foster 1980). Exactly this happened to Wadge's and Isaacs' study which illustrated the risks the Soufrière Hill Volcano has for Montserrat, but the local decision-makers did not pay any attention to it. And once the volcano had resumed activity, the study was initially not presented to the public as the government did not want to cause apprehension in the population.

Scenarios, apart from heightening risk awareness and illustrating possible alternatives, also have a further application. Over the past decade they have gained importance, also in the *International Decade for Natural Disaster Reduction* (IDNDR), in the context of vulnerability. Scenarios concentrate less on the potential disaster as such than on the analysis of the prevailing social, political and economic relations in a system and the illustration of alternative development options. The dynamics of the society in question is put to the fore, also technological innovation, economic growth, population movements, land and resource utilisation. All these factors decisively determine the society's capacities to cope with the damages caused by a disaster. Scenarios can help to reveal potential weaknesses in a societal system. Such weaknesses may potentially lead to a higher vulnerability and it has to be the objective to diminish them as far as possible (Blaikie et al. 1994; Varley 1994; Möller and Clausen 1993).

As we have seen, scenarios have quite a tradition in hazard management and their main application is in hazard prevention and preparation. As descriptive pictures they have primarily two functions in this context: (1) the simulation of different disasters and (2) the vulnerability check in different societies. Scenarios are generally designed by experts and are later on used to transmit the awareness decision-makers and/or the population have of potential disasters, their consequences and the dynamics of their vulnerability. The opportunity to use scenarios as a tool for reconstruction after a disaster has so far hardly received attention, the work of Rubin (1987), Dobler (1988) and Geipel et al. (1988) being the exceptions. They explicitly refer to the potentials of reconstruction scenarios. However, these studies still adhere to the traditional view that scenarios are designed by experts and at high administrative levels and then applied to initiate structural improvements in the scope of central reconstruction processes. But the design of scenarios as alternative perspectives for reconstruction is the responsibility of those who will in future be living with and in this reconstructed world. Reconstruction processes call for an active participation of the affected population, a finding which leads us back to the questions asked in Chapter 3:

1. Does the scenario method as a discursive strategic planning method function in a hazard or disaster situation?
2. Is the scenario method a suitable tool to mobilise the capacities of the affected people so that their needs and priorities can be integrated into the reconstruction planning?

When trying to find an answer to the above questions, it appears sensible to briefly recall the potentials of the scenario method and to examine whether these potentials also apply in a disaster situation. The following comparison will serve this purpose.

General Potentials of the Scenario Method	Relevance in a Disaster Situation
The scenario method facilitates the handling of complexity since the basic relations of the system, central variables and possible cause-and-effect chains are identified. Complexity is reduced to its individual components and is thus easier to comprehend.	Complexity is extremely high in a disaster situation. At the same time instant decisions have to be made in order to satisfy short-term needs and secure long-term interests. Using the scenario as a tool to grasp complexity can result in a better understanding of the whole system and structuring of the problem. Implications of different reconstruction measures on the regional and social structure can also be clearly illustrated.
The scenario method encourages the constructive and creative dealing with uncertainty. It supports a systematic preparation for alternative futures and emphasises the continuous reflection of values, objectives and visions.	Any human being faces great uncertainty when trying to cope with a disaster. The *lebenswelt* has to be reconstituted and it is in this context extremely important to question the values, objectives and visions. The previous structures can in this way be analysed and new ideas be tried. Integrating unexpected events facilitates the mental examination of other disasters.
The scenario method encourages the consideration of alternatives and the development of flexible strategies.	All strategies to cope with a disaster have implications for the future. Reconstruction planning should therefore be considered an opportunity to reduce vulnerability. Analysing reconstruction alternatives and developing flexible strategies can support this process. The horizon is broadened and flexibility and innovation in the decision-making process are enhanced.
Designing scenarios in the scope of participative learning processes improves communication and cooperation between the participants. Prepared scenarios are as a literary description of alternative futures easily comprehensible and an excellent basis for discussion.	The effective coping with disasters requires cooperation between decision-makers and population. Designing scenarios in a participative planning process ensures that the needs and priorities of the local population are given more consideration. This in turn guarantees a greater acceptance of the strategies and more mutual confidence. Scenarios are, furthermore, easily comprehensible so that the reconstruction objectives can be discussed on a large scale.

This comparison suggests that the scenario method challenges our creativity and skills particularly in disaster situations. It also appears to consider the sustainability of reconstruction. These two hypotheses shall be examined and possibly verified in practice in the following chapter.

Let us build a new future. It's never going to be the same as the clock tower in old Plymouth. But we can dream. And then we can make it reality (Allison).

6 Living on the Edge of a Volcano – Scenarios for Montserrat

The volcanic disaster is not only to be understood as a development constraint for Montserrat, but also as an impulse. It offers the exceptional opportunity to consciously encourage structural transformations. Vulnerability can be reduced, rise given to the hopes of the affected, their capacities and abilities fostered and their resilience improved. It should be the aim to involve a great many different people in reconstruction planning, purposefully utilising their ideas and skills. Reconstruction planning should be organised as a process, teaching the people step-by-step how to manage complexity and uncertainty with respect to their long-term interests and priorities. An approximation to this goal will be carried out in this paper in a two-fold approach: a scenario project in Hamburg and later on a scenario workshop on Montserrat. Despite their different approaches, which will be explained below, both operations had the same objective: contributing to improving the long-term options for Montserrat during and after the volcanic crisis.

6.1 A First Approach – Pre-Testing the Scenario Method

Between November 1995 and June 1996 a scenario project was carried out at the Institute of Geography of the University of Hamburg. The following report concentrates on the interpretation of the findings obtained in the course of four scenario workshops, each lasting several days. The scenario workshops are best described as social experiments during which the participants, in a step-by-step process, learn to put their acquired knowledge into practice, errors never being excluded. The organisers' main intention was to test and practice the scenario method and to examine its individual phases – namely the analysis, prognosis and implementation phase (cf. Chapter 3.4). It was our priority to develop and test an appropriate method for the planned scenario workshop on Montserrat.

Following von Reibnitz (1992), the three phases are differentiated in various steps with the following objectives.

1. *Task analysis:* It is the purpose of this step to define and analyse the objective. All relevant information is gathered and evaluated by the people involved in the design of the scenarios. Prevailing models, objectives and strategies are reflected upon, also the potentials and pitfalls of the system under examination. The participants will become acquainted with the system, understand it better and evolve a mutual awareness for the problems. Also, the task has to be clearly defined at this stage. This means that the subject matter and the time scales for the scenarios have to be agreed upon.

2. *Analysis of influences:* During this step, factors are determined which exert influence over the evolution of the system. These variables are gathered, grouped and analysed according to their structure and function. It is important that all factors are clearly disconnected from each other in order to avoid any overlap. The main objective of this step is the analysis of the interactions between the variables and/or the factors and the spheres of influence. The evaluation of these interactions can provide findings about the dynamics of the system.

3. *Trend projections:* The variables which have been determined serve as a basis for the third step. Neutral and descriptive determinants, the so-called descriptors, are now selected by the group. They describe the current and future developments as precisely as possible. Once the actual state of the descriptors has been depicted, projections are made for all descriptors. Usually, a distinction is made between clear and alternative projections. Uncritical descriptors are those that the group sees as clear and predictable, whilst critical descriptors are subject to alternative development options.

4. *Clustering alternatives:* The various development options assumed for the critical descriptors now have to be tied in consistent, logical clusters without internal contradiction. These clusters depict raw scenarios. In the case that the participants simply discuss whether certain descriptors can be combined or not, we talk about an intuitive method. Alternatively, a manual or computer-aided correlation matrix (cf. Chapter 6.1.2) can be used to allocate and evaluate the development options. Finally, the group selects those raw scenarios that show the greatest internal consistency and also represent different development options.

5. *Interpretation of the raw scenarios:* The raw scenarios selected in step four now have to receive a design to become expressive pictures of alternative futures. The results of all previous steps, particularly those of the analysis of influences, have to be considered. Various scenarios are discussed and interpreted, and the group then selects a minimum of two scenarios. They have to be plausible and free of contradiction, but should possibly represent completely different options.

6. *Consequences:* In this next step the potential opportunities and pitfalls of the developments depicted in the scenarios are determined. The pros and contras are to be discussed, evaluated and appropriate measures suggested.

7. *Analysis of disruptive events:* Disruptive events have so far been excluded. These are events that suddenly occur without having been anticipated. They have far-reaching implications and often influence or alter the system and its evolution (cf. Chapter 3.1). This step of the scenario method is designed to examine the effects potential disruptive events may have, and to prepare preventative and reactive measures.
8. *Scenario transfer:* This final step represents the point of intersection between the pure creation of a scenario and its subsequent planning and management practice. In reality, this step already goes beyond the scope of pure scenario-design. The opportunities and risks which have been ascertained for the system now have to be effectively integrated into planning and management. As in step four, the integration process can either be carried out intuitively or structurally. The former method is based on the experience gathered during the design of the scenario whilst the latter develops systematic strategies for the descriptors. Von Reibnitz (1992) has, in this context, recommended the use of a control strategy which is based on all alternative scenarios. Finally, a system for the observation of the associated surroundings also forms part of this step.

Since the scenario project in Hamburg was carried out with the specific intention of practising the scenario method, the following report will solely refer to the first five steps actually carried out and not include the practical implementation of the scenarios. The procedure in part closely follows von Reibnitz (1992), but has in other parts been modified, as the detailed description will show. The report is structured in each individual *task* and the *evaluation* of the respective results. The evaluation considers both aspects as regards contents and communication, and decision processes between the participants.

The participants were members of staff and students of the Institute of Geography, and representatives of the private and public sector. It was our intention to have a heterogeneous group of participants in order to obtain new perspectives. In detail, the group comprised five scientists, three students, two representatives of the media and information technology sector and one representative of the Urban Planning Department, Hamburg. Some of the eleven participants already had experience with the scenario method, others not. Five participants had already stayed on Montserrat (some for a few weeks, some for months), whereas others knew the island from films or publications. The participants were also selected according to their occupation, previous experience, sex and age. The age varied between 27 and 52 years and the ratio of women to men was 5:6.

The *metaplan technique* was used for all workshops. It is a specific process that takes advantage of four basic principles – visualisation, speech, interaction and planned progress. It uses these principles in order to evoke an interactive group process which is guided by the moderator. This technique is highly suitable for the organisation of group-oriented communication and decision processes that deal with systems of medium or high complexity. The moderator uses questions and tasks to motivate interactions between the participants. He also organises and controls the individual steps. The dramatic character of the meeting is attained through a constant alteration between questions and activities or individual group tasks and plenum discussions. This dramaturgy plays an important role in the metaplan technique (Mehrmann 1994).

The metaplan technique requires each participant to be not only speaker but also visualiser of his contributions. Visualisation is deliberately taken advantage of as a communicative aid. Ideas and thoughts are noted down on moderation cards and are readily retrievable at any time. Visualised statements are also a helpful communicative tool since the matter, rather than the personal opinion, is put to the fore. Boards, stickers, arrows, circles, board markers and paper and file cards in various sizes and colours serve as material for the visualisation. All of the participants' senses are addressed. Spoken words, visualised impressions and active intervals enhance their memory and receptiveness. Group work is carried out on the condition that fairness and tolerance are extended to each other, and cooperative behaviour is encouraged (Mehrmann 1994).

6.1.1 Phase I – Tasks and Influences

The complex system Montserrat, as the object of the scenario project, has already been analysed and evaluated in Chapter 4.1. All participants received a copy of this information. The first meeting was primarily set up to mobilise the participants to broaden their horizons with regard to problems, and to stimulate their fantasies. The subject matters that were to be treated in the scenarios were defined by the group during that first meeting.

Task analysis

First of all, *abstract utopias* were drawn up, based on the method of the German *Szenarienwerkstatt des Instituts für Arbeit und Technik* (1991). An abstract utopia is entirely detached from existing tendencies in the course of development. It is a free and imaginary mental experiment without any realistic reference. The abstract utopias were the participants' path to unrestricted fantasies, wishes and fears. All participants were requested to experiment with their ideas and to leave reality behind them. The decisive aspect in this situation was that no pressure be exerted and that all people involved were totally prepared to have and express new, crazy and unthinkable thoughts. Fantasising was the way to open up the corridors into potential futures. The abstract utopias were to encourage free association and facilitate an inductive systematisation and structuring of the tasks.

The intention of and question behind each step was explained and illustrated with examples. Each participant was requested to draw up several abstract utopias. They received five cards on which they could note the basic idea of their abstract utopia and some explanatory notes. All participants were left to themselves for about 30 minutes. The cards were then recollected, shuffled and read aloud by the moderator. Afterwards they were stuck up on boards, the whole group deciding on the order of the cards by their contextual compatibility. Each person was allowed to comment for no longer than one minute, and it was generally not allowed to discuss or criticise. However, the other participants could ask questions when the meaning of an utopia had not been understood.

Evaluation: When the first phase was initiated, some participants questioned its necessity and remarked that they found it rather unusual to approach a matter without a definite and precise question. They clearly backed off since they wanted to adhere to their usual working method. The blockade was removed when the participants were requested to allow their creativity and fantasy to take their course. Once the function of the cards had been explained, the participants were prepared to act. It seemed of importance that the task was anonymous and group discussion prevented. The time given to develop and briefly explain their abstract utopias appeared too short to most of the participants; in short their fantasies and experiments had not yet fully unfolded. Harmony returned when the suggestions were read out and later allocated on the boards. The participants had no hesitation in talking about the utopias presented, restricted themselves to brief statements and did not interrupt each other. Lastly, all participants were actively involved. A mutual and creative group process had been realised. New ideas came up all the time, both with regards to Montserrat and its future development and also with reference to the methodical approach. None of the participants attracted attention through specifically dominant or passive behaviour. The group was highly motivated.

As a result of the first working step we had obtained 47 cards with utopias. Some ideas and topics overlapped, some were unique. A brief analysis of the abstract utopias revealed that the interests of the participants were relatively homogeneous. Some ideas regarding the social or economic structure were predominant, e.g. the right to veto for citizens, cultural integrity, social justice or sustainable economic development options for Montserrat, such as the cultivation of orchids, tobacco plantations or ecologically tolerable technologies (cf. Table 6.1). Disaster utopias were hardly designed, but some individual expectations and desires were expressed. Some participants in this context criticised the ideas and expectations of outsiders in that they could not simply be transferred to Montserrat. After a lengthy discussion the group decided that everybody should explicitly express their own wishes and ideas since they would, in effect, anyhow influence the entire scenario process.

In the next step the abstract utopias were assessed as to their *common ground* and *differences*. The participants were asked to tie clusters and give them generic terms. These names would later on provide the basis for the subject-related scenarios. The clusters also had to be evaluated and rated in importance. The participants had to give points to those clusters that they personally assessed as significant for the creation of the various scenarios. Multi-point questioning as an instrument of the metaplan technique was at this stage applied. Each participant was given eight sticky points and had five minutes to think about their personal rating and write it down. A maximum of three points was to be given to each main cluster or generic term. The group process was again anonymous. After the decisions had been made individually, the participants were requested to allocate their points to prepared fields on the boards.

Evaluation: The majority instantly realised the purpose of this task. There was only one person who did not want to differentiate any further. He feared that the entirety of the system would be lost and with it the abstract utopia. The others argued that they meant to illustrate the basic structures. The participants had a lengthy discussion about the contradiction between approaching the system in its entirety and the necessity to make the system's complexity comprehensible. The latter was finally given preference. The abstract utopias were allocated without any further problems. The subsequent discussion about the generic terms began to drag. The clusters which referred to social or economic utopias were discussed most. Those which had before been described as real abstract utopias in the course of the discussion lost more and more significance. The participants expressed that further discussion about the interactions between the clusters did not make sense since "everything was linked to everything else". Clear limitations were now requested. The multi-point question was received positively. The result is illustrated in Table 6.1, once again clearly showing the homogeneity of the group. Clusters referring to terms such as power relations, economy (in general and with specific reference to Montserrat) and environment management were of significance whilst the clusters *communication*, *work* and *paradise* did not gain any points.

In the next step, different *subjects* and *leitmotifs* were selected from those clusters which had been rated significant. The leitmotifs were as recurring themes necessary to determine the dominant development principles of each scenario, i.e. they facilitated orientation as regards contents. The participants were requested to formulate the subjects and leitmotifs in the form of vivid pictures. These should take the utopias up, but appear realistic in the medium

Table 6.1. Specification of the tasks

Cluster	Points allocated	Choice of some utopias
Power relations	19	Right of veto for the citizens; enforcing the individual power of judgement; cooperative republic; eliminating hierarchies; system-oriented thinking; island of social justice; participation
Communication	–	Problems of the information society; communication channels
Culture	8	Phasing out of the orientation towards the mother country; confident and communicative cooperation of free individuals; cultural integrity; travelling with solar planes to give a feeling of freedom; moderate amount of work; a lot of knowledge and pleasure
Economy, in general	13	Diversified, integrated and autonomous economy; economic niches; subsistence; national cooperation; sustainable development
Work	–	Satisfactory work; enjoyable work
Ideas for economic development	16	Centre for volcanology and disaster prevention; sustainable tourism; tobacco plantations and cigar manufacturing; the island as a museum; wind park; green technologies; cultivation of orchids; biologic gardening; research island; booming tourist island
Management of the environment	16	Controlling the resources; sound environment; no emissions; ecological refugees
Paradise	–	Sound earth, the world as paradise

or long-term (Institut für Arbeit und Technik 1991). The whole group participated in this step. It was requested to determine the number of scenarios and their time frame. The discussion dragged more and more since the participants' concentration and mental mobility had visibly diminished. It also proved difficult to reach any consensus at this point. It was therefore decided to form small teams of two to three people. The teams had an hour to formulate the subjects and leitmotifs, and their results were later on to be presented before the plenum.

Evaluation: The tension which had built up in the larger group now eased. The receptiveness rose visibly, probably due to the fact that the work of each group was later to be presented before the plenum. The presentation then showed that all participants had clearly understood the task. Each group presented its results briefly and precisely. The subjects and leitmotifs which had been worked out varied to a far greater extent than the previous discussion had led us to expect. The subjects of significant importance were *sustainable development* and *democratisation*. The evaluation of the clusters thus had a strong influence on the evolution of a common appreciation of the problems. As Table 6.2 shows, the group agreed on four different themes which already had the character of vivid titles. The first theme, called *Small Caribbean jaguar*, focused on economic growth, the second one, named *Business as usual*, was determined by the continuation of the previous development trends. The other two themes both concentrated on the island's sustainable development, but placed different emphasis on the citizens' role and participation. A time frame of 20 years was determined for all four scenarios.

As Table 6.2 suggests, Arras' (1987) concept was used to distinguish the different scenario types. Arras rejects terms such as trend, alternative or contrast scenario when typologising scenarios. He believes that such terms evoke rash evaluations as regards their contents. Arras therefore recommends distinguishing *static scenarios* from *scenarios of continuity*. A static scenario describes the state striven for at a certain point of time. A scenario of continuity, in contrast, emphasises the process of possible alterations and transformations

Table 6.2. Determining the themes, leitmotifs and scenario types

Theme	Leitmotif	Szenariotype
Small Caribbean jaguar	Coca-Cola company; expanding economic development; determined by an élite; minimised participation; export-oriented economic ideas	State
Business as usual	Continuation of the present state	Flux
Sustainable development from below	"As the Caribbean used to be"; ecological utilisation; maximised participation; equal rights; niche economy and diversification	State
Sustainable development from above	Eco and ethnotourism; resource management; technology (environment, communication); strict ecological regulations; environmental policies from above; orientation towards an ecological world market	State

over a certain period of time. The participants opted for this differentiation which seemed more useful to them. The practice showed, however, that a complete distinction cannot be realised. All scenarios describe a future state and objectives whilst at the same time addressing the means, processes and developments necessary to realise this state.

At the end of the first meeting the participants were given an opportunity to evaluate the method and the results. All participants referred positively to the results since a relatively broad scope of development options had been worked out in the course of the meeting. Those who knew Montserrat quite well were surprised about the amount of new and innovative ideas and perspectives often offered by those who knew comparatively little about the island. Asked about the method, most of the participants positively assessed the alternation between working in smaller groups and the whole plenum. It was also mentioned that the group as a whole had shown a creative and harmonic communicative behaviour. The opinions about the individual tasks varied. Part of the group thought the search for abstract utopias was effective and a refreshing beginning, at the same time teaching that creativity, fantasy and experiments are decisive factors when designing scenarios. Other participants argued to the contrary and said they did not consider this first step of any relevance and also felt it was too time-consuming. The different opinions triggered off a lengthy discussion. The final outcome was a mutual agreement that each individual step and task is necessary in order to reach a common consensus and to inspire the cooperative attitude of the group.

Analysis of Influences

The main task of the second workshop was the detailed analysis of the complex system Montserrat. Since the four basic subsystems had already been defined (Chapter 4.1), it was not necessary to determine the spheres of influence. The task was limited to the identification of the significant factors within the four subsystems. The structure and function of these factors was then to be analysed at a later stage. The participants had to briefly define each factor and describe its actual condition and its significance in the design of the scenario. A correlation matrix served as a tool for the analysis of the interactions between the individual variables. The dynamics and set up of the system were in this way to be filtered out and the participants' understanding of the system enhanced.

Discussions were soon provoked once again. When the first step, *the identification of the variables*, was approached, it became apparent that the factors identified were different with regards to their structures and functions. The group finally decided to distinguish between endogenous and exogenous factors, respectively dependent and autonomous ones. The endogenous and dependent factors were described as those which exert influence on the system from within and which can be directed to a certain extent. These factors were decisive for the later analysis. The participants defined all those factors as autonomous or exogenous which influenced the system from the exterior, for example global developments or sudden, unexpected events. These types of factors were not yet to be integrated into the analysis, but noted down so that they could possibly be used later on when analysing disruptive events. The group also decided to limit its analysis to the time before the onset of the volcanic crisis.

Evaluation: Once the basic procedure had been agreed upon, 42 variables were identified. It became apparent that the definitions in themselves often illustrated overlapping factors. Communicative problems between the participants also surged since they all had their own perceptions. The discussion made slow progress and was extremely difficult, but absolutely necessary in order to reach a common understanding about the significance of the variables and develop a mutually acceptable definition. Finally, the 42 factors were further aggregated so that a consensus was reached about the following 28 variables.

1. Environmental system

1.1 Waste and emissions: products of anthropogenic origin which were not produced purposely and therefore have to be further processed or disposed of.

1.2 Natural resources: all marine and terrestrial resources and their current utilisation and potentials. Also comprises the variety of flora and fauna and the biotopes.

1.3 Quality of soils: the quality of the soil is influenced by various physical and anthropogenic factors. Reduction of its natural functions through erosion, condensation and modifications in its composition.

1.4 Water management: supply of water to the population and the related management which comprises the provision of water and the necessary transport and allocation procedures.

1.5 Land utilisation management: Land utilisation and planning by the various categories, i.e. villages, infrastructure, recreation and conservation. Also comprises the spatial allocation of the population.

2. Social system

2.1 Population: natural dynamics and age structure.

2.2 Occupational qualification: training and education as well as skills.

2.3 Social security: social benefits paid by the state and the family, including remittances from expats.

2.4 System of values: general values of the population regarding central subjects, such as culture, family, politics, planning, the economy or the environment.

2.5 Social and family structures: the social structure comprises the relations between social groups and their possibilities to exert influence at local level. The family structure refers to the family network and the size and composition of a family.

2.6 Cultural infrastructure: comprises cultural and recreational facilities, such as sports fields, green areas and restaurants/pubs.

2.7 Emigration: emigration by its causes (push/pull factors) and the number of emigrants (including the social effects).

2.8 Immigration: immigration by its causes (push/pull factors) and the number of immigrants (including the social effects).

3. Political system

3.1 Extent of self-determination: possibilities and rights of political self-determination the local government has under the influence of the British mother country and the *Organization of Eastern Caribbean States* (self-determination vs. heteronomy).

3.2 Foreign aid: the entire funds transferred by the providing countries (finances, consulting etc.).

3.3 National budget: balance of the government's finances and planned public investments.

3.4 Legislation: comprises local legislation (taxes, restraints of competition, industrial laws; conservation laws) and legislative in the scope of the OECS.

3.5 Development strategies and planning: objectives set by the local and British governments and the OECS, including the associated strategies and plans.

3.6 Participation and political culture: political consciousness and political self-determination of the population and the influence of the various interest groups (unions, NGOs etc.).

4. Economic system

4.1 Gross national product: the total value of the goods and services produced by the residents of a nation during a specified period (i.e. a year).

4.2 Infrastructure: equipping a place with supply and disposal facilities, public transport and roads, communication, health and education facilities.

4.3 Economic structure: the structure of the economic system by sectors.

4.4 Labour market: situation on the labour market according to demand and supply, unemployment rate, individual behaviour and employment policies.

4.5 Investments of the private sector: readiness to invest and behaviour of the private sector.

4.6 Distribution of property: land, capital and productive property by classes, quantity, quality and sectors.

4.7 Technological development: technological standard and development potentials.

4.8 Trade: internal and external trade with land, capital, means of production and consumer goods.

Once the key factors had been identified and defined for all subsystems, the potential *interactions between the variables* were ascertained with the correlation matrix (cf. Table 6.3). A correlation matrix is a methodical instrument for the systematic description of all potential interactions between a certain number of factors and their intensity. This task was again carried out intuitively. The potential direct effect of one factor on another was discussed for each pair of variables. It was vital to ensure that the direct influence was not falsified by indirect effects. The extent of influence was determined by a simple rating (0 = no effect, 1 = some effect, 2 = modest effect, 3 = strong effect) and noted down in the respective fields of the matrix.

Evaluation: The whole process was time-consuming since the group's consensus had to be obtained for each rating. It took several days and dragged on, but numerous new findings about the system were gained in the course of the discussions. In the beginning, every participant was clearly convinced of his own individual assumptions about the interactions. However, other assumptions were expressed in the course of the group discussions. Often, individual participants successfully managed to convince the rest of the group that it had neglected a decisive interaction. This enticed the group to reexamine all assumptions and to modify them in case of need. The process was on the whole very productive, despite or even because of the time-consuming discussions. The group, step-by-step, developed a common understanding of the system's complexity. Once all 784 interactions had been discussed and all fields in the matrix been filled in, the group was proud of its achievements. All participants expressed their astonishment, saying that they had not believed they could cope with the initially "insuperable pile of interactions". They put the success down to the productive and creative cooperation within the group and also expressed their interest in the results of the analysis.

The sums of the lines and columns in the correlation matrix are the basis for the interpretation. The *active sum* (AS) is that of the lines and represents the intensity of influence one factor has on the others. The sum derived from the columns is the *passive sum* (PS) and represents the intensity of direct influence all other factors have on one individual one. Multiplying or dividing the active and passive sum results in values for the analysis of the interactions. The so-called *Q-value* is obtained as quotient of AS/PSx100 and the so-called *P-value* is a product of AS and PS. These calculations facilitate the following categorisation of the influence factors and reveal their implications on the dynamics of the system (von Reibnitz 1992).

1. Active variable ⇨ highest Q-value, exerts strongest influence on the other factors; is itself hardly influenced at all; these factors have an above-average active sum and a below-average passive sum.
2. Passive variable ⇨ lowest Q-value, exerts least influence on the other factors; is itself most strongly influenced; these factors have an above-average passive sum and a below average active sum.

Table 6.3. Correlation matrix for the complex system Montserrat

from \ to	1	2	3	4	5	6	7	8	9	10	11	12	13	14	15	16	17	18	19	20	21	22	23	24	25	26	27	28	AS	AS/ PS*100 (Q-value)
1 A Waste management		2	1	0	0	1	0	0	0	1	0	0	0	0	0	1	2	3	2	3	1	2	1	1	1	0	1	0	23	50.0
2 A Natural resources	0		3	3	0	3	0	0	2	2	2	1	1	2	0	1	2	1	3	2	3	2	3	3	3	1	1	1	45	76.3
3 A Quality of soils	0	3		3	0	2	1	0	1	1	0	0	1	0	0	1	1	2	2	2	1	1	1	1	1	1	1	1	28	68.3
4 A Water management	1	3	2		2	3	0	0	0	2	1	1	0	0	1	1	2	2	2	1	1	3	2	1	1	1	1	1	35	66.0
5 A Energy supply	3	2	1	1		2	0	1	1	1	0	2	0	1	3	2	1	1	2	1	3	3	2	1	2	1	3	3	43	100.0
6 A Land management	3	3	3	3	2		1	0	1	0	1	3	0	1	1	2	1	3	3	1	1	3	2	1	3	2	1	1	46	90.2
7 B Population development	3	3	3	3	3	3		1	2	2	3	1	3	3	0	1	2	1	2	0	2	1	1	2	1	1	0	2	49	233.3
8 B Professional skills	0	1	1	1	1	0	2		3	3	2	1	3	2	2	1	2	1	2	3	3	1	3	3	2	1	3	2	49	122.5
9 B Social security	0	0	0	0	0	1	2	1		3	3	1	1	1	3	1	3	2	2	1	1	1	1	1	0	1	0	1	31	75.6
10 B Values	3	3	2	2	3	1	2	3	3		3	3	2	1	3	0	1	2	3	3	1	1	1	1	2	2	1	2	54	93.1
11 B Social and family structure	1	1	0	0	1	1	3	2	3	3		3	3	2	1	0	1	0	1	3	2	1	0	2	1	3	1	1	40	100.0
12 B Cultural infrastructure	0	1	0	1	1	2	0	1	0	3	2		1	1	0	0	0	1	2	2	1	2	2	2	2	0	0	1	28	65.1
13 B Emigration	1	1	1	1	1	0	3	3	2	1	3	0		0	1	1	1	0	2	1	3	0	1	1	0	0	2	1	31	93.9
14 B Immigration	2	2	1	2	2	1	3	1	0	2	1	2	0		1	1	2	0	1	1	2	1	1	2	1	1	1	1	35	94.5
15 C Self-determination	0	1	0	1	1	2	0	1	3	3	0	0	1	1		3	3	3	3	3	3	3	2	2	2	0	2	3	46	117.9
16 C Foreign aid	3	2	1	2	1	2	1	2	1	3	0	1	2	3	3		3	1	3	1	3	3	2	2	3	0	3	0	51	150.0
17 C National budget	3	2	0	3	0	3	0	3	3	0	1	3	2	0	0	0		3	3	0	3	3	2	3	0	0	0	0	40	80.0
18 C Legislation	1	3	0	3	0	3	0	2	2	1	0	0	0	3	3	0	3		3	3	2	3	2	2	3	3	0	2	47	109.3
19 C Development planning	3	3	3	3	3	3	1	2	1	2	0	3	1	2	3	3	3	3		2	1	3	3	3	3	1	3	3	64	97.0
20 C Participation, political culture	0	0	0	0	0	0	0	0	0	2	3	3	0	0	3	1	3	1	3		0	3	3	2	0	0	0	0	27	62.3
21 D GDP	3	3	3	3	3	1	0	3	3	3	3	3	3	3	3	3	3	1	3	0		1	3	3	3	1	3	3	67	121.8
22 D Infrastructure	3	3	3	3	3	3	0	0	0	3	0	3	0	3	0	3	3	0	3	0	3		3	1	3	1	1	2	50	92.6
23 D Economic structure	2	3	2	3	3	3	0	2	0	3	1	1	3	3	1	2	1	3	3	1	3	3		3	3	1	3	3	59	109.3
24 D Labour market	2	3	2	2	2	3	1	3	3	3	3	2	3	3	1	1	2	2	3	3	3	0	3		3	2	2	2	62	124.0
25 D Private investment	3	3	3	3	3	3	0	2	2	2	0	1	2	2	2	3	2	2	3	1	3	2	3	3		3	3	3	62	121.6
26 D Distribution of properties	0	3	3	2	2	3	0	3	3	3	3	1	1	0	0	0	1	3	3	3	2	2	2	2	3		1	1	50	166.7
27 D Technological development	3	3	3	3	3	1	1	3	0	3	3	1	0	0	1	2	1	0	2	0	1	3	2	0	2	0		3	44	110.0
28 D Trade	3	2	0	2	3	1	0	1	2	3	2	3	0	0	3	0	1	2	2	2	3	3	3	2	3	3	3		52	120.9
PS	46	59	41	53	43	51	21	40	41	58	40	43	33	37	39	34	50	43	66	43	55	54	54	50	51	30	40	43	1258	
PS * AS (P-value)	1058	2655	1148	1855	1849	2346	1029	1960	1271	3132	1600	1204	1023	1295	1794	1734	2000	2021	4224	1161	3685	2700	3186	3100	3162	1500	1760	2236		

0 = no influence, 1 = low influence, 2 = moderate influence, 3 = strong influence; AS = active sum, PS = passive sum

3. Critical variable ⇨ highest P-value; exerts strong influence and is itself also strongly influenced; these factors have above-average active and passive sums.
4. Idle variable ⇨ lowest P-value; exerts weak influence and is itself hardly influenced; elements with below-average active and passive sum.

The matrix can also be interpreted graphically (cf. Figure 6.1). The active and passive sums of the respective factors are positioned in the grid system. The maximum possible active and passive sums are marked on the respective axis of the coordinate system, and the area for the position of all other factors is determined. All factors together add up to the average active and passive sums. These serve to divide the area into four fields. The factors can then, according to their active or passive sum, be allocated to one of the fields. All variables which have been positioned above the 45° line are to be considered relatively active. They show a ratio of active to passive sum which is larger than one (Götze 1991).

Field I ⇨ Area with active variables which can exert strong influence on the system.
Field II ⇨ Area with ambivalent variables which can exert just as much influence on the system as it in turn exerts on them.
Field III ⇨ Area with buffering or idle variables which can hardly influence the system and vice versa.
Field IV ⇨ Area with passive variables which can be strongly influenced but exert hardly any influence on the system itself.

The combined analysis of the matrix and the grid shows very clear results for the system Montserrat. Of the selected variables, the factor *development strategies and planning* has the highest P-value as the product of adding the active and passive sums. This factor is therefore the one to be most closely involved in the system and represents a key variable. The factor with the highest Q-value is the *population development*, thus being the most active element of the system. The system is in absolute terms most strongly influenced by the factors *gross national product*, *labour market* and *investments of the private sector*. The grid also shows all other relatively active factors positioned above the 45° line. These include *energy*, *occupational qualification*, *social and family structure*, *emigration*, *degree of self-determination*, *foreign aid*, *legislation*, *economic structure*, *distribution of property*, *technological development* and *trade*. Some of the critical variables are *natural resources*, *land management*, *system of values* and *infrastructure*.

These results generally confirm the contents of Chapter 4.1. Many variables which have been presented in Chapter 4.1 as constraints have proven to be either active or critical variables (e.g. *limited resources, population, occupational qualification, British dependency* etc.). But the correlation analysis also surprised the participants. Some factors which had originally been evaluated as central to the system's dynamics turned out to be quite insignificant. The factor *participation and political culture* attracted much attention since it was

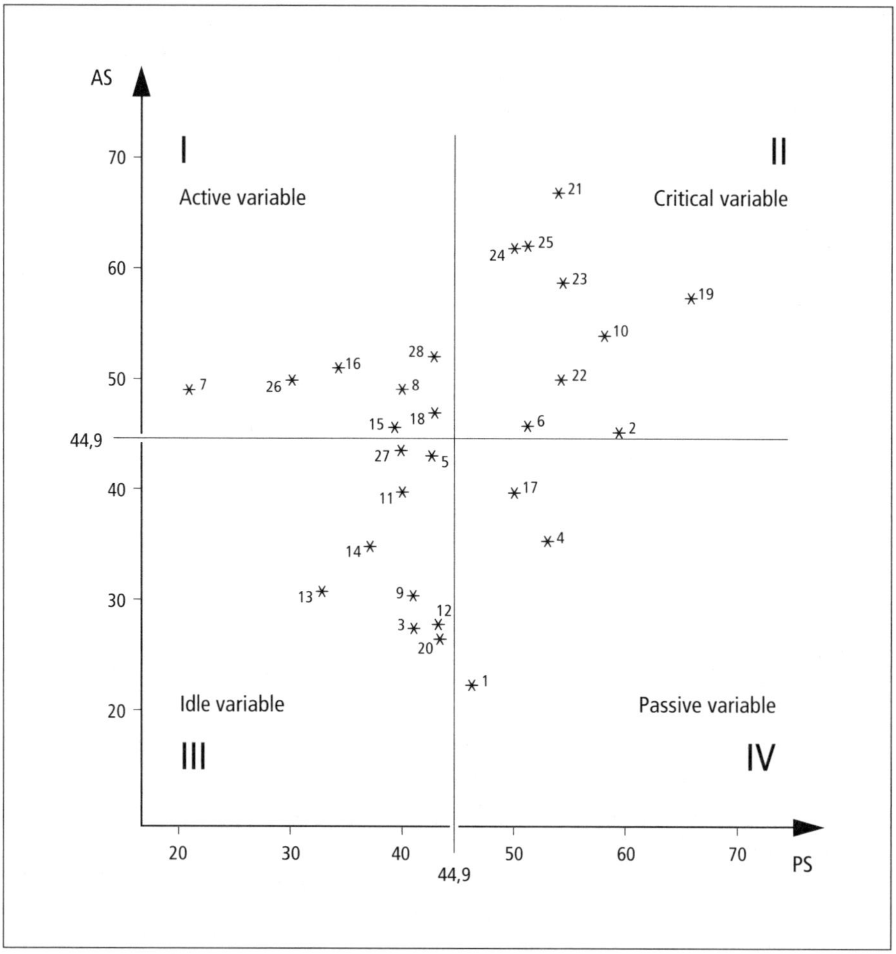

Figure 6.1. Grid for the complex system Montserrat

determined an idle variable in the correlation matrix. Many of the participants, however, recalled that the participation of the citizens had been strongly emphasised in the initial phase. It was uttered that the original intentions had been lost in the course of continuous aggregation. Doubts arose amongst the group and it seemed necessary to review the basic statements derived from the analysis. On this basis, two laws of von Reibnitz' (1992) system dynamics were introduced. The first says that the strongest intensifying effect is usually exerted by the active or critical variables. The second law says that passive and buffering variables depend on all other factors of the system, particularly on the active and critical variables. This implies that idle and passive variables are also important for the system's dynamics, even though indirectly via their interactions with active or critical factors. When attempting to design a scenario or strategy, it is therefore important to analyse which active or critical variables exert most influence. In practice, the analysis of all interactions realised by active variables and amounting to the value '3' revealed that especially *participation and political culture* prevailed quite frequently. Other factors such as the *extent of self-determination*, *legislation* or *distribution of property*, for example, exert influence on this variable at the value '3'. These three factors would consequently have to be changed if the participation of the citizens in planning and decision-making processes was to be enforced.

Particularly in the beginning, the analysis of the variables turned out to be quite complicated. This can be put down to the following factors.

1. The participants encountered difficulty in dealing with complexity, initially even refusing any approximation. In the course of the process conflicts arose because some participants intended to analyse the system in detail whilst others considered it more useful to limit the analysis to the variables determined in order to describe complexity.
2. In view of the task of discussing 784 interactions, many participants first shied away. The task seemed too time-consuming and most of them believed that it could not be realised. To the contrary, the group should concentrate on the problems it considered important.
3. When the interactions between the variables were discussed, quantitative and qualitative elements overlapped and this regularly caused confusion. The participants were also critical of the fact that the interactions were first ascertained with a communicative process but then later documented in form of quantitative numbers.
4. In view of the different perspectives and assumptions of the participants it was often difficult to find a consensus.

All these problems were faced and solved in the course of the practical process. The group always managed to find mutually acceptable solutions. The fear of complexity disappeared once the discussions showed that it was exactly the wide range of perspectives that played a decisive role. Team work and the use of the correlation matrix ensured that interactions were taken into consideration which would surely have been neglected with other methods. The analysis of the influences opened up new perspectives and the group gained a common understanding of Montserrat's complexity. This decisively influenced the further process, i.e. the design of the scenarios. In summary, the analysis was a significant step and in its course all participants had the opportunity to question their assumptions, become acquainted with the perspectives of the others and get a feeling for the complexity of the system, even though the understanding gained is certainly specific to this one group.

6.1.2 Phase II – Trend Projections and Clustering Alternatives

The factors determined in the first phase now served to ascertain the relevant *descriptors*. According to von Reibnitz (1992) a descriptor is a descriptive variable used to make statements as to the condition and potential development of a system. Each descriptor had to be defined objectively and without it overlapping with other descriptors. In the light of the intuitive group-oriented approach it had to be ensured that no more than 16 descriptors were ascertained. The participants therefore concentrated on those variables with potential power for the future developments of the system, i.e. on the critical and active variables. Table 6.4 summarises all descriptors which were selected on this basis.

Evaluation: The group effectively and productively discussed and determined the descriptors. Minor problems were incurred at the beginning of this phase since the group had difficulty in transforming qualitative variables, (e.g. system of values), into neutral, descriptive determinants. This evoked a lively discussion, again necessary in order to clarify the task and procedure within the group. Compared with the previous workshops consensus was reached far more rapidly. The group was well accustomed to the working techniques and showed professional conduct with the scenario method. The further aggregation went off smoothly and the group was able to give detailed and comprehensible reasons why the descriptors had been selected. The experience of the previous workshops and the step-by-step approximation to the tasks seemed to foster the mutual learning process which illustrated not only the advantages and disadvantages of the scenario method, but also the specific characteristics of the complex system Montserrat.

Table 6.4. Determining the descriptors

Variable (active / critical)	Descriptors
Natural resources	Handling the natural resources
Land management	Not applicable*
Population development	Population development
Occupational qualification	Training and education
System of values	1. Family structures 2. Cultural values
Degree of self-determination	1. Caribbean integration 2. Influence of Britain
Foreign aid	1. Amount of funds transferred 2. Structure of the funds transferred
Legislation	Influence through legislation
Development strategies and planning	Not applicable*
Gross national product	Growth rate of the GNP
Infrastructure	Not applicable*
Economic sectors	1. Tourism 2. Offshore banking
Labour market	Vacancies available
Investments of the private sector	Investments of the private sector
Distribution of property	Not applicable*
Trade	Balance of trade

* 'Not applicable' here means that the respective factor was excluded from further analysis since it seemed either not appropriate or not relevant.

Once the descriptors had been determined, their actual condition had to be described. This step was closely related to the analysis outlined in Chapter 4.1 and will therefore at this point not be considered in detail. Based on the actual conditions (before the volcanic crisis), the participants were now requested to use their experience, knowledge and fantasy in order to design potential development options for the descriptors, i.e. to work out *trend projections*. The group decided to base their projections on two periods of time – projections until the year 2006 and alternatively 2016. The times were set believing that assumption for some developments are more predictable over a shorter period of time whereas other developments can only be assessed after a longer period, as for example 20 years. Table 6.5 was used as a tool. Each descriptor was projected into the relevant time, and the participants always had to ask whether they were able to expect a certain development or whether there was uncertainty as to its future. Should the second case be encountered, alternative developments had to be shown and given reasons for. Definite, uncritical descriptors were in this way distinguished from alternative, critical descriptors. If the development was the same for both periods of time, they were not noted again. The group was, in this step, not obliged to find a consensus.

Evaluation: The group was able to immediately approach the task and it did not encounter any great difficulty in its realisation. However, it took a whole day to make the projections and to give reasons for them. Often four or five alternatives were suggested for individual descriptors. The participants expressed that these alternatives could in many cases have been differentiated even further but they did not want to waste time by going into too much detail. The fact that alternative developments were suggested for all descriptors was quite surprising. There were no uncritical descriptors at all. The trend projections were on the whole very realistic since improbable and unexpected events were still excluded from the reflection. Some participants were disappointed about the lack of fantasy and the rather strict adherence to the existing patterns. But the procedure provoked certain factors to be thought about which would otherwise have been left out of consideration. The participants had furthermore expected that the clustering of alternatives would contribute more to understanding the scope of alternative developments. However, despite some disappointments and criticism, it can on the whole be said that the participants were generally satisfied with the working steps and approved of the concentrated and harmonic cooperation. The overall identification of the group with the procedures and results was positive.

The next task was to tie the alternative developments in consistent *integral clusters of assumptions (raw scenarios)*. The combination was supposed to reveal logical and rational fundamental scenario structures. Once again, a group discussion was used to accomplish the task. The alternative development options were intuitively clustered on a flip chart. The group gradually combined the assumption for one descriptor with those of other descriptors and examined whether it was possible that they both occurred.

Table 6.5. Trend projections for the descriptors (1/3)

Descriptor/present state	Projection 2006	Reasons	Projection 2016	Reasons
1) Treatment of natural resources: moderate exploitation, moderate environmental consciousness	A: greater exploitation B: somewhat reduced exploitation C: reduced exploitation	Government depends on tourism, construction industry Small projects are promoted, somewhat growing environmental consciousness Environmental education and management is a major issue		
2) Population development: increase balanced out by emigration	A: remains the same B: increasing	Trend has remained the same for 200 years Immigration increasing (Canada, Caribbean)		Emigration decreasing and immigration increasing
3) Education: British school system, relatively good in comparison to the Caribbean standards, very little continuing education after secondary school	A: better B: worse C: remains the same/ unchanged	Promotion of the Technical College and the UWI section on the island, AUC opens up for Montserratians, adult education, further training in evening schools Brain-drain, lack of funds, teacher training and curriculum remain as they are		Skilled labour stay and apply their know-how, England promotes education more strongly
4) Family structure: both Caribbean family networks and nuclear families	A: remains the same/ unchanged		B: Nuclear families and singles increase at the cost of the networks C: Stronger network structure	Increasing wealth, Americanization, Individualization, achievement orientation, career orientation Decreasing wealth or more solidarity
5) Cultural values: a mixture of tradtional orientation and influence of globalization	A: greater return to tradition B: greater orientation towards USA, Japan and Europe C: remains the same	Greater cultural consciousness in the Caribbean, increased self-confidence, emigration experience, tourism influence, fear of globalization Communication, media, tourism, migration Processes of globalization and tradition have same value		
6) Caribbean integration: OECS, CARICOM, ACS	A: greater East Caribbean integration B: greater fragmentation	OECS promotes further integration increase in inter-island conflicts	C: greater integration in the whole region D: no integration	Political, economic and cultural integration, one political union, one currency Concept of integration is not relevant anymore
7) Influence of UK: colonial administration, offshore control, stronger planning control and development	A: increasing B: decreasing	Political scandals, financial hardship of the UK, burocratization and political control. Financial hardship of the UK, stable Offshore finance situation, autonomy movement, efforts of 1994 bare fruits (administrative reform), economically more independent	C: no self-administration D: independence E: associated state	Changed international political situation. Greater influence of the USA, France and the Netherlands. Changes in basic laws. Internal restructuring, UK's pulling out of its colonies, successful autonomy movement See D, and security interests of UK and Montserrat protected

Table 6.5. Trend projections for the descriptors (2/3)

Descriptor / present state	Projection 2006	Reasons	Projection 2016	Reasons
8) Amount of transfer: high transfer, 70-80% of the state budget by means of transfers	A: unchanged	No structural reform, no economic development		
	B: decreasing	Donor countries shift their support to other regions, political/economic development, conflicts, successful structural reform, financial hardship of donor countries		
9) Structure of transfer: concentration on construction and infrastructure, social welfare and education, state budget; little promotion of economy and environmental protection	A: unchanged	Tourism still a major interest, infrastructural measures and education to attract foreign investment		
	B: concentrated, but on new areas	From conventional to ecological tourism, stronger environmental protection, promotion of fisheries and agriculture		
	C: diversified	Many areas are promoted to the same extent. Economic, social and ecological promotion have the same value		
	D: expansive in new markets	College, windpark, media, high-tech		
10) Influence by laws: moderate to strong depending on the sector	A: stronger	Foreign political pressure, influenced by the international economy, Carribbean, USA, drugs, international environmental protection agreements		British influence decreasing
	B: weaker	Liberal policies, general deregulation		
	C: unchanged	Reform of various areas one after the other		
11) GNP growth rate: 3.2% (1980-92)	A: growing	Economic growth and transfer		
	B: dropping	Regression, crisis, international reasons, reduction of transfers		
	C: unchanged	Stagnation, transfer frozen at the current level. Growth of the local economy is halted.		
12) Trade balance: very negative, 10% export of total trade volume 1993	A: Zero export	The Products are consumed by the local population, no export production		
	B: negative trade balance with a growing export percentage	Niche production for exports		

Table 6.5. Trend projections for the descriptors (3/3)

Descriptor/present state	Projection 2006	Reasons	Projection 2016	Reasons
13) Tourism: weakly developed, about growthing, particularly villatourism, cruise ships	A: growing market sector	Concentration of the economy on tourism development, expansion of airports and ports, diversification of the tourism product, construction of hotels, villa tourism reduces, mobility of the tourists increases (island hopping), prices for plane tickets drop (cheap flights), competition islands are overcrowded, tourists demand more ecologically and cultural oriented holidays, competition advantages.		
	B: stagnation	No increase in the number of beds, infrastructure not expanded, turn down of image campaigns, diversification does not work, competition with other islands	C: dropping	Flight tourism out, Carribbean out
14) Offshore: decrease after the scandal in 1989, now low	A: growing	Government invests in the Online registration of Offshore companies, P.O. Box offshore, new niches, new companies, more money, trust-building state regulation measures		
	B: meaningless as a sector	Decrease, strong competition, more scandals, others sectors are seen as economically more important offshore is given up due to lack of profits.		Generally less offshore, loses in global importance
15) Work market: average, concentrates on tourism, construction and public sector.	A: stronger	By means of growth, labor unions, state, employers, cooperation, promotion programs, large infrastructure projects.		
	B: weaker	By means of recession, lay-offs, growth in capitalistic competition relations, periphery is left out of economic globalization, reduced transfer, public sector reform.		
	C: unchanged	Continues to concentrate on tourism, construction and the public sector		
16) Investment by the private sector: low to average	A: higher	Increase in enterpreneurship, good investment policies, flow of international capital, cheap labour, booming tourism, niche expansion, transportation infrastructure.		
	B: low	Investments don't lead to profit, no incentives, distance to profitable markets too great, lack of enterpreneurial drive.		

Evaluation: Altogether six clusters of alternatives were determined, four of them closely resembling the initial themes and leitmotifs. For this reason the group decided to focus on these four alternatives, which are outlined in Table 6.6.

It was remarkable how quickly the participants realised that the combinations were limited when checked regarding their consistency. Some of the participants tried to combine those assumptions which they considered desirable. But they soon realised that these combinations did not form consistent and integral clusters. Even the initial leitmotifs could not be linked in the way

Table 6.6. Clusters of alternatives

Descriptor	Alternative 1	Alternative 2	Alternative 3	Alternative 4
Natural resources	increased exploitation	increased exploitation	slightly reduced exploitation	reduced exploitation
Population development	increasing	remaining constant	remaining constant	increasing
Training/Education	improving	remaining of the same quality	improving	improving
Family structures	core families and singles	remaining the same	closer network	core families and singles
Cultural values	orientation towards the States, Europe etc.	maintaining existent values	stronger emphasis on traditional values	orientation towards the States, Europe etc.
Caribbean integration	integration in the entire Caribbean region	ntegration in the Eastern Caribbean region	no integration	integration in the entire Caribbean region
Influence of UK	increasing	no self-government	associated state	no self-government
Amounts transferred	constant	constant	decreasing	constant
Structure of transfers	expansive into new areas	remaining the same	diversified	concentrated, but shifts
Legislation	more	same	less	more
GNP	rising	constant	constant	constant
Trade balance	negative, with rising exports	negative, solely imports	negative, solely imports	negative, with rising exports
Tourism	growing	growing	growing	growing
Offshore-finances	growing	growing	without significance	growing
Labour market/vacancies	improving	remaining the same	improving	worsening
Investments of the private sector	more	more	less	more

that the group had imagined. On the whole, clustering the alternatives was the most interesting step since the results of the previous tasks now formed a unit which was not produced arbitrarily as the group had feared. When checking the consistency, the results of the analysis of influences was constantly referred to. The participants were, at the end of this third workshop, convinced of the scenario method. They now recognised that every single step had been necessary. The raw scenarios that they had developed were evaluated as useful and comprehensible development alternatives for Montserrat. These alternative development options now had to be examined in order to show precise options for the future.

6.1.3
Phase III – Scenario Interpretation

The four raw scenarios now had to be built up to become meaningful alternative scenarios. Since it would have been too time-consuming to involve the whole group and would not have led to any new results, the moderator of the workshops carried this task out. She undertook the in-depth interpretation of the raw scenarios and *designed the scenarios.* When interpreting the raw scenarios, the results of the analysis of influences were again closely referred to. Furthermore, some global scenarios, prognoses and statistics were also involved in order to examine the results of the workshops in light of other facts and findings. It took some weeks to design the scenarios. Each scenario was built up in the same way and presented on about 12 pages. Once finalised, the four scenarios were presented to the group in a further workshop. The group was now requested to analyse and comment on the scenarios and to modify them when this was deemed necessary.

Before approaching this task, the group had to visualise the scenarios. This was done for two reasons. On the one hand, the participants were to be reminded of the contents of the raw scenarios and, on the other hand, the *visualisation of the scenarios* was carried out as preparatory work for their presentation on Montserrat at a later stage. The group was split in two and each team had the task of graphically depicting two scenarios. The first team used pictures to describe the main aspects of each scenario. The other group decided to draw up thematic maps which showed the main spatial implications of each scenario. It was later discussed which method was preferred and the whole group mutually agreed that, depending on the target group, both methods had their advantages and that a combination of both might also be a good solution.

Evaluation: The group now had the opportunity to examine the scenarios the moderator had designed. All participants were on the whole very satisfied with the contents of the scenarios and commented that all major aspects worked out previously had been taken into consideration. It appeared to be a problem that the scenarios were not strictly structured, and the narrative style was also criticised. The scenarios were exciting and logical stories about Montserrat's potential futures, but a comparison between the four scenarios was found difficult. After a lengthy discussion the group, nevertheless, decided to keep the narrative style, but it was also agreed that a booklet had to be prepared for the presentation of the scenarios on Montserrat.

This booklet with the title "*Future is possible*" had two principal goals. First, the necessity to discuss Montserrat's future despite or particularly because of the volcanic crisis was to be illustrated and second, the principal aspects of the four scenarios were to be clearly explained and supplemented by pictures. In order to facilitate the comparison of the four scenarios, all of them were closely related to the four subsystems, though with a different emphasis. The booklet is presented on the following pages. The text was written by the moderator whilst the rest of the group graphically illustrated the most characteristic aspects.

The scenarios presented in the booklet are naturally determined by the group's own values, goals and perspectives. The participants had attempted to consider the perspectives of the Montserratians and integrate the results of the interviews, but the scenarios were at the end of the day designed by people who lived in a totally different *lebenswelt* and were not threatened by a volcano. The group in Hamburg was, for these reasons, convinced that the Montserratians would develop completely different scenarios. That is to say that the scenarios worked out in Hamburg and outlined on the following pages are in no case representative or complete. And this was not their intention. They were, on the contrary, designed to encourage the affected population on Montserrat to carry on and to confront the situation and to design their own scenarios and strategies for the reconstruction phase.

Future is possible ...

The small Caribbean Jaguar

Economy

In the year 2020 the political actors proclaim a high level of economic growth by promoting exports and the acquisition of direct foreign investments. This became possible through prior infrastructural achievements by the British motherland, a strong promotion of human capital as well as strong market-economy oriented state intervention as major elements of the structural reforms introduced around the turn of the century. The opening of attractive construction sites in northern and eastern Montserrat, the newly built industrial park on Spanish Point as well as a large number of investment incentives in the area of taxes and customs led to a flood of foreign capital and know-how. Especially production-oriented and software companies established subsidiaries in Montserrat. The modern telecommunications provide a new network of services in Internet. They also led to a revival of the offshore financial sector. Large infrastructural projects such as the extension of the airport and the piers in the capital, as well as the building of a number of hotels led to a boom in construction. The economic prosperity on the island is accompanied by a clearly improved employment situation. The number of immigrants from the Caricom countries is growing. But with it also grows the dependence on global economic conditions.

Politics

Along with the various investment incentives, educational policies were an important measure in national structural reform. Higher priority is also given to the technological subjects in primary and secondary schools. The national college and the University of the West Indies contribute to a high standard of qualification from the Montserratians. The close cooperation, particularly between the British motherland and the Caribbean integration partners have an impact on political action. Future-oriented planning measures and the coordination of economic and political activities over the internet allow for strategic and commonly-devised actions. The state transfers are concentrated especially on the development of new technologies. In general, the neoliberal politics promotes private investments.

Human Welfare

The economic prosperity on the island led to a decrease in emigration while the immigration from the neighboring Caribbean islands and North America increased. The north-american lifestyle has an important influence on the culture and way of life of the Monserratians. They start talking about "coca-colazation"; the middle class in particular is Americanized to a large scale. It uses the material wealth for the acquisition of luxurious homes. The average 3-person household size is clearly a reduction. The Caribbean large family with its specific network structure does not exist anymore. Career-orientation, wealth, achievement and the global connection are important values in the modern society. On the other hand, the cost of living and real estate have risen. The sick, the disabled and other social outsiders have a hard time in this achievement-oriented society. The state cannot provide any relief because the economic reform has led to a total neglecting of social issues. This situation is made worse by the increasing conflicts among the local population and the immigrants.

Ecology

The growing number of tourists and the growing population which accompanied the economic boom have increased the pressure for the use of natural resources and the creation of an infrastructure. Drinking water is partially imported because the ground water is contaminated and water-purifying facilities are not well-equipped. Chemical residue and waste still seep into the air and the soil. The emission rates set by the government are too low and also the state efforts to integrate protected areas in the national area usage program fail. Environmental problems continue to increase.

Future is possible ...

2 Sustainable Development - from below

Economy

Montserrat's economy is diversified in 2020. A relatively strong mix of economic sectors with a close connection between them has been established. Producers and consumers communicate directly about the desired goods. Short production processes dominate, also in the food chain and for consumer goods. The product selection has a regional orientation. Common-market orientation means money. The traditional sectors such as agriculture and fishing have experienced changes. Today environmentally-sound farming dominates. Measures for the preservation of nature such as tree-planting, caring for walkways and parks are the special jobs of the farmers. These activities contribute greatly to the preservation of an attractive landscape. Small-scale tourism is integrated into the regional economic activities. Trade and business profit from the tourists who are particularly interested in culture and nature. The tourists enjoy the beautiful nature, the close contact to the population, and the creative activities on historical plantation and the good care of the national parks, crafts and art workshops. Large touristic projects are missing. The innovative way of doing business, which primarily strives for the preservation of nature and culture is accompanied by active enterpreneur-ship, a low portion of foreign invest-ments and modest economic growth.

Politics

The relatively low importance of state support leads to an increasing tendency for self-responsibility and public participation. Politics take place today in network structures in which the state depends on the cooperation and participation of citizens. The moderation of social processes has priority. Citizen forums, discussion and workgroups influence political action, which focuses particularly on social and environmental compatibility. The mobilization of a number of societal powers results in a growing level of self-confidence and a strive for autonomy among the citizens. Today Montserrat enjoys military protection as an associated state of Great Britain; otherwise it is completely independent from the former motherland.

Human Welfare

Environmental education and the learning of alternative, resource-friendly means of production play a very important role. Family structures and a lively neighbor-hood are influencing elements in society. Private capital today is no longer used for the acquisition of luxury and consumer goods but rather serves the mutual assistance in the active communities. The social capital is thus very large. The return to tradition, family and culture have strengthened the self-help-potential and self-confidence of the people. The decreasing wealth is balanced out by a commitment to solidarity. Many emigrants return to Montserrat in order to put in their contribution to preservation of an environmentally and socially-sustainable way of life. The global influence on national culture disappears along with it.

Ecology

National environmental education and environmental campaigns, as well as the successful work of the National Trust have turned the Monserratians attention to environmental and developmental issues and the problems associated with them. Many of them, particularly the up and coming leaders, use this knowledge actively in both professional and private areas of life. The dangers of an uncontrolled economic growth are known to a great extent. Environmental policies and actions become the center of interest. Next to a widespread amount of environmental laws, which contain two national parks and various protected areas, many communities are actively involved in environmental protection activities. The caring for protected areas, the supervision of rare species as well as the creation of car-sharing-systems, recycling, beach-cleaning and environmental education of the tourists are all a part of it. The degradation of natural resources has clearly decreased.

Future is possible ...

Business as usual

Economy

By 2020 Montserrat's economy is dominated by tourism and tourism-related industries such as construction, trade and transportation. The state investments are concentrated on the touristic infrastructure. Along with the extension of the airport and the harbor in the capital, an increasing number of attractive construction sites on the water and in the western parts of the island are opening up. In this way, foreign investors are enticed to build further villas, apartment houses and small hotels. The construction sector blooms correspondingly. The airport extension and global trends in tourist demands are responsible for a strong growth in tourist arrivals and tourist spending. Particularly in the areas of villa and ecotourism Montserrat is experiencing a boom in growth. But the competition with the neighboring Caribbean islands and many Pacific island states increases. The online-registration of companies and online-consulting of private individuals leads to a revival of the offshore financial sector. The local employment effects nevertheless remain low. Agricultural production today has an increasingly disappearing importance. Export production of agricultural products has disappeared altogether. Over 90% of the food is imported. Fisheries reduce themselves down to sport-fishing for the tourists.

Politics

Long-term planning and consistent politics have become extremely difficult in Montserrat due to cyclic changes in government from voting period to voting period. Due to the short legislation periods, the political leaders have little opportunity to experience the long-term consequences of their political decisions. Political action is widespread reactive and sector-oriented. A necessary coordination of measures is missing. The dependence on transfers from Great Britain is still large. The political independence seems to be far away from economic independence. The OECS integration efforts are successful in countering this. The eastern Caribbean monetary system offers a stabile framework for the goods, services and capital movement between the member states.

Human Welfare

Continued emigration causes a reduction of the quantity of qualified working power. The population of Montserrat is overaged. On the other hand, the amount of work-immigrants from the Caricom countries has increased. The percentage of foreigners is about 25%. The educational level of the Montserratians is relatively good but the skills and focus of school education is not adapted enough to the market conditions and future challenges. The high level of unemployment among the younger population in particular leads to growing problems on the work market. The brain-drain continues. In general, the income development differs strongly and leads to a separation of social worlds. Family networks and remittances can provide some relief. But the conflicts between the low-wage workers from the Caribbean and the local population continue to grow.

Ecology

The tourist-oriented infrastructure development takes place at the cost of other forms of land use. Many areas which were intended for agricultural production or nature preservation according to the area usage program are being used for tourist purposes. Today, the beaches and beautiful valleys of Montserrat are in private ownership and there is little access for the public. The pressure on natural resources is increasing in the west of the island. Particularly ground water reserves are seen as in great danger. The garbage disposal also causes problems. Furthermore, a conflict-free connection between the growing market ecotourism and the necessities of nature protection can not be created. In the protected areas, investment mainly concentrates on infrastructural development at the cost of a detailed guidance and education of visitors. The small number of park rangers can only provide visitor information, resource monitoring and management to a small extent. The increasing threat of danger to the domestic flora and fauna makes a revision of the national environmental laws necessary soon.

Future is possible ...

4 Sustainable Development - from above

Economy

The influence of law making on local economy is strong. It focuses on environmental compatibility of the economy. This is determined today from emission and norms, energy taxes, environmental impact assessments, growth management and fees for using non-regenerable resources. The provision of qualified labor and technological development, which is strongly oriented on the use of domestic resources and the solution of country-specific problems, are at the center of economic policies in 2020. Modern technologies such as solar and wind energies, ecotourism, agriculture and fisheries make up the economic structure. This development could be promoted through foreign direct investment and close cooperation between international consultant agencies. Government consultation for farmers and fishers is the result of the cooperation. In the meantime, small and midsize firms have established themselves in Montserrat which can react relatively flexible to economic changes due to their various niche markets. The state-promoted touristic nature preservation has contributed to the creation of attractive environmental tourism programs. In the area of furniture production and electronic micro production, strong importance is put on the longevity and reparability of the products produced in order to avoid waste. Some companies concentrate on technical environmental protection and waste recycling and storage.

Politics

The promotion of environmental quality by means of an environmentally responsible market economy is the major goal of politics. The government supervises the environment and its pollution with poisons, in order not to risk the developmental options for following generations. Independent controllers as well as the national Ministry of Environment examine a number of developmental projects. The Ministry of Environment even has right of veto regarding environmental questions. Along with the governmental regulation and surveillance, the strong lobbying on regional and international levels has contributed to the successful environmental and economics policies. The national and regional sea and area usage are coordinated in the framework of eastern Caribbean integration. The focus of governmental investment in environmental technologies is supported by the British motherland and other countries from the western world.

Human Welfare

The educational policies have led to an increase amount of well-qualified individuals, who can be relatively easily integrated into the work market. This is not the case for less qualified individuals, who either are pushed into informal working situations or become unemployed. The gaps within private income distribution are becoming bigger. The high amounts of subventions in the environmental and technological areas have resulted in a widespread neglect of social issues. The participation of citizens is low; their social security is insufficient. The communication, economic, and financial structures have led to a dependency and orientation on the outside. The individualization and isolation of many Montserratians is growing. The high level of state responsibility limits the motivation and autonomy of the mass population.

Ecology

The strict environmental expenditures, the widespread resource management as well as the consistent environmental education contribute strongly towards the relatively intact natural environment. Furthermore, environmental "taboo-zones" as well as the large number of protected areas and national parks guarantee the protection of important ecosystems. Limiting the number of visitors as well as the work of the national park administration have been decisive in the development of an environmentally-adapted tourism. Park rangers manage and survey the resources as well as the activities of the visitors. Behavior rules when hiking, bike riding or bird watching are generally accepted and understood as a contribution to the preservation of the environment. Economic usage is adapted to environmental concerns to a large extent. The increase in the use of public transportation has resulted in a clear reduction of ground and air pollution. The new electric buses make a strong contribution.

6.1.4 Concluding Evaluation

The scenario project in Hamburg proved that a group of people can use the scenario method as a tool to gain a mutual understanding of a system's complexity. The logical sequence of the individual tasks has proven to be decisive since it firstly facilitates a reflection of the values, goals and activities and secondly facilitates the understanding of the factors and spheres of influence. The continuous exchange of information enhances the participants' impression that their own ideas and suggestions are taken into consideration. Open discussions are the basis of an extensive learning process within the group. The participants are encouraged to think about alternatives, broaden their awareness and deepen their understanding of the system.

This pre-test of the scenario method also illustrated that it is at any stage possible to come to a consensus within the group, even though there might initially be numerous distinct values, interests and goals. This finding refers to both the methodical implementation of each task and to the final agreement on various alternative futures for Montserrat. The participants were extremely motivated to act collectively and they showed considerable openness towards other perspectives. The participants, on the whole, identified themselves with the results of each task, though these results were often subject to the conflict between the desire to grasp the system in its entirety, and the need to develop a mutual understanding about the system's complexity. It also has to be kept in mind that the findings can at any time be surpassed by the dynamics of change.

For exactly these reasons, the scenario method basically requires a constant repetition of the processes, either with one and the same group or with different participants. Decisive is, above all, the permanent reflection and modification of the values, goals and activities. This, as outlined in Chapter 3, requires the following abilities: resilience, creativity, tolerance towards errors and multiple perspectives. As the workshops show, communicative and decision-making processes can actually be improved with the scenario method. The participants proved to be very creative and were able to consider various different perspectives. It was characteristic of the group that it experimented with ideas and fantasies and constantly learned from experience.

Since the actual volcanic crisis was largely excluded from the reflections and a final analysis did not take place, it can at this point not be determined whether the resilience of the participants was encouraged. In the course of the scenario project in Hamburg the participants did not have to prove whether they were able to accept unexpected events and the uncertainty of a disaster situation and it was not ascertained whether they were able to adapt flexibly to changing conditions. A test with disruptive events showed that the group even had considerable difficulty to mentally cope with imaginary, sudden events. When the implications of the actual volcanic crisis on Montserrat were analysed, no major problems were encountered, but the group commented that this was a rather frustrating and paralysing procedure.

In summary the project in Hamburg confirmed most of the potentials of the scenario method, particularly with regards to the creative dealing with complexity and the experiments with alternative development options. It became apparent that the potentials of the method, as outlined by Arras (1989), can be improved when not only the results but also the communication and decision processes of the group are analysed and considered. Only one question remains open – can the scenario method contribute to coping with uncertainty? The answer to this question will now be sought. The individual tasks of the scenario method will be accomplished with another group on Montserrat at the time of the ongoing crisis.

6.2 Second Approximation – The Scenario Method in a Hazard Situation

The results of the scenario project in Hamburg were presented to various public representatives on Montserrat in the autumn of 1996. Most of them were quite interested, above all the *Physical Planning Unit.* It recognised the potential benefits of the scenario method for the organisation of a sustainable and participative planning process and decided to practise it with a larger group. A two-day workshop was planned and carried out on the island in January 1997. It was financed by the *University of the West Indies' Centre for Environment and Development* (UWICED) as part of a *World Bank / Commonwealth Secretariat* transfer. 23 representatives of the public and private sector, schools and NGOs attended the workshop.

Participants:

Dawn Bramble	Surveyor, Land and Survey Department
Lennox Brown	Managing Director, Montserrat Electricity Services Ltd.
Stephen Brown	Technical Officer, Montserrat Water Authority
Roselyn Cassel	Director, National Development Foundation
Tom Dawkins	Owner 'Paradise Shirts', representing private sector interests
Neville Dublin	Manager Customer Services, Cable & Wireless
Margaret Dyer Howe	General Manager, Montserrat Water Authority
Lachlan Dyett	Student, MSS, Salem Campus
Natalie Edgecombe	Geographical Information System (GIS) Dept., Physical Planning Unit
Sarita Francis	President, Montserrat National Trust
Claude Gerald	Director, Department of Agriculture
Franklyn Greenaway	Physical Planner, Physical Planning Unit
Alan Gunne–Jones	Chief Physical Planer, Physical Planning Unit
Trevor Howe	Environmental Health Officer, Environmental Health Department

Cindy Johnson	Student, MSS, Salem Campus
Franklyn Margetson	President, Montserrat Chamber of Commerce and Industry
Carole Osborne	Hotel Manager representing the Montserrat Tourist Board
Colin Riley	Geography Teacher, MSS, Salem Campus
Carl Roberts	Managing Director, Cable & Wireless
Henrietta Ryan	President, Rotaract (Youth Organisation of the Rotary Club)
Denis Silcott	Student, MSS, Salem Campus
George Skeritt	Representative, Montserrat Volcano Observatory
Peter White	Director, Land Development Authority

Again, a multi-step group procedure was to be carried out in order to design scenarios for Montserrat. However, the intention was another this time since the group was dealing with a real situation and not carrying out an ivory-towered test. Focal points of the project were the reflection of the existing development plans, an increase in the awareness of the problems and the practical implementation of the scenarios with appropriate strategies. The following five tasks were closely related to the overall goals:

1. The potential and ability of the participants to draw up alternative pictures of a desirable future had to be activated.
2. The participants had to identify themselves with the problems, and express their individual desires and needs.
3. Priorities had to be set for planning, and the existing development plans and strategies (those worked out before and during the crisis) had to be examined.
4. The scenario method was to be tested in an actual hazard situation, especially with a view to the mechanisms between the affected population and the decision-makers.
5. In-depth discussions were to be held about how sustainable development could be realised on Montserrat.

Running a scenario project over some months was quickly dismissed since the practice demanded a rather urgent consensus about the future development of the northern region. Most of the participants were also employed and could, at best, commit themselves for a few days. The experience gathered with the pre-test in Hamburg was taken advantage of and a tighter procedure opted for, i.e. certain tasks were selected (cf. Chapter 6.2.1 to 6.2.4). Because of the larger number of participants two moderators were in charge of the group sessions, in turn running the discussions, taking minutes and organising matters.

Introductory Phase and Presentation of the Booklet

The goals and procedures of the workshop were briefly presented and the participants were then given the opportunity to introduce themselves and express their personal interests and expectations. Because of the heterogeneous composition of the group, the interests naturally varied considerably. Some emphasised the implications of the actual crisis on Montserrat's future, others were concerned with the general development of distinct sectors, such as education, the economy, environment, politics or sports. The expectations of the participants were on the whole similar. Most of them saw the workshop as an experience and learning process. They recognised it to be a unique opportunity to examine alternative development options for Montserrat. Many participants also expressed the need to exchange the various ideas that existed about the development of the north and to mutually decide on a sustainable strategy. Some of the comments are exemplarily listed below:

- "I feel responsible for the youth of our country and therefore I'd like to look for some directions."
- "I'd love the idea to start from scratch and sometimes wished us back to Columbus' times. We could then take into consideration our mistakes and make sensible decisions."
- "We live on a small island with precious resources. Sustainable development is vital and planning needs to consider this."
- "I'm interested in the views of other private sector representatives and I'd like to share ideas about what should be done in the north."
- "I would like to learn what can happen to us, what we see for the future and its impact on our crisis situation."

In the course of this introductory phase the general objectives and the procedures of the scenario method were explained. The scenarios that had been designed during the workshop in Hamburg were presented. The intention was to give the participants a general impression about the way scenarios are constructed. The booklet with a brief summary was then handed out to the participants and they were given some time to study the contents and evaluate it with a questionnaire.

When asked "*Which future do you like best*", the majority favoured the scenario called *Sustainable Development from below.* 16 of the participants preferred this scenario (no. 2), four chose the *Small Caribbean Jaguar* (no. 1). Two participants selected the scenario *Sustainable Development from above* (no. 4) and one participant opted for *Business as usual* (no. 3). However, the next answer showed a more differentiated picture since the participants could well imagine living in one or the other future. It was also ascertained which scenarios received worse ratings from the majority of the participants. Table 6.7 summarises the answers. Attention should be paid to the fact that the participants often selected more than one category.

Table 6.7. Evaluation of the different scenarios

Answer	Scenario 1	Scenario 2	Scenario 3	Scenario 4
Yes, I really like this future	2	11	–	–
Yes, I could live in this future	5	10	4	2
No, I don't like this future	7	2	5	8
No, I really don't like this future	7	–	8	1
Undecided	1	1	3	9

The most disputed scenario was the *Small Caribbean Jaguar*, with 14 negative and seven positive evaluations. It was noticeable that particularly the younger participants, between 18 and 34 years of age, preferred this scenario as the best option. The second scenario *Sustainable Development from below* was clearly rated positively (21 participants). Only one participant rejected this scenario. The scenario which received the worst rating envisaged following the previous path of development, i.e. *Business as usual.* Sustainable development directed by the élite was evaluated negatively by nine participants, a further nine remained indecisive, and only two participants could imagine living in such a future.

The participants were also given the opportunity to note any wishes for modifications on the questionnaire. There was quite a clear tendency to integrate positive aspects from other scenarios into the favoured one. Those advocating the *Small Caribbean Jaguar* for example requested socio-economic modifications, such as lower costs of living, certain control measures for the immigration of foreigners and management strategies for the protection of natural resources. The people supporting the scenario *Sustainable development from below* mainly called for modifications with respect to the economic development. Some wished to encourage ecotourism, offshore banking and a more ecological agriculture, whilst others, in contrast, evaluated tourism as critical and preferred to support modern communication technologies. Parallels to the workshop in Hamburg could easily be drawn. The participants of that first scenario project in Germany also wished to combine their expectations, but later on they realised that this often proved impossible, especially at the stage where the alternatives are clustered and examined with respect to their consistency.

When asked about the vulnerability of the scenarios to hazards, the answers given varied. Seven participants chose the *Small Caribbean Jaguar* as the least vulnerable scenario. They argued that it provided enough capital and know how for preventative and reactive measures. Four participants expressed their concern about exactly this scenario, saying that it was extremely vulnerable since especially the marginal groups had hardly any alternative, and the strict economic orientation with extensive infrastructural measures conditioned vast damage in the case of a disaster. 13 participants opted for the second scenario, giving the following reasons: "*A sustainable development approach in a holistic manner putting people first would be less vulnerable in the long-term. Investment in education and resource management is absolutely vital to mitigate hazards. These are very important aspects for the future of post-volcano Montserrat*". In view of the actual crisis it was no surprise that twelve participants considered the *Business as usual* scenario as very vulnerable.

The participants generally regarded the mental examination of the scenarios developed in Hamburg as an important step for the entire working process. The reasons given varied: (1) the scenarios conveyed an impression about what was expected of the participants in the course of the workshop, (2) the scenarios illustrated different alternatives for Montserrat about which the participants wanted to reflect intensively, and (3) the scenarios motivated the participants to take an active part in Montserrat's future in a time of crisis and to consider its options, especially in view of the goals of reconstruction.

6.2.1
Phase I – Utopias and Visions

Just as in Hamburg, this phase was designed to formulate *abstract utopias* in order to enter the world of fantasy, desires and fears and to learn to express them. In the light of the acute volcanic crisis this mental experiment seemed to be of considerable significance. It gave the participants the opportunity to dream, fantasise and hope and free themselves of the paralysing effects the volcano had on their daily lives. Again, the multi-questioning method of the metaplan technique was applied. Since the number of participants was rather large and would have produced too many cards, the group was split into teams of two. The teams were requested to gather ideas by a simple brain storming procedure. Afterwards they had to decide on four abstract utopias. The cards were then collected and read aloud by the moderators.

Evaluation: As Figure 6.2 shows, different fantasies and pictures were accumulated in this phase. The participants expressed the ideas and values that seemed most important to them, e.g. humanity, solidarity, self-determination, ecologically sound activities, peace or harmony. These aspects were either described in detail or formed part of overall goals. They referred to both global and local developments, such as the current globalisation process or the return of other Montserratians from abroad. Even individual wishes, such as owning a yacht or becoming a professional golfer, were expressed and supplemented the utopias depicted in Hamburg by a wide spectrum of ideas and pictures. The majority of the visions did not refer to the actual development trends of the island, i.e. they were real utopias. In the light of the volcanic crisis it was impressive to see that the participants had no problems in experimenting with their wishes and dreams. This phase proved to be extremely valuable since the island's future had turned from something rather hopeless into something worthwhile discussing, influencing and shaping.

In the further course of the workshop the participants were requested to determine and discuss the *common ground, differences and interactions* between the abstract utopias. As before in Hamburg, clusters were tied and given generic terms. The participants quite quickly accomplished the task and agreed on the following eight clusters: *Education, Transportation, Economy, Health, Politics, Social, Environment, Dreams.* These eight clusters were surpassed by the cluster which expressed the desire most urgently put forward by all participants: the *cessation of volcanic activity*. The clusters revealed that the

Figure 6.2. Abstract utopias and cluster generation (1/3)

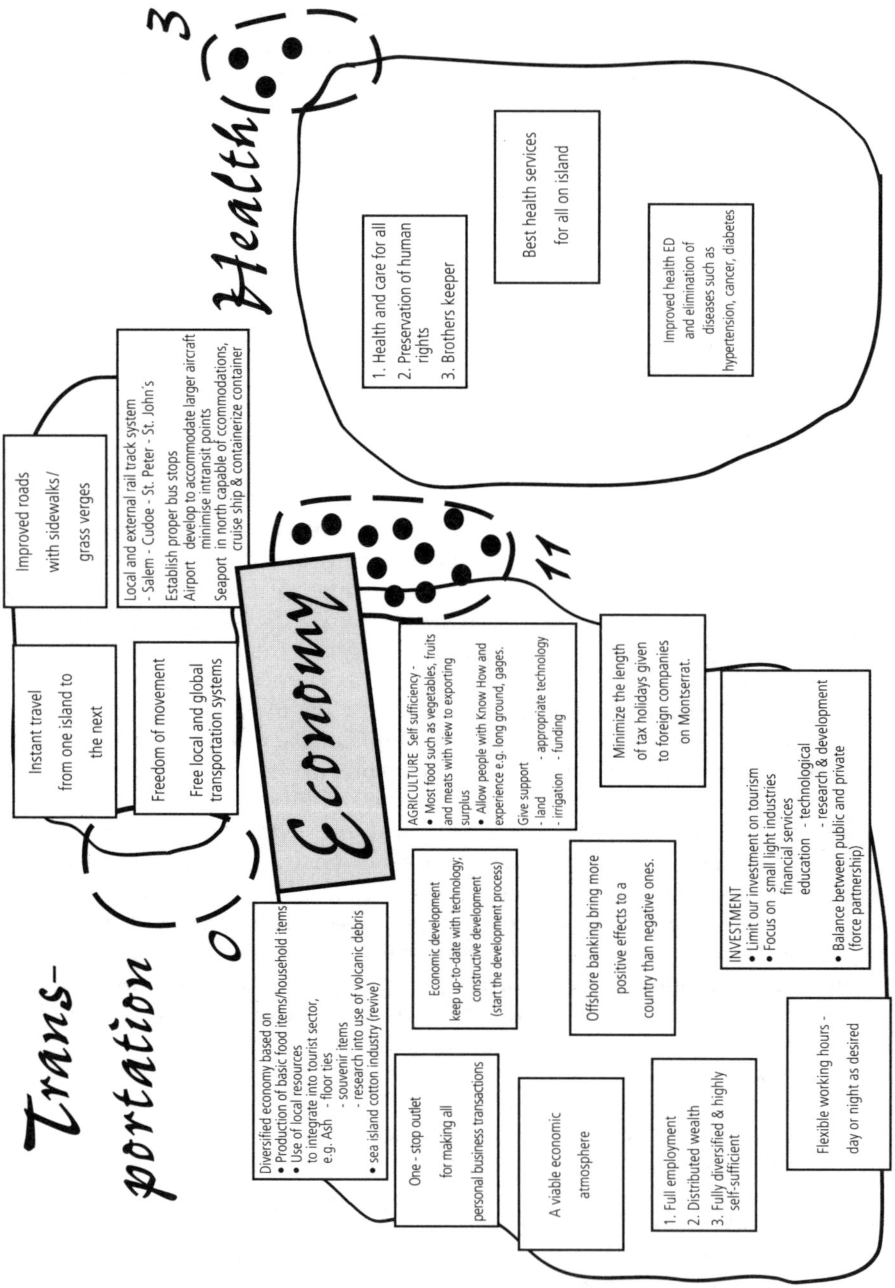

Figure 6.2. Abstract utopias and cluster generation (2/3)

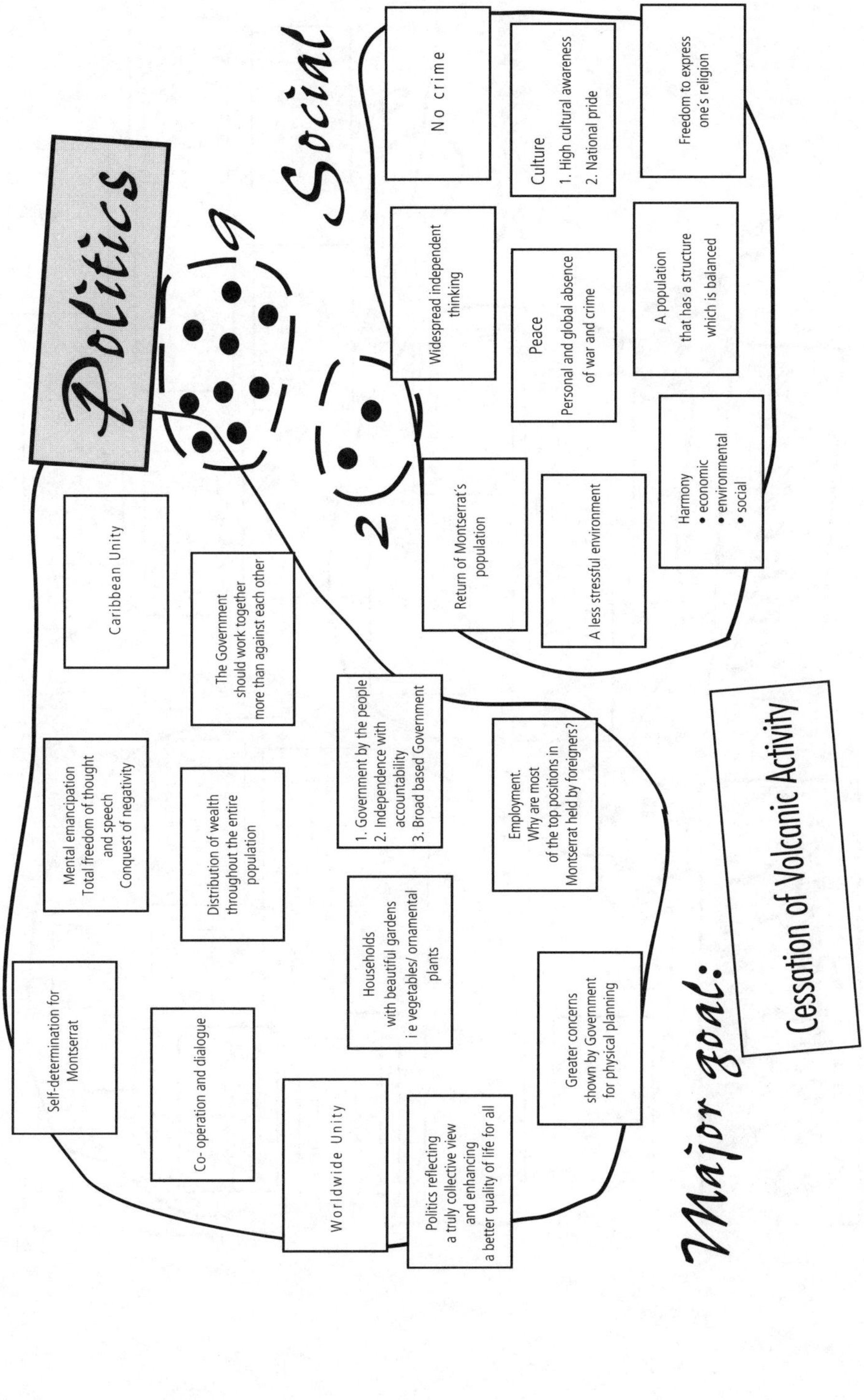

Figure 6.2. Abstract utopias and cluster generation (3/3)

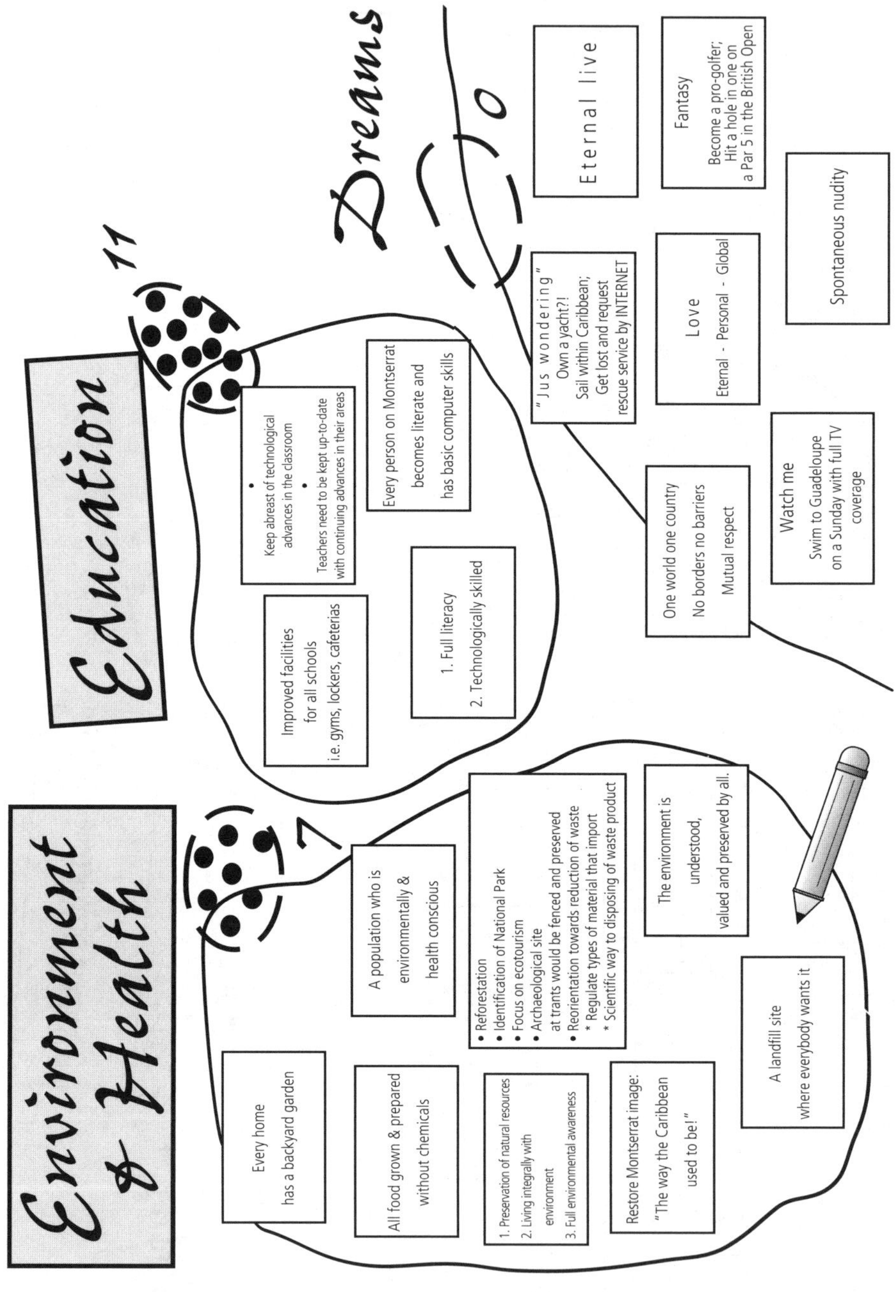

majority of the utopias referred to political, economic and, to a slightly lesser extent, social and ecological questions.

The multi-point questioning technique was applied to *evaluate the clusters. Education* and *Economy* were allocated eleven points and were thus highest-ranking in priority. They were followed by *Politics* (nine points), *Environment* (seven points), *Health* (three points) and *Social* aspects (two points). The remaining clusters *Transportation* and *Dreams* did not receive any points. It was surprising that the cluster *Education* had received so many points since it had hardly been considered before. On the other hand, the cluster *Social* which had initially contained numerous abstract utopias, at this stage turned out to be of little significance. As had been experienced in Hamburg, this task provoked that concrete ideas came to the fore. The group now had to restrict itself to the most important clusters. The participants selected *Education, Politics* and *Economy* and also insisted on grouping *Health* and *Environment* since they were of the opinion that both had to be addressed as a unit. This request represented an interesting difference in the awareness of the two groups. Whilst the participants in Hamburg had hardly linked individual health and environment, the Montserratian group insisted on their close relation. Another difference was detected in the course of this task. Contrary to the workshop in Hamburg, the social system here was represented exclusively by *Education.* The task was on the whole accomplished without any major problems. The multi-point questioning technique found general approval since the participants had the impression that their own opinions were taken into consideration.

6.2.2 Phase II – Draft Scenarios

Central objective of this second phase was the *construction of alternative draft scenarios* for Montserrat for a period extending over the next 25 years. Based on the experience with the workshop in Hamburg, the group was directly split into teams of three. The teams were asked to use the four clusters and agree on a descriptive title and general contents for the scenarios which were later to be designed in detail. Once this task had been accomplished, the teams once again sat down for a brain storming session in order to collect pictures and scenes to describe the different options for the island's future. Amongst others, the team members asked themselves the following questions: "What could Montserrat's future look like?", "What could it look like in the regional or global context?", "What would the Montserratians themselves think and do?", "Which decisions could the local politicians make for the future development?" etc. Since the task was designed to obtain alternatives, each team was asked to propose at least two different titles and development options. The results were to be presented to the plenum at a later stage.

Evaluation: It took the teams about an hour to produce their extremely constructive results. All participants had well understood the task and lively discussions contributed to the positive outcome. All teams were active and creative and, despite the occasional difficulty in finding a consensus, they all strictly adhered to the goals defined. They were, once again, highly motivated and wished to present and discuss their results before the plenum. Posters were used to illustrate the goals, models or fundamental ideas for the draft scenarios (Figures 6.3 and 6.4). Each team chose a representative who presented and explained the results. Some of the general statements are summarised below.

The first team presented a draft called *The Cohesive Society.* It pursued twelve goals, shown in Figure 6.3 (corresponding to the team's own). The team focused on a society in which all sectors are closely related and which maximises the use of its potentials. Their second title, *An enlightened Society where high Standards of Education and Dialogue are the Norm,* described a future in which education and research were just as important as the participation of the citizens in planning and decision processes (cf. Figure 6.4). This draft also concentrated on families and a common system of values for the entire society. On the whole, the team emphasised social aspects, even though these had been quite insignificant in the previous evaluation of the clusters.

The second team presented a draft titled *Harmonious Montserrat* (cf. Figure 6.5). The picture that accompanied this title showed a future in which mankind has recognised the significance of the natural environment and their own health. Accordingly, the team focused on the improvement and sustainable management of the ecosystem, early environmental education and an improved health system. Their second draft, called *Indigenous Montserrat,* concentrated on the island's economic development, thereby emphasising the mobilisation of local knowledge and skills. Central aims were a diversification of the agricultural production, an integrative production system and the promotion of local craftsmanship and modern communication technologies. As before, this team also concentrated on one cluster each.

The third team, which was set up of students, drew up a draft titled *Back to the Basics* (Figure 6.7). A modern ecological agriculture here formed the basis of a coherent economy and society. In *Back to the Basics* people would not move away from the land and exports would not outweigh imports. This would allow more money to circulate in the country which again would result in a high standard of living, a drop in brain-drain and a less corrupt society. The team's alternative dealt with the political autonomy of the island and was called *A look into the future – Montserrat as an independent Nation.* The team explained that a successful economic development on the basis of a modern agriculture, tourism and offshore banking was the prerequisite for political independence. This draft also emphasised improved education and conservation. As figure 6.8 suggests, all four clusters were addressed by this draft.

The fourth team also referred to Montserrat's political independence as the title clearly shows: *An independent (self-determined) Montserrat.* The draft is presented in figure 6.9 and the team has clearly shown its principal vision for all four clusters. The team members had also prepared suggestions for the realisation of their vision. Their second draft, *Montserrat as an integral Part of the Mother Country (UK)* (Figure 6.10) was also based on the political situation, however, represented the other option. It referred to Martinique and Guadeloupe as *Departements d'autre mer* and suggested that Montserrat became an integral part of Britain. The team realised that the result would mean a stronger position of power for the mother country and had already discussed

Figure 6.3. The cohesive society

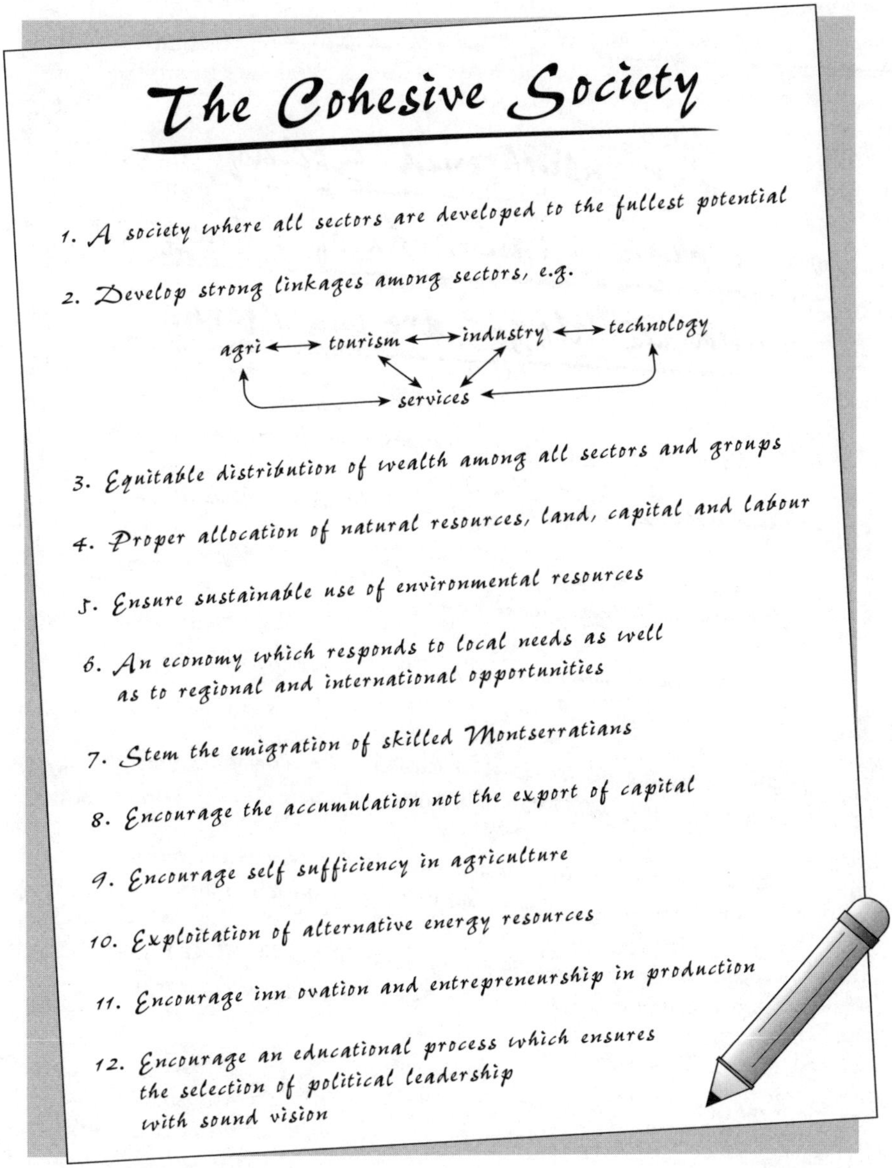

Figure 6.4. An enlightened society where high standards of education and dialogue are the norm

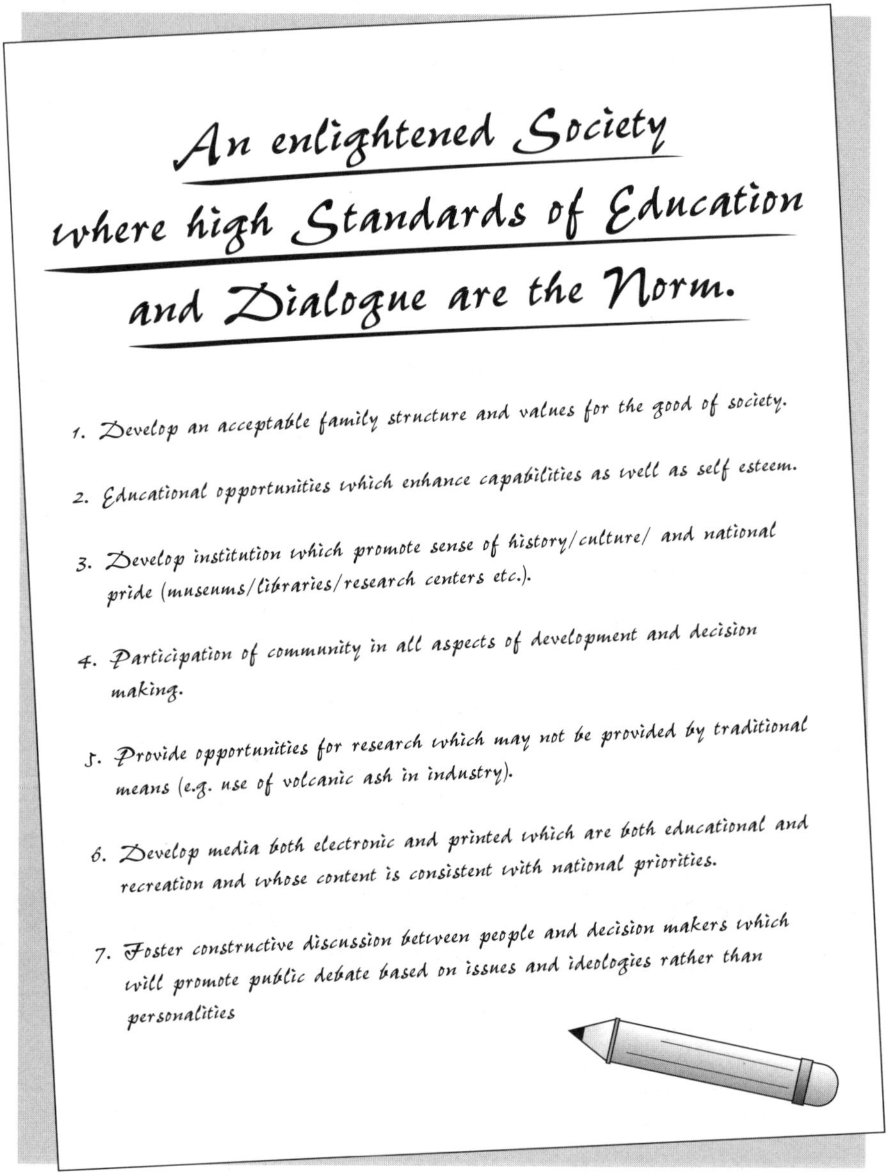

Figure 6.5. Harmonious Montserrat

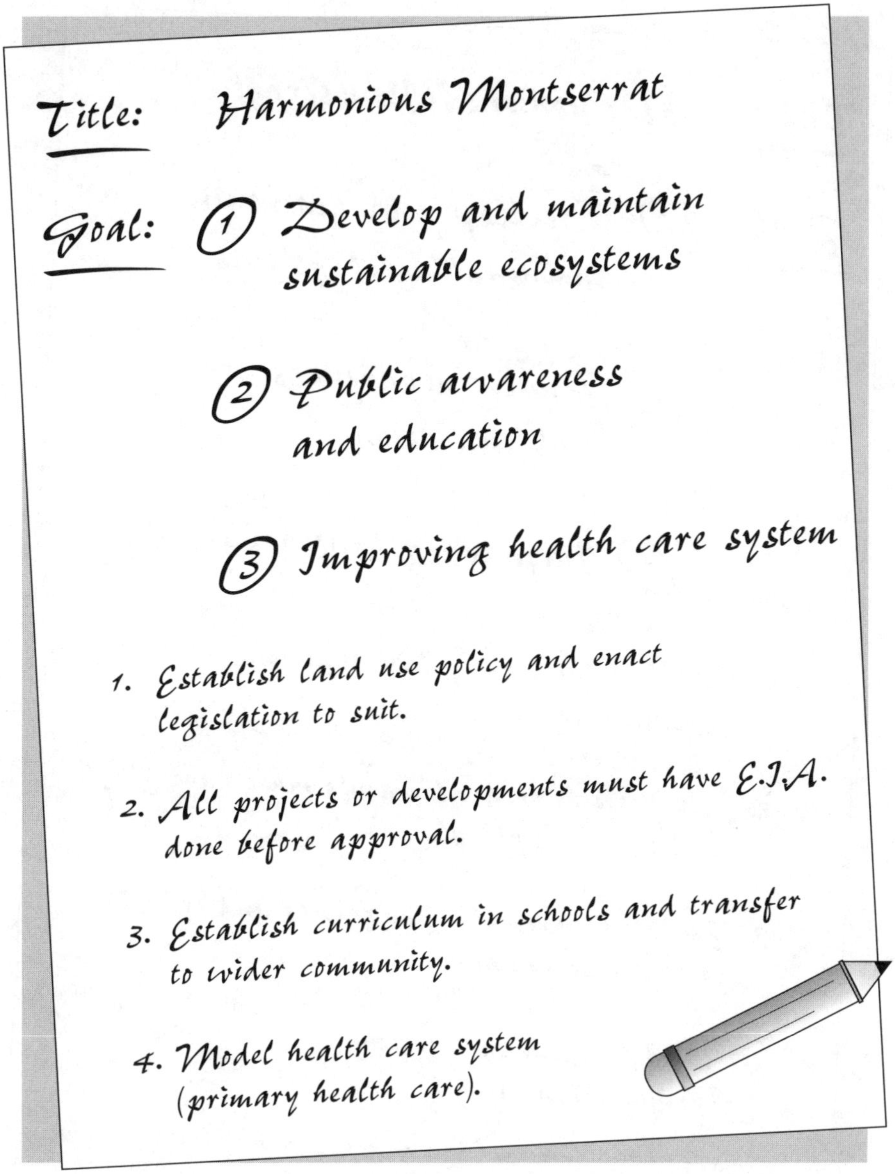

Figure 6.6. Indigenous Montserrat

Figure 6.7. Back to basics

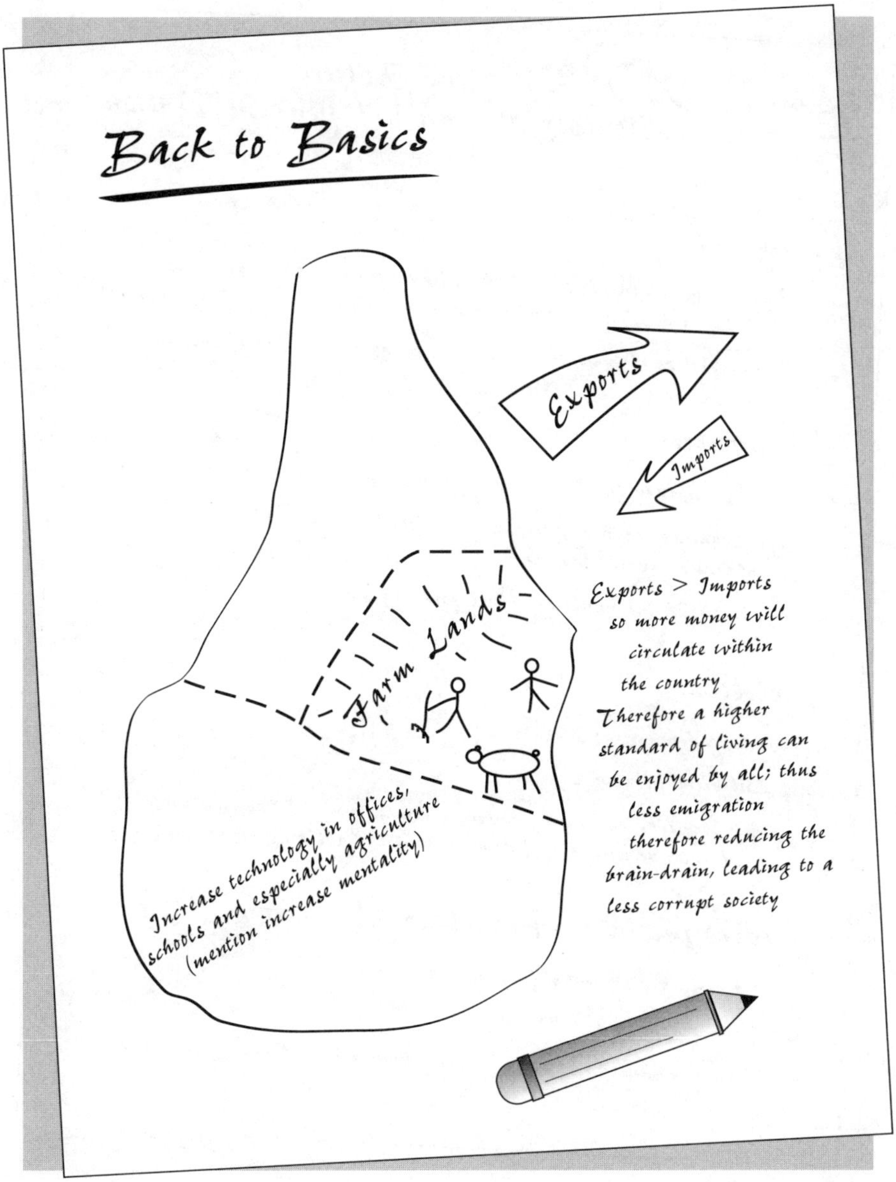

Figure 6.8. A look into the future – Montserrat an independent nation

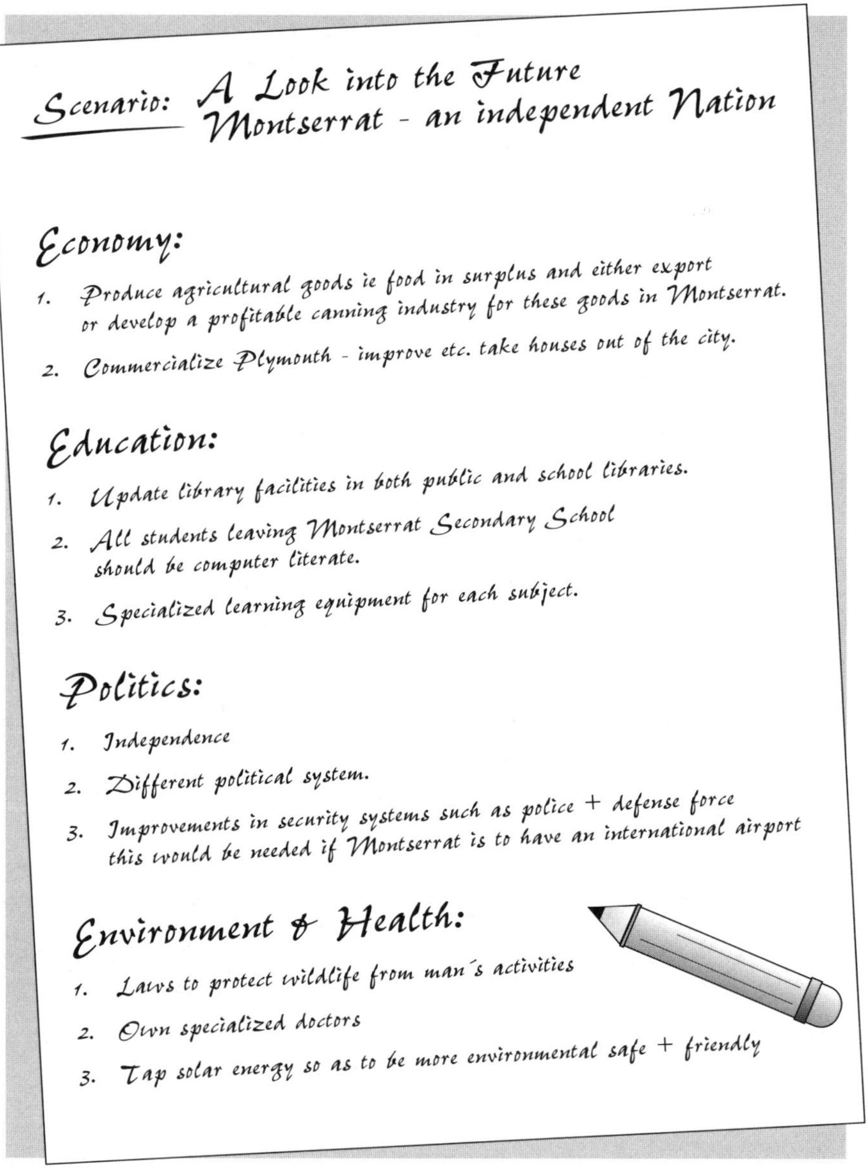

Figure 6.9. An independent (self-determined) Montserrat

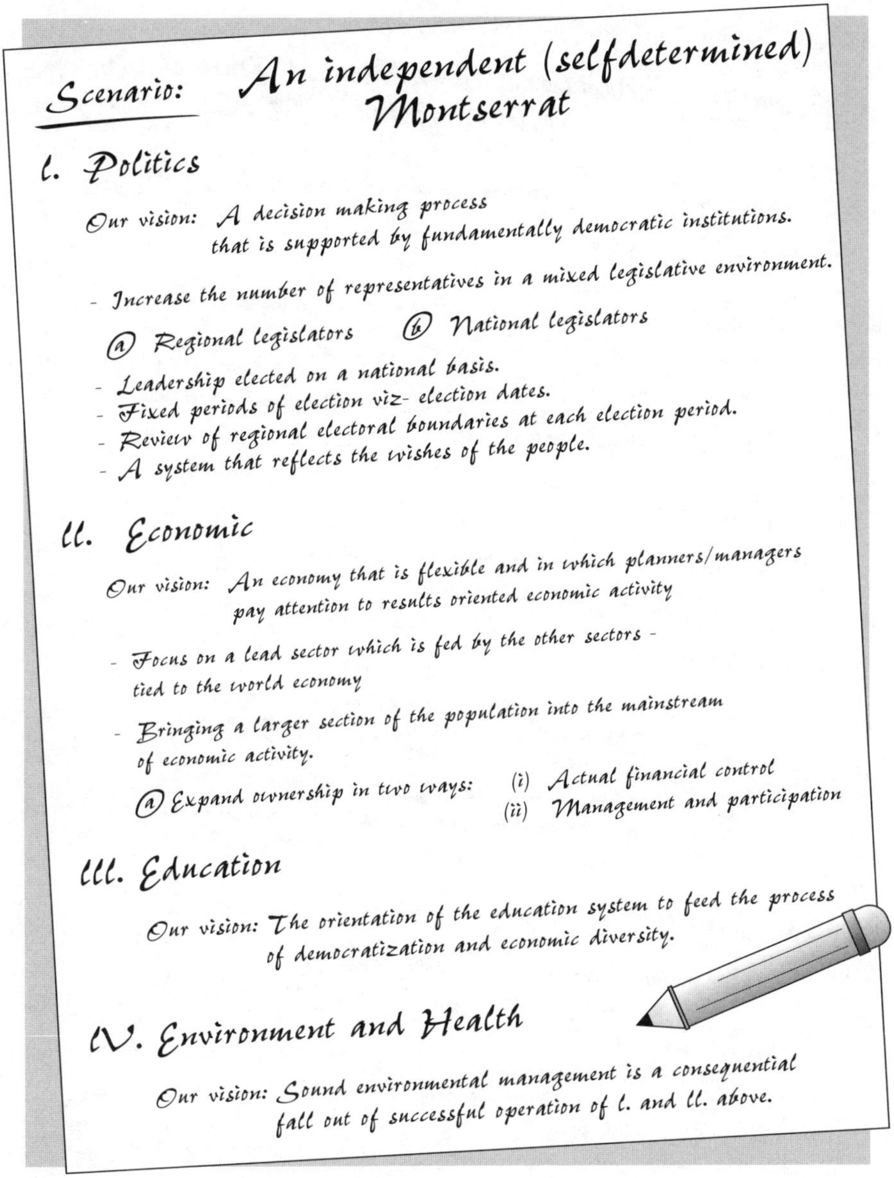

Scenario: An independent (selfdetermined) Montserrat

I. Politics

Our vision: A decision making process that is supported by fundamentally democratic institutions.

- Increase the number of representatives in a mixed legislative environment.

(a) Regional legislators (b) National legislators

- Leadership elected on a national basis.
- Fixed periods of election viz- election dates.
- Review of regional electoral boundaries at each election period.
- A system that reflects the wishes of the people.

II. Economic

Our vision: An economy that is flexible and in which planners/managers pay attention to results oriented economic activity

- Focus on a lead sector which is fed by the other sectors - tied to the world economy
- Bringing a larger section of the population into the mainstream of economic activity.

(a) Expand ownership in two ways: (i) Actual financial control (ii) Management and participation

III. Education

Our vision: The orientation of the education system to feed the process of democratization and economic diversity.

IV. Environment and Health

Our vision: Sound environmental management is a consequential fall out of successful operation of I. and II. above.

Figure 6.10. Montserrat is an integral part of the Mother Country (UK)

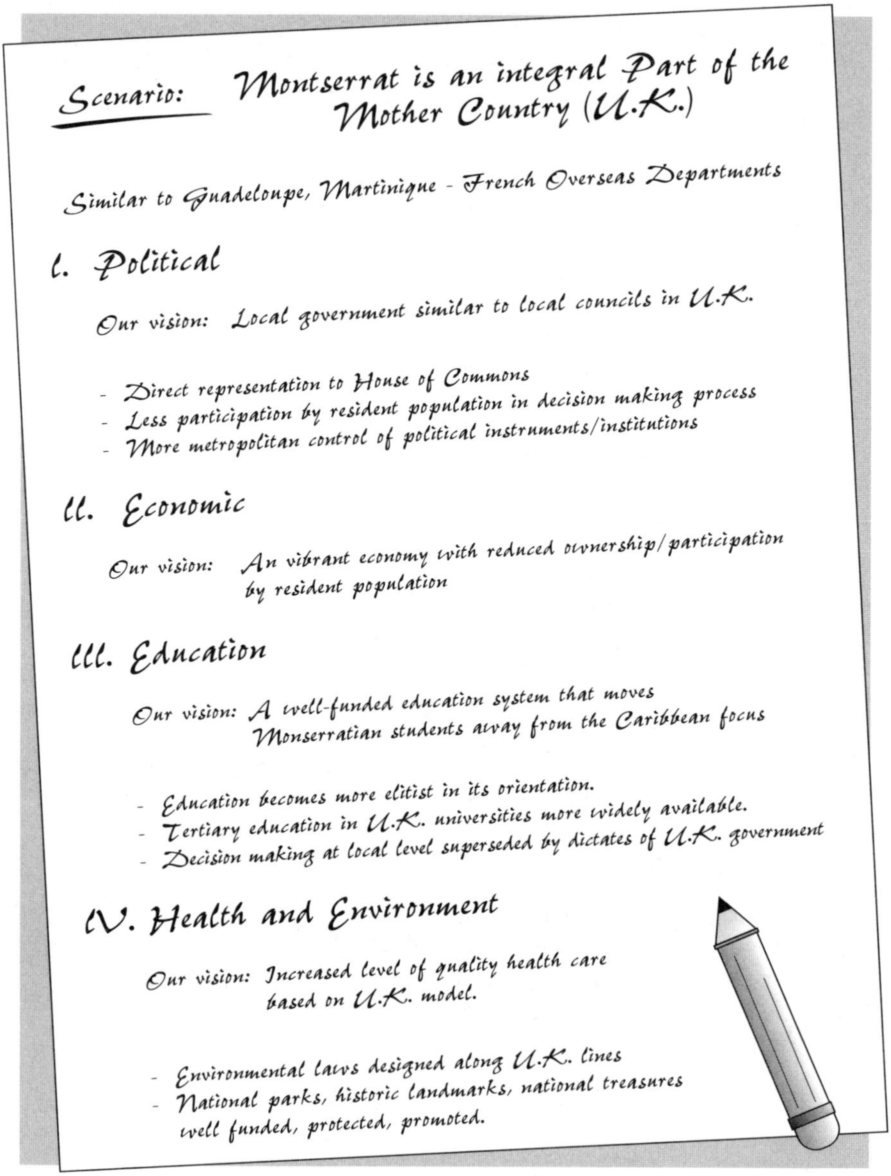

Scenario: Montserrat is an integral Part of the Mother Country (U.K.)

Similar to Guadeloupe, Martinique - French Overseas Departments

I. Political

Our vision: Local government similar to local councils in U.K.

- Direct representation to House of Commons
- Less participation by resident population in decision making process
- More metropolitan control of political instruments/institutions

II. Economic

Our vision: An vibrant economy with reduced ownership/participation by resident population

III. Education

Our vision: A well-funded education system that moves Monserratian students away from the Caribbean focus

- Education becomes more elitist in its orientation.
- Tertiary education in U.K. universities more widely available.
- Decision making at local level superseded by dictates of U.K. government

IV. Health and Environment

Our vision: Increased level of quality health care based on U.K. model.

- Environmental laws designed along U.K. lines
- National parks, historic landmarks, national treasures well funded, protected, promoted.

Figure 6.11. Environmentally sound Montserrat

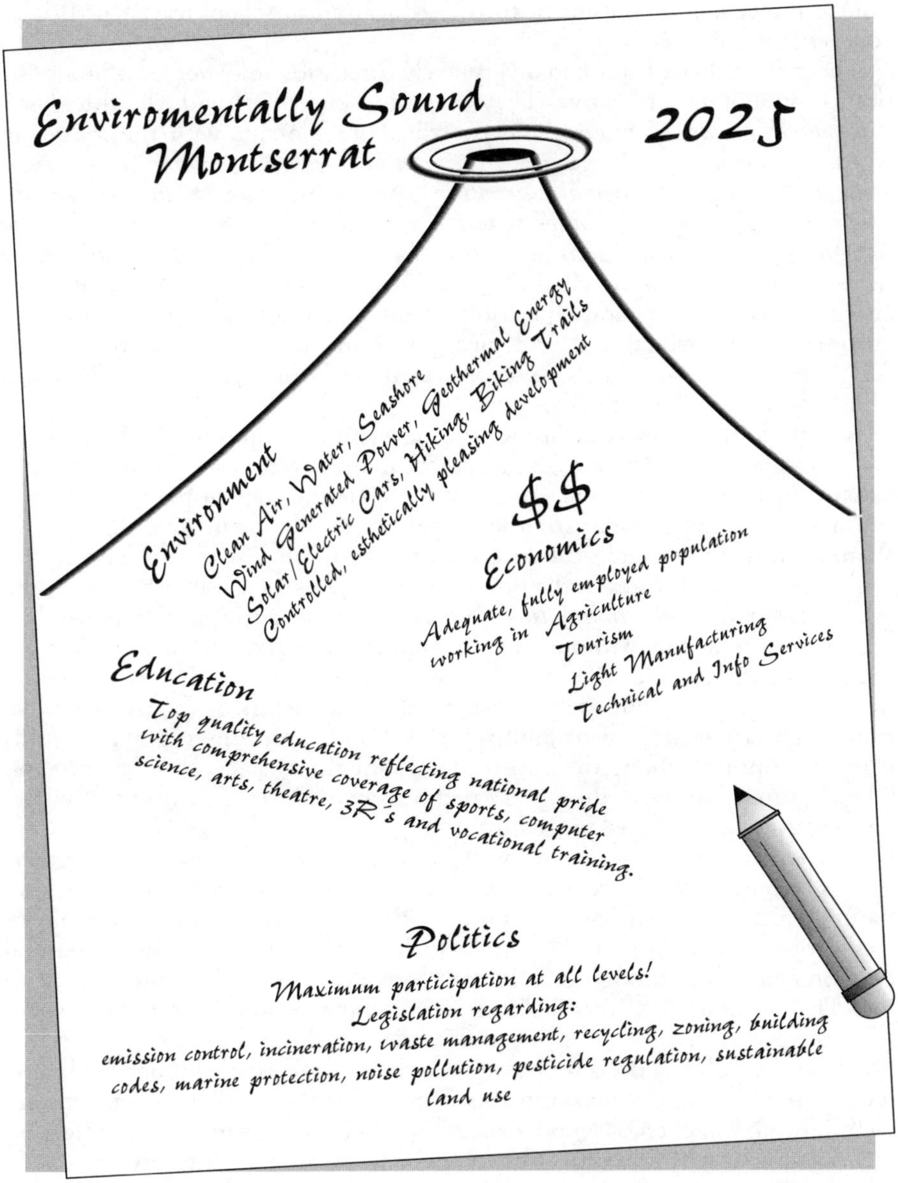

a number of implications, such as a possibly lower Montserratian contribution to the economic output or further education of school leavers at British colleges and universities.

The fifth and last team had a very lively discussion and were not in a position to present an alternative. They presented their draft with the title *Environmentally sound Montserrat* and described its contents with the following words: "*It would be a Montserrat in a kind of Greenpeace fashion, where environmental issues outweigh all the other aspects. Important is an aesthetically pleasing environment. All other sectors – education, economy, politics – would have to reflect different aspects supporting the vision of an environmentally sound Montserrat.*" The group called for alternative energies (e.g. solar and wind energy), wished to establish a comprehensive conservation legislation and encourage a diversified and ecologically sustainable economic development. All these requests and goals were projected onto one symbol: the Soufrière Hills Volcano (cf. Figure 6.11).

As the draft scenarios illustrate, the development options for Montserrat vary considerably. However, some subject matters were to a greater or lesser extent addressed by all teams. Amongst these are the state of the natural environment, the sustainability of the economic development and the societal framework including political decision processes. Since scenarios have to describe alternative futures and not simple variations of a subject matter, the group favoured a *restriction to two alternatives.* These, again, in themselves had to be consistent and without any contradiction. A compromise had to be made since the time for the workshop was running out and it was, therefore, not possible to carry out the analysis of influence factors or the trend projections. The group decided to jointly proceed and determine the most significant assumptions about the future development in a step-by-step process. These assumptions were then to be tied into two consistent clusters (raw scenarios), using an intuitive procedure.

Evaluation: The first raw scenario, called *Proud Alliouagana,* was everybody's favourite (cf. Figure 6.12). As its name implies, the scenario focuses on nature, culture, knowledge and abilities of the local population. Central objective of the scenario is an economy which adapts to the needs of the natural environment. Even though it is tightly controlled, it is at the same time flexible. This economy is diversified and concentrates on the local resources. An integrative resource management and modern technologies also play an important role. Political decision processes are to a large extent influenced by the local population, the scenario in this context suggesting that the citizens are well-informed and enjoy good education. Political culture, communication technologies and Montserrat's history and culture are taught to proud citizens who have learned to be self-confident. Land utilisation plans, building regulations and flexible management processes take the sustainable utilisation of resources into consideration and lessen the society's vulnerability.

The second raw scenario, called *Sterling Montserrat,* represented the conscious alternative (cf. Figure 6.13). In view of the volcanic crisis the participants believed this option to be more probable. The name symbolises that it focuses on the economic development of the island. The economic profit is expressed in Pound Sterling and the raw scenario sees Montserrat as an inte-

Figure 6.12. Proud Alliouagana

Proud Alliouagana

People oriented, Self- determined, Cohesive & Indigenous, Environmentally sound economy

Economy:

* Diversified economy, where all sectors are developed at the fullest potential.
* Strong focus on modern technology.
* Proper allocation of natural resources (Music, Agriculture, Land, Capital, Labour etc..).
* Craft Development: Using indigenous resources & knowledge.
* Bringing a large section of the population into the mainstream of economic activity.

Politics:

* A new political system which takes into consideration what we would like to achieve in topic of economy, see above.
* Focusing on ideologies rather than personalities.
* Decision making process that is supported by fundamentally democratic institutions.
* Participation of community in all aspects of development & decisionmaking.

Education: available, relevant, independent, comprehensive.

* Educational opportunities which enhance capability & self-esteem.
* Top quality education reflecting national pride with comprehensive coverage of sport, computer science, arts, theatre, music...
* The orientation of the education system has to feed the process of democratisation and economic diversity.
* Minimum education standard.
* Develop Institutions which promote sense of history & culture.
* Ensure widespread comprehensive political education.

Environment & Health:

* Land allocation & physical planning.
* Sustainable use of resources.
* Development in harmony with nature.
* Clean air, water, seashore.
* Wind generated power, geothermal energy, solar/electric cars
* Hiking, biking trails.
* Controlled, aesthetically pleasant development.
* Comprehensive health care system.
* Legislation regarding: emission control, incineration, waste management, recycling, zoning, building codes, marine protection, noise pollution, pesticide regulation, sustainable land use.
* Public awareness.
* Curative health and counselling.
* Enhancing indigenous medicine.

Figure 6.13. Sterling Montserrat

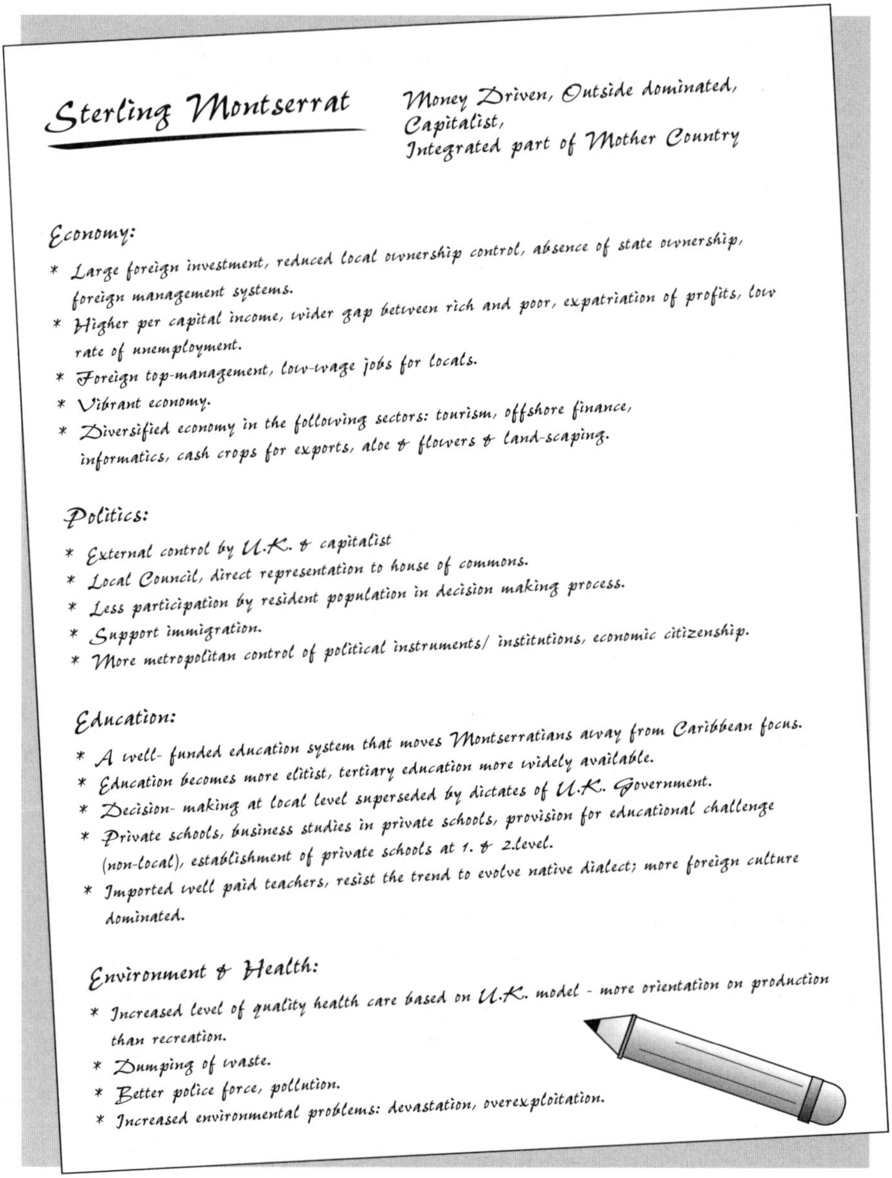

Sterling Montserrat

Money Driven, Outside dominated,
Capitalist,
Integrated part of Mother Country

Economy:

* Large foreign investment, reduced local ownership control, absence of state ownership, foreign management systems.
* Higher per capital income, wider gap between rich and poor, expatriation of profits, low rate of unemployment.
* Foreign top-management, low-wage jobs for locals.
* Vibrant economy.
* Diversified economy in the following sectors: tourism, offshore finance, informatics, cash crops for exports, aloe & flowers & land-scaping.

Politics:

* External control by U.K. & capitalist
* Local Council, direct representation to house of commons.
* Less participation by resident population in decision making process.
* Support immigration.
* More metropolitan control of political instruments/ institutions, economic citizenship.

Education:

* A well- funded education system that moves Montserratians away from Caribbean focus.
* Education becomes more elitist, tertiary education more widely available.
* Decision- making at local level superseded by dictates of U.K. Government.
* Private schools, business studies in private schools, provision for educational challenge (non-local), establishment of private schools at 1. & 2.level.
* Imported well paid teachers, resist the trend to evolve native dialect; more foreign culture dominated.

Environment & Health:

* Increased level of quality health care based on U.K. model - more orientation on production than recreation.
* Dumping of waste.
* Better police force, pollution.
* Increased environmental problems: devastation, overexploitation.

gral part of its British mother country. The island is directly represented in the British *House of Commons*, the participation of the local population, nevertheless, being low. All sectors, i.e. economy, education, culture and politics, are dominated by Britain, and the Montserratians represent European rather than Caribbean values. The economy booms and is diversified, but social and ecological matters are on the whole neglected. Increasing environmental problems and social disparities aggravate the island's vulnerability which is in the short term met by large capital transfers and modern technologies.

Even though the procedures were simplified in comparison to the project in Hamburg, this process was again time-consuming as the group had to find a consensus for each matter. It was furthermore difficult to guarantee the consistency of the raw scenarios and at the same time make sure that they were different. Some of the participants initially refused to consider the alternative *Sterling Montserrat.* They did not regard this scenario desirable. However, in the end the group came to the decision that this alternative should also be worked out as it differed most from *Proud Alliouagana* and also seemed most realistic. The entire working process was, despite all the lengthy discussions and sometimes difficult decision processes, very productive. Numerous perspectives were involved and most participants were able to identify themselves with the results. All participants were aware that both raw scenarios had to be checked and analysed further, especially regarding their consistency, before advancing to designing proper scenarios. The group, nevertheless, was of the opinion that the raw scenarios already facilitated the understanding of alternative development options, opportunities and risks and enabled the discussion of different planning strategies in a heterogeneous group, the latter being a task of the next phase.

6.2.3
Phase III – Options and Strategies

It was a major concern of the scenario workshop to give the participants an indication of how to put a scenario into practice. It was also the *Physical Planning Unit's* intention to ascertain the participants' priorities for the development of the northern region. This implied that the plans and strategies the unit had worked out before and during the crisis needed to be examined and adopted where necessary. All participants were therefore given the opportunity to *develop their own strategies.* The original intention was to develop strategies for both raw scenarios and then summarise them, according to von Reibnitz' (1992) pattern, to one major strategy. But due to the lack of time, the group decided to exclusively develop strategies for its favourite scenario *Proud Alliouagana.* Four teams were built, each concentrating on one of the main clusters (*Education, Politics, Economy, Health and Environment*). The teams had one hour to develop their strategies and to discuss them. As before, the results were later to be presented to the plenum. A table had been specifically designed for this presentation, and the participants were requested to allocate their ideas to the following columns: *challenges, goals, short and long-term measures,* and *actors.*

Evaluation: The first team addressed the cluster *Politics* (cf. Table 6.8). As the table suggests, a new political system which also takes Montserrat's economic development into consideration represents the major challenge. The four main political goals were stronger participation of the citizens in political decision processes, better representation in political committees, national elections for all political representatives and a stronger focus on political programmes and ideologies rather than personalities. Main actors were, according to this team, the local and British governments, local interest groups such as the unions, the Chamber of Commerce and various citizen's initiatives. The initiation of discussions at community level was suggested as a short-term measure. To this end the political conditions and participation options for the citizens were to be documented and published in a comprehensible form. The discussion processes were furthermore to be improved by the interest groups and a political reform committee. The team envisaged a new constitution in the short term and for the long term a new political system that finally emphasised programmes and ideologies instead of individual personalities.

The second team referred to the cluster *Economy* and distinguished the following challenges: exploiting the potential benefits of all economic sectors, focusing on modern technologies, efficient allocation of the natural resources and promoting crafts and culture (cf. Table 6.9). Central to the economic development were agriculture, tourism and other services, such as offshore finances and communication technologies. These goals were closely related to modern technologies which were the second challenge. Here, computer literacy was one major goal. In the opinion of the group the efficient allocation of natural resources requires a detailed survey, based upon which adapted land utilisation and other long-term strategic plans can be developed. The team also defined goals for the promotion of craftsmanship and culture, e.g. income-generating activities should encourage self-employment and the cultural heritage should be preserved. A number of short and long-term measures were presented for all categories, e.g. market research, incentives to private businessmen, provision of internet and other technological access and national endeavours for scientific conferences, exhibitions and awards.

The third team addressed the cluster *Health and Environment*. Table 6.10 shows the suggestions this team made. It saw life in harmony with nature as a major challenge, requiring the coordination and effective control of all development projects. The *Physical Planning Unit* was to become the main actor in this context. But the citizens also have their stake in the development and their activities clearly determined the sustainability of any development. They are as such requested to draw up their own regulations for a life in harmony with nature. The group saw an urgent need to properly collect and dispose of the waste. Extensive recycling was one of the long-term measures proposed. When the team addressed the subject of health care, a lengthy and lively discussion was provoked. The team had initially planned free health care for all citizens but soon realised that this could not be financed. It then agreed on compulsory health insurance. Traditional medicine like teas, herbs etc. was to be integrated, requiring communication and the transfer of knowledge from one generation to the next. For this reason the integration of the elder population in community projects was determined as one measure. When it came

Table 6.8. Strategies for the sector politics

Challenges	Goals	Actors	Measures short term	Measures long term
New political system which takes into consideration the economic development of Montserrat	i Increased participation in political process	i Community groups ii Labour Unions iii Chamber of Commerce iv Politicians v Individuals	i Community discussion and debate ii Research, writing & publishing by social scientists iii Publishing political systems options for consideration	Referendum on political options
	ii Increased representation in political system	i Political directorate ii Pressure groups * Community Groups * Labour Unions * Chamber of Commerce iii UK Government	i Petition the UK Government ii Decision by UK Government	New constitution
	iii Leadership elected at the national level A Head of state B Head of goverment	i Political directorate ii Pressure groups	i Create a constitution reform committee ii Decision by UK Government	New constitution
	iv Focusing on ideologies rather than personalities	i Political directorate ii Pressure groups iii Individuals	i Public discussion on political ideologies	A non-political party system

Table 6.9. Strategies for the sector economy

Challenges	Goals	Actors	Measures short term	Measures long term
Maximum sector development	Agriculture development Tourism development Services, Financial, Commercial, Industrial	Ministry of Agricul./Farmers Department/Board Tourism, Hoteliers, Real Estate Agency, Tour Operator, Airlines, Financial institutions, Ministry of Finance, Private sector	Assessment of products & resources Enhance product promotion Train providers of tourism services Introduce enabling legislation Provide incentives for local & foreign entrepreneurs and investors	Increase efficency and stability of production Refine/upgrade plant and product Modern financial centers, Stock market, Mortgage financial company Dev. & maintain institution for promotion of investment opportunities
Focus on modern technology	Increase computer literacy Adopt or use appropiate teachers Increase technological education & training	Ministry of Education Private Sector/Individuals Utility companies, Government & private sector Schools, Technical college, business	Expand concession on importation of computers, provide awards of recognition for technology assessment Enhance curriculum & teacher training and upgrade labs, science, fine arts, computers, music	Computer centre National showcase or exhibition of technology Widespread access to Internet
Proper allocation of natural resources	Device inventory of resources i.e. land, capital, human resources	Ministries: Agriculture, Finance, Labour, Stats Private sector, L.D.A. Educational institution, i.e. National, Regional & International	Agree on land use policies Land acquisition Develop skills inventory	Integrate land use Data base for example GIS Maintain forward planning of land use beyond invested use
Craft/ Cultural development	Create income generating activities-self employment Preserve cultural heritage	Government, small business, craftsmen, artists Generell population as consuments	Craft training & outlet/ market upgrade end product Identify potential Craft trainers local & regional Enhance industrial activity to provide more employment opportunities for women	Craft & buy local exhibition-competition Research use of indigenous material Provide further incentives in construction industry

Table 6.10. Strategies for the sector health and environment

Challenges	Goals	Actors	Measures short term	Measures long term
Development in harmony with nature	Proper co-ordination of effective control of development projects	Physical Planning Unit		Full-time recycling
	Proper waste management/recycling	Environmental health officers, householders, businessmen	Proper collection and disposal	
Comprehensive health care systems	Mandatory health insurance schemes	Government & employers		Legislation
	Adequate & capable medical practitioners	Ministry of Health	Import expertise to train nationals	Scholarships
Curative health & counselling	Reduction /Elimination of physiological diseases	Health educators & Nutrition officers		Education about proper diets
Enhancing indigenous medicines	Successful transfer of knowledge from one generation to another	Older knowledgeable people	Talk about them, discover, experiment	Write a book or pamphlets
Legislation regarding emission, recycling, zoning, building codes, marine protection, noise pollution, pesticide regulation	Proper legislation/ taxation to protect the environment	Parliament	Taxation (punative) to prevent importation of non-recyclable products	Legislation

Table 6.11. Strategies for the sector education

Challenges	Goals	Actors	Measures short term	Measures long term
Educational opportunities which enhance capability & self esteem & national pride	Expose citizens to a variety of disciplines which would help to build self esteem and national pride	Educational department, sports department, cultural department, as well as media & private sectors	Increase motivation in the classroom Develop programmes to promote national pride Equip institutions with resources (capital, man power) to develop programmes such as sports, arts and culture	Introduce students to a wider range of subjects in the formal education Have more practical work in subjects in the non-formal system
Comprehensive computer literacy	Ensure that a larger percentage of the population is computer literate	Educational department, schools, computer unit and private sector	A comprehensive computer educational programme for all secondary students	A comprehensive computer educational programme for all
The orientation of the educational system has to ensure widespread comprehensive political education	Ensure that every citizen understands the political process	Educational department, media, schools, parliamentary representatives, youth services and community groups	Introduction of civics into the school curriculum	Continued discussions using the media and civic groups
Develop an educational programme which would promote the use of natural resources	Introduce opportunities where people can be educated to use natural resources to create economic diversity	Educational department, media, schools, youth services and community groups	Introduction of economic education to the school curriculum and in small business sector	Continued discussions using the media and civic groups

to legislation, many ideas were suggested, e.g. taxes on plastics articles or deposits on glass bottles etc. The team expressed its apprehension about the fact that the responsibility was in most cases passed to public institutions, but hoped that the people would in a step-by-step process become more and more conscious about their own responsibility.

The fourth group dealt with the cluster *Education*. The challenges it proposed are shown in Table 6.11 and include better education in computer technology, environment and resource management and politics. The team suggested initiating education programmes to enhance national identity, building up comprehensive computer training for all secondary students, ensuring that a broader range of subjects were taught at school, encouraging continuous discussions in the media and civic groups and forming a closer cooperation between the schools and the private sector in order to guarantee an education which would meet the requirements of the labour market. This would also ensure that the use of Montserrat's resources could be optimised and an even more diversified economy established. The team suggested the following major actors: the *Ministry of Finance, Economic Development and Sports*, the *Ministry of Education, Health and Community Services*, the media, the private sector and youth and civic groups. The team also emphasised that it was necessary to establish a computer unit, ensuring that every Montserratian had access to a computer and the internet and the opportunity to obtain the required skills.

The teamwork proved to be very productive and went on for about two hours. The participants did not encounter any difficulties in accomplishing the task. The teams led very lively, sometimes even heated debates. The tables can only show extracts of the results since a vast number of goals and measures were suggested in the course of the entire working process. It was remarkable that, once the participants had presented the results of their teamwork, they still expressed other ideas. This phase of the workshop once again demonstrated clearly that it was not only the results that were important, but also the learning process that prompted the participants to approach the subject matter further. The participants had become active planners, able to jointly determine Montserrat's future. The overall objective, i.e. understanding how the future of the island and particularly its northern region could be designed and organised, was achieved.

The participants furthermore expressed that dealing with the goals, measures and actors of each sector had helped them understand complexity of the system and the necessary decisions and activities. They were interested in carrying out an analysis of the influential factors so that they might better understand and determine their interactions. The participants were quite astonished with the parallel results of the individual teams. Almost all of them had addressed participative planning methods, computer literacy as a goal of education, foresighted integrative development plans, economic diversity and flexibility, and the sustainable utilisation of resources. The team that had addressed the economic sector, for example, called for a survey of the natural resources and an efficient education in computer technology. These aspects were also mentioned by the teams that had referred to education, and health and environment. This and other examples illustrate that the sectors and measures interact closely. As a last task the measures were evaluated with the multi-point questioning technique. The following priorities were ascertained.

1. a new political constitution
2. an integrative waste management
3. models for discussion and participation on community level
4. a broadened curriculum at the schools
5. comprehensive education in computer technology for the entire population
6. establishing a computer unit
7. grants for education/training

6.2.4 Final Evaluation

How can we now rate the success of the scenario method in an ongoing hazard situation? First of all, we have to pay attention to the fact that the results derived from the scenario workshop do not represent reliable knowledge that allows a control of the environment. Instead, the participants experience a new conscience about the constitution of their *lebenswelt*, a world which the volcano has driven into complete disorder. The scenarios facilitate the discussion of extremely complicated problems via descriptive drafts. They revive the feeling that the world surrounding the affected can be reorganised and they contribute to developing strategies and measures in an acute hazard situation. The participants proved that they were able to develop alternative options for Montserrat's future, even though the workshop was carried out at a time when the Soufrière Hills Volcano was extremely active. This even reinforced the group's confidence in its ability to effectively react to the crisis.

The communication and decision processes between the participants play an important role in this context. The group was large and heterogeneously composed but, nevertheless, worked extremely effectively and constructively. The participants presented numerous ideas and visions for Montserrat's future and developed innovative strategies. In the course of the different phases and tasks the participants developed a common awareness about the problematic nature of the situation. The workshop also improved their ability to 'think in alternatives'. On the one hand, the group reflected on the prevailing values, goals and activities. On the other hand, the workshop enhanced the group's willingness to mentally approach less favoured alternatives, such as the scenario *Sterling Montserrat.*

The workshop also illustrated that not only was the creativity of the participants enhanced, but also their resilience, which proved to be extremely high. The participants constructively approached uncertainty by developing new and positive visions of their future. Especially the scenario *Proud Alliouagana* challenged the abilities of the participants who demonstrated a keen interest to develop further strategies for the practical implementation of the scenario. This new and clear vision conveyed to the participants that the crisis could alternatively be understood as an impulse for innovative transformation processes.

It was also characteristic of the workshop that all tasks were carried out without major problems. The participants always grasped the intention of the individual working steps which might, in part, be related to the previous testing of the scenario method in Hamburg. A simplified and time-efficient procedure had been developed and this was easily understood by all of the participants. It has to be mentioned, however, that the discussions were often hemmed by a lack of time. Many subject matters could not be debated in detail and some tasks, which would surely have resulted in new findings, could not be carried out. Amongst these were the analysis of influences and the clustering of alternatives. Both tasks would certainly have further enhanced the participants' awareness and understanding of the complexity of the system. Accomplishing these tasks would also have further ensured the internal consistency of the scenarios.

The comparative analysis as regards contents of the scenarios developed in Hamburg and Montserrat shows that the different groups actually suggested different alternatives. The participants of the workshop held in Montserrat were challenged to develop alternatives other than those initially presented to them. They certainly achieved this goal, not least with the *Sterling Montserrat* scenario. However, the fundamental goals of both groups were, astonishingly enough, very similar. Both strive for a sustainable development and a strong participation of the citizens. The reasons for this may vary. Living in harmony with nature and participation in planning and politics is a desirable goal to many of us. It might also be of significance that the scenarios designed in Hamburg were influenced by the results of the interviews earlier conducted on Montserrat. Conversely, the booklet with the summary of the Hamburg scenarios may have influenced the participants of the workshop on Montserrat.

On the whole the workshop confirmed the overall long-term development goals the Montserratians had earlier expressed in the interviews. The *Proud Alliouagana* scenario illustrates the interest in a sustainable development of the island. Common goals are the protection of the natural resources, an integrative economic development, modern education and political reforms. Both the Montserratians interviewed and those participating in the workshop expressed urgent needs for modifications in all these sectors. This was also confirmed by the strategic reflections. The scenario workshop can, to some extent, therefore, be considered a societal warning system, revealing problems and goals that are perceived by the population. The members of the *Physical Planning Unit,* for example, commented that the workshop had greatly contributed to their getting a clearer impression about the actual needs and desires of the local population. They also emphasised that using scenarios facilitated the determination of the long-term goals of reconstruction.

The processes which were decisively initiated by the scenario workshop, in summary, correspond to the advantages of participative reconstruction as outlined by Maskrey (1994). The advantages shall at this point briefly be recapitulated.

1. The problems, needs and priorities of the population are clearly structured and articulated.
2. The potentials of reflection, communication and learning are enhanced. Flexible and creative solutions to problems are found.
3. Visions are developed as to the society's future.
4. The existing resources are mobilised.
5. The strategies proposed usually correspond to the local situation.
6. Bureaucratic constraints and corruption are reduced.
7. Resilience is enhanced.

The goals initially set were also to a large extent realised. The workshop did not only mobilise the abilities and emotional capacities of the participants, but also offered them the opportunity to introduce their competence for a responsible and sustainable shaping of the reconstruction process. The direct implementation of the results, in the sense of actively integrating them into the reports and plans of the political administration or realising the strategies, can, therefore, not be the sole yardstick of success. The mental implementation of the knowledge is certainly just as important as some of the following statements show. They all originate from the final review at the end of the two-day workshop on Montserrat.

- "A magnificent exercise which stimulates positive thinking. I would love to do more with that."
- "This should be a model for our representatives."
- "I'm very tired and it really needs a lot of effort, but the type of decision-making we experienced here should be moved to the real decision-makers."
- "I will use the scenario-method for many aspects in my life. I do not need a crystal ball anymore."
- "I will be waiting the rest of my life that some of our visions become true."
- "The workshop enabled me to think positive of the current crisis and help others to see a future for Montserrat after the crisis."
- "We should do similar exercises with other communities and groups and then we should put all our ideas together into a main strategy and test it for the whole country."
- "It is our responsibility now to put the scenarios onto paper and take them to a bigger forum. We also need to send the visions to our representatives."

These statements illustrate that the participants clearly call for a broad presentation of the scenarios and strategies, and that further workshops should be conducted with other groups. They strive to find a common and broad consensus for a sustainable development of their island, including the respective implications for reconstruction. The workshop was thus not the end of the work, but the beginning of it all. The real work, i.e. the implementation of the results, still has to commence and, in fact, in two directions: (1) as task for the decision-makers who have to implement the results in their planning and management strategies, and (2) as task for each individual participant to integrate the experience in their daily lives.

Many of the workshop's participants have actually started to pursue their tasks. For example, an employee of *Cable & Wireless* has since the workshop permanently emphasised the necessity to grant every citizen access to the internet. In the course of the workshop he had made this a task for himself and is now pursuing it. Other participants published statements in the Montserrat Newsgroup or national newspaper, expressing that it was vital to develop a common vision about Montserrat's future, particularly in the time of crisis (cf. Textbox 6.1). And the public representatives also developed visions for the reconstruction process, as will be shown in the concluding chapter. The scenario workshop, in effect, equally motivated all participants, no matter whether member of the public or private sector, student or scientist to put the results into practice. For this reason alone it has to be important to conduct further workshops with other participants, or at least discuss the scenarios with as many other people as is possible. A snow ball effect could thus be initiated which would, in the sense of Oakley (1991), enable the population to participate in the shaping of a desirable future.

Textbox 6.1. Peter B. White, Extract of Statement Published in the Montserrat Reporter on 11 March 1997

> ... What needs to be done by the authorities is to narrow the range of options based on a vision of where it is perceived Montserrat should be heading. This does not necessarily mean an actual detailed plan for the future. What it means is defining boundaries, developing goals and finding ways to facilitate the achievement of these goals. Specific plans can and must follow, but a vision of what we hope to achieve will go a long way to calm fears even if the uncertainty is not immediately eliminated... How many of us are ready to live with such a vision? How many are ready for us to make decisions for ourselves in our best interest?

In order to be safer, we may have to be more willing to acknowledge how unsafe life really is (Parfit).

7 "Living with the Unexpected" – Perspectives for a Constructive Dealing with Complexity and Uncertainty

Complexity and uncertainty determine the planning and management of our interactions with the environment. Planning and management strategies can therefore only be successful when we accept complexity and uncertainty. This calls for an openness towards new, sudden and unexpected developments just as well as for our readiness to accept variations, chaos and disorder as challenge, enabling us to act in a non-linear world. It is decisive that our awareness, decisions and activities are flexible. We all have to adapt to continuous changes and modifications by constantly changing ourselves. We have to learn to cope with complexity and uncertainty by trying to shape our future according to our (personal) interests and goals. Such an approach requires planning and management methods which guarantee a continuous exchange of information, flexible communication and determination mechanisms, an active observation of the environment and a permanent reflection and modification of our values, objectives and activities.

In this paper the considerations concentrated on the question in how far scenario planning serves as a point of reference when dealing with complexity and uncertainty. To find an answer, the different steps of the scenario method were practised with two completely different groups. The first group mainly consisted of scientists and students who critically tested the method within the scope of a scenario project that was run over a few months. The other group was composed of students, planners and managers who expressed their personal wishes and goals for planning during and after the volcanic crisis on Montserrat. In both cases it became apparent that the scenario method contributed greatly to overcoming mental barriers, to illustrate complex developments and to prepare the ground for a range of possible decisions and activities designed to shape the future. Both groups constructed alternative scenarios for the island, and the second group also suggested strategies and measures for putting its favourite scenario into practice.

The implementation and realisation of a scenario is the transition point between pure planning and management. This does not mean, however, that planning at this point ceases to be of importance. To the contrary, in view of complexity and uncertainty, planning and management have to melt to an integrative unit. This is the way to ensure that our interactions with the environment can at any time be questioned, tested and adapted whenever deemed necessary. The scenario method can be an important tool for these processes. Its specific value is to be sought in its potentiality to vividly illustrate possible alternative development options and their respective consequences. The options can thus be understood and evaluated by as many people as possible (Stiens 1995). Nevertheless, the scenario method should be accompanied by other planning and management methods. A sensible combination of various methods has to be called for since it improves our adaptability to constantly changing conditions and facilitates that we can constructively approach complexity and uncertainty (Mitchell 1997; Stiens 1996; Georgantzas and Acar 1995; Merry 1995; Mittelstaedt 1994; Peters 1994).

But what exactly is a sensible combination of methods and what are its objectives? Which approaches are there that should be integrated? Which difficulties are to be coped with? Based on the research results presented in this publication and Mitchell's (1997) reflections, the following principles can be set up for a constructive dealing with complexity and uncertainty.

1. grasping complexity
2. developing a vision
3. activating a process
4. defining a plan
5. implementing and monitoring

These principles shall in the conclusion be explained in detail. The first three principles facilitate a summary of the research results and place special emphasis on the potentials of the scenario method. The other two principles point at further areas of activity which are to be understood as a continuation of this work. This is to say, they focus on what is to come after the scenarios have been constructed. The implications for a sustainable reconstruction of Montserrat shall be the centre of attention when addressing this subject matter.

1. Grasping complexity

We will never be able to entirely understand complexity. Any attempt to illustrate the complexity of a system, as for example with a computer model or a technical method, in itself constitutes a reduction of the phenomenon. It is, nevertheless, important to gain a certain feeling for complexity. It is not until we have managed to improve our understanding of the interactions between the systems society and environment and until we have been able to grasp their complexity that we can consciously manage our interactions with the environment. This is especially true for resource and hazard management. Chapters 2.2 and 2.3 have clearly shown that both approaches have for a long time focused on a simple cause-and-effect model. The environment was to be controlled by regulating and preventative measures. Central aim was a balancing stabilisation of any system. However, the discussion in Chapter 2.1.1 has shown that natural and social systems are open, non-linear and far-from-equilibrium systems. Management strategies which strive for a balance are consequently not suitable for complex systems as these are based on an disorderly state. Having realised this, many scientists and managers are currently endeavouring to improve their approaches towards the interaction human-environment by applying integrative methods of analysis and adaptive management strategies (Mitchell 1997; Blaikie et al. 1994; Comfort 1994; Varley 1994; Shackley et al. 1996; Funtowicz and Ravetz 1994; Dovers and Handmer 1992; Palm 1990; Mitchell et al. 1989).

Chapter 4 focused on the interaction human-environment on Montserrat. Even though this island is comparatively small with its entire surface of only 102 sq km (about 39 sq mi), it would be wrong to assume that it was for this reason less complex. When approaching Montserrat, the methodical challenge here, too, is to gain a better understanding about the complexity of this insular system. Within the scope of this paper it has repeatedly been attempted to accomplish this task. First of all Palm's (1990) structuralistic approach was applied. As we have seen, it aims at differentiating the basic correlations within the system, i.e. those of the gatekeepers and the individuals. It proved to be particularly valuable to follow this method and to subsequently observe and analyse the interactions human-environment and individual-society from different perspectives. It became apparent that the genesis, structure and function of the system Montserrat in the sense of the sensitivity to initial conditions decisively depends on specific components, all of them being characteristic for small islands. Amongst these characteristics are the scarcity of space and resources, the small market, an unproportional significance of the export sector, a small number of inhabitants, emigration and the decisive influence certain individuals have.

Viewing the system from different perspectives is, in summary, extremely important in order to gain a better understanding of a complex system. But changing perspectives does not only imply the integration of different perceptions, interests, goals and space and time scales when analysing a complex system. Changing perspectives also means that different states of the system have to be analysed. This is the reason why the complex system Montserrat has not only been approached in the times of an apparent secure and normal situation, but also during a disorderly and uncertain situation, i.e. during the volcanic crisis. As a result, it became apparent that unpredictable, sudden events and fluctuations had in 'normal' times to a large extent been excluded from plans and activities. This includes the gatekeepers just as well as to the local population. The analysis of the interests and objectives showed that both groups strove for an imaginary state of balance. They desired a secure and orderly life in harmony with nature; at the same time they hoped for economic growth, which, at the end of the day, does constitute a change.

These basic objectives on the whole hardly changed during the volcanic crisis. Yet, the crisis forced the people to constantly adapt to new, extreme events, whereby time, space and the resources of the island became extremely critical factors. The conflicting interests increased and finally escalated with the disaster in the summer of 1997. All decision processes and activities on the island were obstructed by the discontinuity of the normality, the routine and the order, that is namely by the uncertainty about the further course of volcanic activity. The radical spatial and social transformations caused by the disaster finally did not allow any other option but to re-interpret and re-shape the *lebenswelt* in a step-by-step process.

A re-interpretation and re-shaping of the *lebenswelt* after a disaster is more effective when the previously prevailing natural, social, political and economic correlations within the system are reflected upon and analysed. This can enhance the understanding and subsequent reduction of vulnerability as it is always a result of the dynamic interaction human-environment. The analysis of influences as an important step of scenario planning can in this context be of great help. Chapter 6.1 presented such an analysis of the complex system Montserrat before the volcanic crisis. It showed that the analysis assists to develop a common understanding and a common language for the complexity of a system. Based on this experience, the group in question determined the active and critical variables of the system which, according to them, presented starting points for a possible reduction of Montserrat's vulnerability towards future disasters. To this end alternative decisions and activities were thought through in order to show options for a sustainable development of the island. This methodical approach proved to be very fruitful, particularly when attempting to investigate the local conditions of vulnerability with different participants and to subsequently develop measures for its reduction. The question of how effective such an analysis of influence is at the time of an ongoing disaster and conducted with the people concerned was, however, too complex to deal with in the context of this work. To all appearances it will be just as effective in such a situation. Some practical experience has already been made with similar methods (Anderson and Woodrow 1998).

2. Developing a vision

Chaos, complexity and uncertainty are by no means a reason for standstill. We have to continue our activities and it is of great importance to have a vision about a desirable future – this is important in a twofold way. On the one hand, a vision is something rather stable, since it is always built around central values and interests which continue for a long period of time. A vision gives us a certain orientation in a time of constant changes. It shows us the path of a desired development. On the other hand, a vision is something dynamic. It emphasises the necessity to mobilise our abilities in order to realise future goals. It challenges our creativity, self-reliance and ability to act. A vision is, in summary, at the same time stable and dynamic, and for exactly this reason it gives us the confidence that we are able to react despite of chaos, complexity and uncertainty. "Effective visions must be stable but constantly challenged – and changed at the margins ... The vision must act as a compass in a wild and stormy sea and, like a compass, it loses its values if it's not adjusted to take account of its surroundings ... Effective visions are beacons and controls when all else is up for grabs" (Peters 1987, p. 403).

Scenario planning is next to its relative, the *Zukunftswerkstatt*, one of the most significant methods for the development of visions. A central characteristic is its openness towards the future, drawing up unexpected, sudden and new developments. Uncertainty towards the future is consciously produced in the various working steps, since it is the doubt that allows us to cross borders, to surpass our own imagination, to open our eyes and mind for new perspectives and to induce us to experiment with new ideas. The workshops in Hamburg and Montserrat clearly illustrated that abstract utopias (freely imagined mental experiments outside the realistic context) improve the fantasy and creativity of the participants and encourage them to continue on this path. In this way they developed many illustrating and innovative future visions for Montserrat. These visions took up the ideas of the abstract utopias and at the same time provided a basis for the construction of alternative scenarios.

Visions are of particular importance after a disaster. A vision gives the affected people concerned confidence in the future of their lebensraum and motivates them to act. Furthermore, completely new values and objectives can be worked out on the basis of visions. These can in the sense of the sensitivity to the initial conditions have important consequences. New values and objectives can initiate a conscious structural change during reconstruction so that the society's vulnerability to other disasters can be reduced. Particularly in the time after a disaster the affected people are more open for changes and show a greater preparedness to carry out measures which reduce their vulnerability (Anderson and Woodrow 1998). It is thus of great importance for reconstruction to develop visions for the future of a society. Such a vision can in times of disorder and destruction actually function as a compass which shows the people the direction and gives them an idea of how to reconstruct their *lebenswelt* and make it less vulnerable.

However, in practice, decision-makers usually opt for a resumption of the predisaster state instead of considering the potentialities of a structural change. Reconstruction programmes are usually centralistically organised, dominated by short-term goals and hardly ever visionary (Rubin 1995; Maskrey 1994; Geipel et al. 1988). The affected people are, furthermore, usually perceived as victims of the situation who are not capable of contributing to reconstruction. The scenario workshop on Montserrat, however, clearly showed that the knowledge and abilities of the affected population was quite productive. They proved to be clearly capable of developing perspectives for a visionary (re)construction of the northern island. The joint fantasising and playing around with desires and ideas made them rediscover their ability to shape their own *lebenswelt*. They developed a number of scenarios in the course of the workshop and all of them with their vivid design greatly facilitated the discussion about the objectives of reconstruction. The participants' favourite scenario *Proud Alliouagana* is an excellent example of a vision that works in the above mentioned two-fold way. On the one hand, it encourages the participants' creativity, self-reliance and willingness to act and, on the other hand, it is a relatively stable vision which continues to have an impact for a longer period of time, as the following principle will illustrate.

3. Activating a process

In a society there are always many co-existing interests and objectives. Developing a vision is therefore by no means a task which can only be accomplished by experts or smaller teams. To the contrary, as many people as possible should ideally come together in order to exchange their experience, knowledge and values and to find a common consensus about a desirable future. This, of course, requires that a participative process is activated and mutual fantasising, learning and experimenting encouraged. As we have seen, the scenario method is particularly suitable to activate such a process. It facilitates the sensitisation towards complexity, a reflection about the interests and objectives and the construction of alternative strategies. Furthermore, scenarios are easily comprehensible since they are vivid pictures which can readily be discussed with most people. It is therefore one of the most significant tasks of scenarios to initiate discussions about alternative future options. This is to say that the processes which follow the construction of a scenario are at the end of the day of significant importance.

The scenarios which were designed during the workshop in Hamburg are in this context to be understood as starting point. They activated the discussions and decision processes on Montserrat. The four scenarios were first of all presented to a school class and various ministries and authorities. They received great interest and always triggered off lively discussions about a desirable future of the island and possible strategies for its realisation. The representatives of the *Physical Planning Unit* (PPU) also recognised the potentials of scenarios for finding a consensus about the development and reconstruction programmes during the volcanic crisis. This is why the PPU initiated the scenario workshop on Montserrat. Some local people were selected and requested to develop their own scenarios and strategies. A creative working process evolved during the workshop. The motivation of the participants to present their own planning goals and ideas for Montserrat was extremely high. At the end of the workshop the participants expressed their wish to present their scenarios and strategies to a broader public and to carry out further workshops with other groups. They wished to integrate as many people as possible in the reconstruction process.

A comprehensive report containing the scenarios and strategies was actually prepared after the workshop and distributed to all ministries and authorities on Montserrat. Unfortunately, there was no opportunity to further evaluate the scenarios in order to then present them to a broader public. Other scenario workshops have also not been carried out. The situation on Montserrat is in this respect no exception. Scenario planning has time and time again met resistance in the political-administrative sector (Institut für Landes- und Stadtentwicklungsforschung 1989). One of the reasons is that the ministries tend to stick to their well-established bureaucratic planning and management styles. They also usually argue that the temporal, personal and material efforts for the development of scenarios is too high. It is usually not considered that this argument is just as valid for other methods such as prognoses, computer simulations, cross impact analyses or Delphi-method. A cost-benefit-analysis of all these methods would surely be helpful in order to meet these prejudices.

There is another reason, however, why the work with the scenarios was interrupted on Montserrat: the escalation of the crisis in the summer of 1997. This phase constantly required new decisions and adjustments in order to protect the lives and property of the local population as best as possible. Extremely acute problems had to be coped with and there were neither time nor resources available for scenario workshops. But despite the dramatic events on Montserrat in the summer of 1997, the results of the scenario workshop continued having an effect. In October of the same year another group of people came together in order to discuss *Global Issues* for Montserrat with similar methods. Further consultations with the local and the British government finally determined the central aim of reconstruction for Montserrat. This proposal considered, amongst others, some ideas of the scenario workshop and was titled *The Vision of Montserrat* (see Textbox 7.1).

Textbox 7.1. The Vision of Montserrat.

> The rebuilding of a healthy and wholesome Montserrat, founded upon a thriving modern economy with a friendly, vibrant community in which all of our people through enterprise and initiative can fulfil their hopes in a truly democratic and God fearing society. (Government of Montserrat 1998, p. 13)

4. Defining a plan

Developing a vision and activating a process only make sense when problems can be solved. To this end it can be of use to work out a plan that selects and defines the various strategies and measures necessary to realise the development goals. Similar to a vision, a plan is designed to offer orientation, with the decisive difference that a plan is far more precise and represents something like a link between planning and management. A plan is the basis against which the suitability of decisions and the effects of strategies and measures can repeatedly be checked. On the basis of the previous vision, the following plan was worked out for Montserrat: *Sustainable Development Plan, Montserrat Social and Economic Recovery Programme – A Path to Sustainable Development* (Government of Montserrat 1998). This plan now actually serves as the basic document for a sustainable and integrative reconstruction of the island until the year 2002. It comprises the strategies and measures to guide Montserrat from the crisis to development. The *Sustainable Development Plan* (SDP) is organised by sectors and four main themes were determined to create a link between the sectors: (1) developing the north, (2) efficiency and effectiveness of the public sector, (3) partnership with and promotion of the private sector and (4) protecting the vulnerable and promoting social welfare.

The SDP bases on the assumption that volcanic activity will continue until it will slow down early in the year 2000. "This activity may include periodic explosive phases resulting in ash and pumice falls across the North and further pyroclastic flows and explosions from collapse of the dome, that are likely to affect the current unsafe zone ... The entire unsafe zone will be highly vulnerable until early 2000. From that time onwards the current central zone, if previously evacuated, and Salem could be habitable. As a consequence of volcanic activity, no areas south of Lawyers Mountain would be suitable for any major development over the plan period. The northern zone is considered suitable for all forms of development, but the central zone for residential use only. It is assumed that no off-island evacuation will be necessary" (Government of Montserrat 1998, p. 47).

The island might, according to this view, in the next few years face further explosive phases and pyroclastic flows because of further collapses of the dome. Solely the north is considered safe and suitable for larger development projects. This means that Montserrat's future depends decisively on the built up of an economic and social infrastructure in this safe zone. The development shall in an initial transition phase be guided by the government. It intends to create the basis for the investors' confidence and activities of the private sector: "As Montserrat moves away from crisis management to development, an objective of the Government of Montserrat will be to encourage the private sector to recommence activities that it previously undertook and to also look to new areas to involve the private sector in the delivery of publicly financed services" (Government of Montserrat 1998, p. 14). As before the crisis, the SDP strives for a public sector reform. Especially the financial dependence on Great Britain is eventually to be reduced. The SDP finally also intends to support the most severely affected population. Such support is guaranteed until they can return to self-reliance.

The SDP is a significant document for all of the ministries and authorities since it defines the central strategies and measures for the reconstruction of the island. Many of the proposed strategies also reveal that the gatekeepers have actually recognised the opportunity for a structural transformation. They have the clear intention to reduce the island's vulnerability. It remains to be seen whether these plans will actually be realised. Some factors naturally give rise to doubt, amongst them the adherence to previous development trends, the overall sector-oriented approach and the limited participatory involvement of the local population in the designing process of the plan itself. Even though the plan had been presented to the public in a number of public hearings on Montserrat, Antigua and even London to involve the Montserratians living abroad, their reaction was rather passive. The plan was even hardly discussed by the members of the internet Newsgroup, usually a lively forum. It appears that the plan had already been worked out too much in detail when it was presented to the public, thus hindering the people's motivation to contribute their own ideas. The ministries and authorities practised a rather short-term type of participation which was primarily designed to promote the plan and to receive a general acceptance. It would surely have been far more fruitful and sensible at the point in time to discuss scenarios and to use them as a basis to jointly develop alternative options together with the population. This opportunity has been missed so that the SDP until now remains without any alternative.

Particularly this aspect clearly shows that plans should with respect to chaos, complexity and uncertainty in no case be perceived or treated as a static product. Each decision which is made under a condition of complexity and uncertainty can have undesired consequences. There again, it might also happen that new developments decisions which initially seemed sensible completely useless in the new situation. Plans therefore have to remain modifiable. The options should be kept open for alternative strategies which should be at hand in case the need occurs. This way a strategy which proves to be wrong or unsuitable can in a relatively short time be replaced. The new strategy can then at the given time carefully be introduced and realised. This approach which we have earlier referred to as *Flexing* (see Chapter 2) makes sure that a once defined plan is not by all means adhered to. Innovative development opportunities can thus be integrated far more effectively and potential threats often recognised at an early stage. Defining a plan, in summary, should in any case not be the end of the planning process. The decisive factor is the way the plan is put into practice and depends on the fact whether the constantly changing environmental conditions are actively observed and reacted to.

5. Implementing and monitoring

Implementing means that the strategies and measures specified in the plans are actually realised. In practice, however, there are all too often many obstructions that interfere with an effective implementation. Amongst these obstructions are conflicts about the priority of measures, incompetent planners and managers, a lack of resources, split up organisational structures, corruption, different levels of information, a variety of perceptions and an isolated consideration of partial matters (Mitchell 1997; Wentz 1992; Dörner and Reither 1978). Many people, furthermore, usually show great resistance against changes of any type so that new plans, projects and activities are often seen as a threat and therefore rejected (Chapter 2). This is especially the case when planners and managers are not familiar with the local conditions and do not integrate the population that is finally affected by their projects and activities. For this reason plans are usually implemented far more successfully when they offer suitable local solutions and consider participative approaches, namely take advantage of the population's knowledge and abilities, not only during the initial developing but also later on in the implementation stages. When these potentials are successfully activated, a range of advantages can be realised, amongst them lower expenses, better solutions and, naturally, a far greater acceptance by the local population (Wentz 1992).

As the analysis of the meso level in Chapter 4 showed, some eleven plans, numerous strategies and a number of projects were worked out for Montserrat between 1993 ad 1998. Their successful implementation was above all obstructed by the following factors: they were to a large extent sector-oriented; the access to and level of information were too different; conflicts between the interest groups; adherence to the existing organisational structure; the small pool of decision-makers; despite some endeavours still far too little involvement of the local population in planning and management. One further negative factor is to be mentioned at this point. When the plans, strategies and projects were worked out, foreign teams of experts were often involved as consultants. These experts were in most cases, however, not (sufficiently) familiar with the local conditions. In the light of these findings, it does not surprise that in the interviews the local population uttered its frustration about the failure of the many plans, strategies and projects (cf. Chapter 4).

This scepticism, of course, also extends to the SDP. The population again fears that its successful realisation might also fail because of the above obstructions. Any doubts as to the financial funding of the plan were cleared away in June 1998 when the British Minister of Foreign Aid Clare Short announced the financial framework for the reconstruction of Montserrat up to the year 2001. It provides the considerable sum of £ 75 million. The financial basis for implementing the SDP is therefore given, as Governor Anthony Abbot clearly illustrated: "By August I would hope that we have an agreed SDP which, combined with the knowledge that we have 75 million pounds budget to work with, will enable Montserrat to make great progress. This will provide a framework on which Montserratians can plan for the future. I am confident it will be possible to agree a Country Policy Plan by the Autumn of this year which should clearly identify which projects are committed, which are planned and what might be possible in the long term" (Anthony Abbot, press release of 11 June 1998). This is a good beginning, but the other obstructions for a successful implementation of the SDP still remain.

In this context a possible starting point would be what has been explained in Chapter 5 as *participation as an end.* It is a process which slowly unfolds so that planners and managers can step by step find solutions, measures and appropriate activities. Above all, they can go through this process together with the local population, jointly develop ideas and strategies and test them as they go along. It would surely be sensible to initiate smaller projects in order to quickly come up with results which can, in case of need, also quite easily be modified. Focal point should be a management which is able to learn and to accept mistakes. Characteristic of such a management is the simultaneousness of project and plan, respectively the blending of planning and management. The first practical steps can in this way be lessons for the further development of the activities and plans. An overlapping instead of a strictly sector-oriented approach would also allow that interactions can be taken into consideration and even taken advantage of. Planners and managers should, above all, not function as controlling instances anymore, but as moderators between the different interests, and they should promote public and private activities. Their abilities to cooperate and communicate therefore play an important part (Wentz 1992). Requirements for an effective and successful implementation are, in summary, a changed (self-)appreciation of planners and managers, an integrative approach and willingness to deal with mistakes and to continue learning all the time.

A management can only learn and adapt to mistakes when it is prepared to closely monitor the effects of its measures and the dynamic changes of the environment. Monitoring therefore involves two aspects. On the one hand it examines and, if appropriate, improves the effects of measures, laws, projects etc. and, on the other hand, it determines and investigates natural, political, economic and social changes in order to prepare appropriate reactions. Monitoring thus represents an important feed back link within planning and management processes. Unfortunately, it has so far received little attention in practice. The financial and staff resources allocated to monitoring are usually unproportionally low in comparison to the funds spent on the actual development of plans and projects. Mitchell (1997) sees the reason in a lack of knowledge about the different evaluation methods and, above all, in the planners' and managers' lack of interest to make their work controllable. "It is difficult to conceive how we can learn from experience or mistakes if we do not systematically monitor our initiatives and activities. Yet the reality is that there is often considerable resistance to monitoring. People who have invested professional reputations or egos in policies or plans are often not anxious to see them monitored, if there is significant likelihood that errors will then become public knowledge. This reluctance and inertia in many instances still has to be overcome, and this will not be a minor task" (Mitchell 1997, p. 289).

As a consequence, monitoring should be encouraged. An improved participation of the population would in this context, again, be of benefit. The population could on the one hand force politicians, planners and managers to establish a control and it could, on the other hand, itself become active in monitoring, especially when there is a lack of funds or staff: "In the future, an opportunity and challenge will be to determine how local citizens' groups and non-government organizations can become more involved in monitoring and assessment. Many governments are reducing their activities, and partnerships for monitoring and assessment offer opportunities to maintain capacity for tracking progress, even at a time of fewer financial resources" (Mitchell 1997, p. 279). Since there are different monitoring methods, it would, of course, be necessary to carry out an analysis of the advantages and disadvantages of each method. It should also be remembered that the scenario method itself offers some links to monitoring. The method, for example, usually requires the establishment of a so-called Umfeldbeobachtungssytems (UBS) (*environment observation system*). Especially business management has used this method very successfully. It observes external developments which are considered important to the company and then analyses their potential influence on the critical and active variables of the system. Finally, if deemed necessary, careful appropriate adaptations of the company's strategy are undertaken (von Reibnitz 1992). Once again, the analysis of influences is an inevitable step. It ensures that those variables are identified which are important to the dynamics of the system. These can then be used as indicators for monitoring. The analysis of disruptive events also provides another link to monitoring since it can help to identify potential threats which can then be dealt with in time.

Just listing the five principles clearly indicates that they have to be seen and treated as integrative parts of a unit. This is the only way to ensure that potential opportunities and risks for the future are identified. But, after all, we must not forget that our future will always remain uncertain. This is exactly the reason why we have to encourage our resilience, that is our ability to accept the unexpected and to cope with it. It has to be the most important task with respect to the shaping of our future. Resilience is the normative goal of sustainability, whilst sustainability is the process that eventually leads to resilience. This process has to meet many requirements:

- reduce vulnerability
- improve the quality of life, especially with reference to the environment, education, economy and social health care
- encourage the partnership between single individuals, the private sector, NGOs and governments
- blend planning and management

The participants of the workshop on Montserrat have in the past nine years experienced the devastating effects of hurricanes Hugo and the Soufrière Hills Volcano. They have proven that they are aware of all the above mentioned requirements. Their favourite scenario *Proud Alliouagana* comprises exactly these requirements. This, in effect, means that the opportunity to realise sustainable development on the island is at this point in time better than ever. The population is sensitised to disasters and an uncertain future. Their preparedness to accept measures which are designed to reduce their vulnerability is comparatively high. It is now all down to realising and putting into practice what has been recognised and expressed in the scenario *Proud Alliouagana.* Moreover, the locals' experience with disasters has to be comprehended and taken advantage of. And their knowledge should also be passed on to the following generations: "Truly involved people can do anything" (Peters 1987, p. 286). This is the way we can learn to live with the unexpected.

References

Abler RF (1993) Desiderata for geography. An institutional view from the United States. In: Johnston RJ (ed) The challenge for geography. A changing world – a changing discipline. Blackwell, Oxford/Cambridge, pp 215–238

Adams RLA (1973) Uncertainty in nature. Cognitive dissonance and the perceptual distortion of environmental information – Weather forecasts and New England beach trip decisions. In: Economic Geography 49 (4): 287–297

Afheldt H (ed) (1985) Bilder einer Welt von morgen. Modelle bis 2009. Horst Poller, Stuttgart

Agustoni H (1983) Szenario. Technik oder Flop? In: Management-Zeitschrift 52 (9): 319–321

Albala–Bertrand JM (1992) Natural disasters in Latin America. Economic patterns and performance Rules. London

Albers G (1993) Über den Wandel im Planungsverständnis. In: Raumplanung (61): 97–103

Allen PM (1994) Coherence, chaos and evolution in the social context. In: Futures 26 (6): 583–597

Anderson RM, Woodrow PJ (1998) Rising from the ashes. Development strategies in times of disaster. 2. edn. Westview Press, Boulder, Colorado

Anderson RM, Woodrow PJ (1989) Rising from the ashes. Development strategies in times of disaster. 1. edn. Westview Press, Boulder, Colorado

Anderson A (1985) Creativity and regional development. In: Papers of the Regional Science Association 56: 5–20

Appleyard D (1979) The environment as a social symbol. Within a theory of environmental action and perception. In: Journal of the American Planning Association 45 (2): 143–153

Aptekar L (1991) The psychological process of adjustment to natural disasters. Natural Hazards Research and Applications Information Center, Working Paper 70. University of Colorado, Boulder, Colorado

Arbeitsgemeinschaft NRW (ed) (1988) Werkstattbericht 35. NRW 2020 – Mikroelektronik, Arbeitsmarkt und Gestaltungsmöglichkeiten. Ministerium für Arbeit, Gesundheit und Soziales des Landes Nordrhein–Westfalen, Düsseldorf

Arras H (1989) Szenarien als Instrumente zur Kommunikation. In: Institut für Landes- und Stadtentwicklungsforschung (ed) Szenarien in der Stadtentwicklung. Zum Stand der Diskussion. waz-Druck, Dortmund, pp 27–38

Arras H (1987) Zur Notwendigkeit und Methodik von Szenarien. In: Verwaltungsrundschau (6): 187–195

Aysan YE (1993) Vulnerability assessment. Keynote paper. In: Merriman PA, Browitt CWA (eds) Natural disasters. Protecting vulnerable communities. Thomas Telford, London, pp 1–14

Bak P, Chen K (1991) Self-organized criticality. In: Scientific American. January 1991: 46–53

Baker PL (1992) Chaos, cosmos and fractals. Social change and development in Dominica. Paper presented at the Islands of the World III Conference, 18.–22. May 1992, Nassau

Banathy BH (1993) From evolutionary consciousness to guided evolution. In: World Futures (36): 73–79
Barzey RO (1989) The wrath of hurricane Hugo, September 16–17, 1989. Plymouth, Montserrat
Bass S, Dalal–Clayton B (1995) Small island states and sustainable development. Strategic issues and experience. International Institute for Environment and Development, London
Bates F (1982) Recovery, change and development. A longitudinal Study of the 1976 Guatemalan earthquake. The University of Georgia Press, Athen/London
Bates F, Peacock WG (1993) Living conditions, disasters and development. An approach to cross-cultural comparisons. The University of Georgia Press, Athen/London
Bayrische Rückversicherung (ed) (1993) Risiko ist ein Konstrukt. Wahrnehmungen zur Risikowahrnehmung. Knesebeck, Karlsruhe
Bayrische Rückversicherung (ed) (1987) Gesellschaft und Unsicherheit. Knesebeck, Karlsruhe
Bechmann G (ed) (1994) Risiko und Gesellschaft, Grundlagen und Ergebnisse interdisziplinärer Risikoforschung. Westdeutscher Verlag, Opladen
Beck U (1991) Politik in der Risikogesellschaft. Essays und Analysen. Suhrkamp, Frankfurt
Beller W, D'Ayala P, Hein P (eds) (1990) Sustainable development and environmental management of small islands. Man and the Biosphere Series 5. The Parthenon Publishing Group, Paris
Bennett RJ (1989) Whither models and geography in a post-welfarist world? In: Macmillan B (ed) Remodelling geography. Oxford, p 273–290
Bennett RJ, Chorley RJ (1978) Environmental systems. Philosophy, analysis and control. Methuen, London
Berke PR, Wenger D (1991) Montserrat. Emergency planning, response and recovery related to hurricane Hugo. Texas A&M University, Texas
Berke PR, Roenigk DJ, Kaiser EJ, Burby R (1996) Enhancing plan quality. Evaluating the role of state planning mandate for natural hazard mitigation. In: Journal of Environmental Planning and Management 39 (1): 79–96
Berke PR, Beatley T, Feagin CE (1993) Hurricane Gilbert strikes Jamaica. Linking disaster recovery to development. In: Coastal Management 17 (2): 1–23
Berke PR, Kartez J, Wenger D (1993) Recovery after disaster. Achieving sustainable development, mitigation and equity. In: Disasters 21 (2): 93–109
Berke PR, Beatley T, Feagin CE (1991) Household recovery following hurricane Gilbert. St. James and St. Thomas parishes, Jamaica. Texas A&M University, Texas
Berz G (1994) Versicherungswirtschaft und IDNDR. Gemeinsame Aspekte und Aufgaben. In: Geographische Rundschau 46 (7–8): 429–433
Bird JH (1989) The changing worlds of geography. Clarendon Press, Oxford
Birzer M (1994) Problemlösung durch Dialog. Das Buxtehuder Modell. In: Vierteljahreszeitschrift für Sicherheit und Frieden (4): 154–158
Birzer M (1994) Bürgerbeteiligung im Dialogverfahren. In: Tribüne (132): 74–76
Birzer M, Feindt PH, Spindler EA (1997): Nachhaltige Stadtentwicklung. Konzepte und Projekte. Economica, Bonn
Blaikie PM (1996) New knowledge and rural development. A review of views and practicalities. Paper for the 28th Internationel Geographical Congress the Hague, August 5–10, The Hague
Blaikie PM (1985) The policial economy of soil erosion in developing countries. Longman, London
Blaikie PM, Brookfield H (1987) Land degradation and society. Longman, London
Blaikie PM, Cannon T, Davis I, Wisner B (eds) (1994) At risk. Natural hazards, People's vulnerability and disasters. Routledge, London
Blankenship JR (1990) The wildlife of Montserrat. Plymouth, Montserrat
Blecke U (1978) Plausible Pfade in die Zukunft. In: Manager Magazin (12): 120–125
Bohle HG (1994) Dürrekatastrophen und Hungerkrisen. Sozialwissenschaftliche Perspektiven geographischer Risikoforschung. In: Geographische Rundschau 46 (7–8): 400–407

Boholm A (1996) Risk perception and social anthropolgy. Critique of cultural theory. In: Ethnos 61 (1–2): 64–84

Bolin R (1982) Long-term family recovery from disaster. Institute of Behavioral Sciences, University of Colorado, Boulder, Colorado

Bossel H (1978) Bürgerinitiativen entwerfen die Zukunft. Neue Leitbilder, neue Werte – 30 Szenarien. Fischer, Frankfurt am Main

Brauner C (1990) Das verdrängte Risiko. Können wir Katastrophen verhindern? Herder, Freiburg

Brenner C (1993) Die Berücksichtigung von Unsicherheiten in ausgewählten Bereichen der strategischen Planung. Wirtschaftswissenschaftliche Fakultät, Tübingen

British Geological Survey (ed) (1997) Montserrat. Note by the chief scientific adviser. Nottingham

Brundtland GH (1987) Our common future. Call for action. In: Environmental Conservation 14 (4): 291–294.

Bryant E (1991) Natural hazards. Cambridge University Press, New York

Bucher H, Gatzweiler HP (1993) Die Raumordnungsprognose als Instrument einer aktiven Raumordnungs- und Städtebaupolitik. In: Informationen zur Raumentwicklung (12): 807–817

Buffonge C (ed) (1996) Volcano. A chronicle of Montserrats volcanic experience 1995–1996. Plymouth, Montserrat

Burby R, Cigler B, French S, Kaiser E, Kartez J, Roengik D, Weist D, Whittington D (1991) Sharing environmental risks. How to control government's losses in natural disasters? Westview Press, Boulder, Colorado

Burton I, Kates RW (1964) The perception of natural hazards in resource management. In: Natural Resources Journal 3: 412–441

Burton I, Kates RW, White GF (1993) The Environment as hazard. 2. ed. The Guilford Press, , New York/London

Burton I, Kates RW, White GF (1981) The future of hazard research. A reply to William I. Torry. In: Canadian Geographer 25 (3): 286–289

Cambel AB (1993) Applied chaos theory. A paradigm for complexity. Academic Press, Boston

Capra F (1984) Wendezeit. Bausteine für ein neues Weltbild. Scherz, München

Caribbean Conservation Association, Eastern Caribbean Natural Area Management Project (eds) (1980) Montserrat preliminary data atlas. Kingston, Jamaica.

Castells M (1992) The World has changed – Can planning change? In: Landscape and Urban Planning (22): 73–78

Casti J (1994) Complexification. Explaining a paradoxical world through the science of surprise. Harper & Collins, New York

Casti J (1990) Szenarien der Zukunft: Was Wissenschaftler über die Zukunft wissen können. Klett–Cotta, Stuttgart

Casti J (ed) (1979) Connectivity, complexity, and catastrophe in large-scale systems. John Wiley & Sons, Chichester/New York

Caviedes C (1991) Five hundred years of hurricanes in the Caribbean. Their relationship with global climatic variabilities. In: Geo Journal 23 (4): 301–310

Center of Excellence for Sustainable Development (ed) (1998) The Wingspread principles. A community vision for sustainability. www.sustainable.doe.gov/wingspread2/wingprin.html

City Plan Resource Center (ed) (1996) Choose your future. City Plan Vancouver. Vancouver, British Columbia

Clarke M, Wilson G (1985) The Dynamics of urban spatial structure. The progress of research programme. In: Transactions (10): 427–421

Clausen L, Dombrowsky WR (1983) Einführung in die Soziologie der Katastrophen. Osang, Bonn

Clausen L, Dombrowsky WR, Fenner H (1994) "Die ersten Stunden entscheiden..." – Vom wissenschaftlich begleiteten Katastrophenschutz zur dauerhaften Entwicklung. In: Geographische Rundschau 46 (7–8): 465–466

Clemons EK (1995) Using scenario analysis to manage the strategic risk of reengineering. In: Sloan Management, summer 1995: 61–71

Cohen J, Uphoff N (1979) Feasibility and application of rural development participation. A state of the art paper. University of Cornell, New York

Cohn RC, Klein I (1993) Großgruppen gestalten mit themenzentrierter Interaktion. Ein Weg zur lebendigen Balance zwischen Einzelnen, Aufgaben und Gruppe. Grünewald, Mainz

Collingridge D (1983) Heding and flexing. Two ways of choosing under ignorance. In: Technological Forecasting and Social Change 23 (2): 161–172

Collymore J (1995) Disaster mitigation and cost-benefit-analysis. Conceptual perspectives. In: Barker D, McGregor D (eds) Environment and development in the Caribbean. University of the West Indies Press, Barbados/Jamaica, pp 111–123

Comfort LK (1995) Self organization in disaster response and recovery. The Maharashta, India, earthquake of September 30, 1993. Natural Hazards Research and Applications Information Center, Quick Response Reports No. 74. University of Colorado, Boulder, Colorado

Comfort LK (ed) (1988) Managing disaster. Strategies and policy perspectives. Duke University Press, London

Cook, Robin (1998) A new partnership. Speech to the Dependent Territories Association, London, 4 Februay 1998

Corker I (1986) Montserrat. A resource assessment. Overseas Development Administration, Surrey, England

Covello VT (1983) The perception of technological risks. A literature review. In: Technological Forecasting and Social Change (23): 285–297

Cross JA (1992) Natural hazards within the West Indies. In: Journal of Geography 91 (5): 190–199

Dale, A (1994): Scientific uncertainty in a world of risk. In: Futures 26 (8): 862–867

Daum E (1993) Im Chaos gibt es keine Orientierung. Probleme und Chancen einer geographischen Weltdeutung. In: geographie heute (96): 45–46

D'Ayala PG (1992) Islands at a glance. In: Environmental Management 16 (5): 565–568

Deutsche Forschungsgemeinschaft (ed) (1993) Naturkatastrophen und Katastrophenvorbeugung. Bericht des Wissenschaftlichen Beirats der DFG für das Deutsche Komitee für die IDNDR. VCH Verlagsgesellschaft, Weinheim/Basel/Cambridge

Deutsches IDNDR-Komitee für Katastrophenvorbeugung (ed) (1996) Jahresbericht 1996. Deutsche IDNDR-Reihe. Bonn

Deutsches IDNDR-Komitee für Katastrophenvorbeugung (ed) (1997) Katastrophenvorsorge. Eine humanitär-soziale Pflichtaufgabe für Staat und Gesellschaft. Deutsche IDNDR-Reihe 5. Bonn

Devaney RL (1991) Chaos, fraktals and dynamics. Computer experiments in mathematics. Addison-Wesley, Marlo Park

Development Unit (ed) (1996) Montserrat. Economic report for 1995 and January - June 1996. Plymouth, Montserrat

Diller C (1996) Die Regionalplanung als Mediatorin einer nachhaltigen Entwicklung. In: Raumforschung und Raumordnung 54 (4): 228–234

Dobler R (1980) Regionale Entwicklungschancen nach einer Katastrophe. Ein Beitrag zur Regionalplanung des Friaul. Münchener Geographische Hefte No. 45. Michael Laßleben, Kallmünz

Dörner D, Reither F (1978) Über das Problemlösen in sehr komplexen Realitätsbereichen. In: Zeitschrift für experimentelle und angewandte Psychologie (4): 527–551

Dovers SR, Handmer JW (1992) Uncertainty, sustainability and change. In: Global Environmental Change 2 (4): 262–276

Dow K (1992) Exploring differences in our common future(s). The meaning of vulner ability to global environmental change. In: Geoforum 23 (3): 417–436

Drabek TE (1986) Human system response to disaster. An inventory of sociological findings. Springer, New York/Berlin/Heidelberg

Drabek T, Key W (1984) Conquering disaster. Family recovery and long term consequences. Irvington Publishers, New York

Eastern Caribbean Natural Area Management Project (ed) (1980) Montserrat. Preliminary data atlas. Kingston, Jamaica

Ebert R, Gnad F, Stierand R (1992) Informieren, Erörtern, Verhandeln. Stufen der Bürgerbeteiligung auf regionaler Ebene. In: Raumplanung (57): 89–97

Edmonds B (1997) What is complexity? The philosophy of complexity per se with application to some examples in evolution. http://www.fmb.mmu.ac.u...e/evolcom/evolcom.txt

Eikenberg C (1998) Journalisten-Handbuch zum Katastrophenmanagement. Deutsches IDNDR-Komitee für Katastrophenvorbeugung, Bonn

Eisenführ F, Weber M (1994) Rationales Entscheiden. Springer, New York/Berlin/Heidelberg

Elias N (1988) Über die Zeit. Suhrkamp, Frankfurt am Main

Emel J, Peet R (1989) Resource management and natural hazards. In: Peet R, Thrift N (eds) New models in geography. The political-economy perspective. Unwin Hyman, London, pp 51–73

Emergency Department of the Governor's Office (ed) Montserrat today. Releases from 1995 to 1998. Montserrat

Evers A, Nowotny H (1988) Über den Umgang mit Unsicherheit. Die Entdeckung der Gestaltbarkeit von Gesellschaft. Suhrkamp, Frankfurt am Main

Fanger U, Furtak RK, König W, Sandner G (eds) (1982) Problems of Caribbean development. Regional interaction, international relations and the constraints of small size. Beiträge zur Soziologie und Sozialkunde Lateinamerikas No. 21. W. Fink, München

Faupel CE, Styles SP (1993) Disaster education, household preparedness and stress responses following hurricane Hugo. In: Environment and Behavior 25 (2): 228–249

Feindt PH (1997) Kommunale Demokratie in der Umweltpolitik. Neue Beteiligungsmodelle. In: Aus Politik und Zeitgeschichte (27): 39–46

Fellner A, Gestring N (1988) Szenarien zur Stadtentwicklung. Institut für Landes- und Stadtentwicklungsforschung des Landes Nordrhein-Westfalen, Dortmund

Fergus HA (ed) (1996) Eruption. Montserrat versus volcano. University of the West Indies School of Continuing Studies, Plymouth, Montserrat

Fergus HA (1994) Montserrat. History of a Caribbean colony. Macmillan Press, London/Basingstoke

Fergus HA (1983) Montserrat. Emerald isle of the Caribbean. Macmillan Press, London/Basingstoke

Fergus HA (1975) History of Alliouagana. University Center, Plymouth, Montserrat

Figley CR, McCubbin HI (eds) (1983) Stress and the family. Volume II – Coping with Catastrophe. Brunner/Mazel, New York

Fisch R, Boos M (1990) Vom Umgang mit Komplexität in Organisationen. Konzepte, Fallbeispiele, Strategien. Universitätsverlag, Konstanz

Fishburn PC (1992) Entscheidungstheorie. Die nächsten einhundert Jahre. In: Reckenwald H (ed) Ökonomische Wissenschaft in der Zukunft. Düsseldorf, pp 138–146

Flechtheim OK (1987) Ist die Zukunft noch zu retten? Hoffmann & Campe, Hamburg

Flechtheim OK (1980) Futurum. Beiträge des Instituts für Zukunftsforschung. Minerva Publikationen, München

Foreign and Commonwealth Office (ed) (1994) Background brief. Britain's dependent territories. London

Foreign and Commonwealth Office (ed) (1996) Britan gives 25 pounds to volcanic island. Press release. London

Foster HD (1980) Disaster planning. The preservation of life and property. Springer, New York/Berlin/Heidelberg

Friesema HP, Caporaso J, Goldstein G, Lineberry R, McClery R (1979) Aftermath. Communities after natural disaster. Sage Publications, Beverly Hills, California

Fritsch B (1980) Möglichkeiten und Grenzen der Zukunftsforschung. In: Analysen und Prognosen, July/August: 3–5

Funke J (1992) Wissen über dynamische Systeme. Erwerb, Repräsentation und Anwendung. Lehr- und Forschungstexte Psychologie. Springer, New York/Berlin/ Heidelberg

Funtowicz SO, Ravetz JR (1994) Emergent complex systems. In: Futures 26 (6): 568–582

Funtowicz SO, Ravetz JR (1993) Science for post-normal age. In: Futures 25 (7): 739–755

Funtowicz SO, Ravetz JR (1992) The good, the true and the post-modern. In: Futures 24 (10): 963–976

Funtowicz SO, Ravetz JR (1991) Three types of risk assessment and the emergence of post-normal science. In: Golding D, Krimsky S (eds) Social theories of risk. Praeger, New York

Funtowicz, SO, Ravetz, JR (1990): The emergence of post-normal science. In: Rossi C, Tiezzi E (eds) Ecological physical chemistry. Elsevier, Amsterdam, pp 253– 283

Geipel R (1994) IDNDR und Hazardforschung am Beispiel des Friaul. In: Geographische Rundschau 46 (7–8): 393–399

Geipel R (1992) Naturrisiken. Katastrophenbewältigung im sozialen Umfeld. Wissenschaftliche Buchgesellschaft, Darmstadt

Geipel R (1983) Katastrophe nach der Katastrophe? Ein Vergleich der Erdbebengebiete Friaul und Süditalien. In: Geographische Rundschau 35 (1): 17–26

Geipel R (1980) Il Progetto Friuli. Martin Internazionale, Udine, Italy

Geipel R (1977) Friaul. Sozialgeographische Aspekte einer Erdbebenkatastrophe. Münchener Geographische Hefte No. 40. Michael Laßleben, Regensburg

Geipel R, Härta R, Pohl J (1997) Risiken im Mittelrheinischen Becken. Bericht über ein von der Deutschen Forschungsgemeinschaft gefördertes Projekt. Deutsche IDNDR Reihe No 4. Bonn

Geipel R, Pohl J, Stagl R (eds) (1988) Chancen, Probleme und Konsequenzen des Wiederaufbaus nach einer Katastrophe. Münchener Geographische Hefte No 59. Michael Laßleben, Regensburg

Gell–Mann M (1997) Visions of a sustainable world, Santa Fè, New Mexico

Gell–Mann M (1994) Das Quark und der Jaguar. Vom Einfachen zum Komplexen – Die Suche nach einer neuen Erklärung der Welt. Piper, München/Zürich

Georgantzas N, Acar W (1995) Scenario-driven planning. Learning to manage strategic uncertainty. Quorum Books, Westport

Gessenharter W (1996) Warum neue Beteiligungsmodelle auf kommunaler Ebene? Kommunalpolitik zwischen Globalisierung und Demokratisierung. In: Aus Politik und Zeitgeschichte 96 (50): 3–13

Giampietro M (1994) Using hiearchy theory to explore the concept of sustainable development. In: Futures 26 (6): 616–625

Girvan NP, Simmons DA (eds) (1991) Caribbean ecology and economics. Caribbean Conservation Association, St. Michael, Barbados

Gleick J (1987) Chaos. Making a new science. Penguin Group, New York

Götze U (1991) Szenario-Technik in der strategischen Unternehmensplanung. Deutscher Universitätsverlag, Wiesbaden

Gomez P (1980) Szenarien als Planungshilfen. In: Management-Zeitschrift 49 (9): 416–420

Gornig M (1993) Gesamtwirtschaftliche Trends als Rahmenbedingungen regionaler Entwicklungen. In: Informationen zur Raumentwicklung (12): 819–830

Government Information Unit (ed) (1996) Putting life first in emergency. In: Government in Action 3 (4): 3

Government of Montserrat (ed) (1998) Sustainable development plan. Montserrat social and economic recovery programme – A path to sustainable development. Montserrat

Government of Montserrat (ed) (1996) Government house press statement. Montserrat

Government of Montserrat (ed) (1994) National environmental action plan (NEAP). Montserrat

Government of Montserrat, Emergency Operations Center (eds) (1995) Volcanic emergency plan. An official guide to evacuation proceedures in a volcanic emergency. Montserrat

Government of Montserrat, Government of the United Kingdom (eds) (1994) Country policy plan. Montserrat

Gray GAL (1994) Existing environmental legislation. Strengths, weaknesses and possibilities for enhancement. Montserrat

Gray–Fow MJG (1989) Faith in the future. The role of belief in shaping events. In: Didsbury HF (ed) The future. Opportunity not destiny. Bethesda, pp 283–294

Gutteling JM, Wiegmann O (1996) Exploring risk communication. Kluwer Academic Publishers, Dordrecht/Boston/London

Haas EJ, Kates RW, Bowden MJ (1977) Reconstruction following disaster. MIT Press, Cambridge/London

Habermas J (1988) Die Einheit der Vernunft in der Vielfalt ihrer Stimmen. In: Merkur 42 (467): 1–14

Häußermann H, Siebel W (1989) Szenarien zur Stadtentwicklung. Zwischen hohen Ansprüchen und bescheidener Praxis. In: Institut für Landes- und Stadtentwicklungsforschung des Landes Nordrhein-Westfalen (ed): Szenarien in der Stadtentwicklung. Zum Stand der Diskussion. waz-Druck, Dortmund, pp 13–36

Halfmann J, Japp KP (eds) (1990) Riskante Entscheidungen und Katastrophenpotentiale. Elemente einer soziologischen Risikoforschung. Westdeutscher Verlag, Opladen

Hanf H (1986) Entscheidungslehre. Einführung in Informationsbeschaffung, Planung und Entscheidung unter Unsicherheit. Oldenbourg, München/Wien

Hanisch R, Moßmann P (eds) (1996) Katastrophen und ihre Bewältigung in den Ländern des Südens. Schriften des Deutschen Übersee-Instituts No. 33. Deutsches Übersee-Institut, Hamburg

Hansel C, Lambrecht M (1993) Wo kämen wir hin...? Zur Erstellung von Szenarien. In: Raumplanung (61): 148–154

Hard G (1995) Zwei Versionen der klassischen Geographie – Oder: Wie man Geographietheorien vergleichend bewerten kann. In: Klagenfurter Geographische Schriften (10): 35–51

Harden C, Pulsipher L (1992) Come a nasty gale. The response to hurricane Hugo on the island of Montserrat, West Indies. In: Focus 42 (2): 9–14

Harvey D (1990) Between space and time. Reflections on geographical imagination. In: Annals of the Association of American Geographers 80 (3): 418–434

Harvey D (1969) Explanation in geography. Hodders & Stoughton, London

Hawkins S (1992) Die illustrierte kurze Geschichte der Zeit. Rowohlt, Reinbek

Helbrecht I (1991) Das Ende der Gestaltbarkeit? Zu Funktionswandel und Zukunftsperspektiven räumlicher Planung. Wahrnehmungsgeographische Studien zur Regionalentwicklung No. 10. Bibliotheks- und Informationssystem der Universität Oldenburg, Oldenburg

Herbold R, Wienken R (1993) Experimentelle Technikgestaltung und offene Planung. Strategien zur sozialen Bewältigung von Unsicherheit am Beispiel der Abfallbeseitigung. Wissenschafts- und Technikforschung edn 39. Kleine, Bielefeld

Hermann RE (1982) Disaster planning for local government. Universe Books, New York

Hesse JJ, Zöpel C (eds) (1990) Der Staat der Zukunft. Nomos, Baden-Baden

Hesse JJ, Kreibach R, Zöpel C (eds) (1989) Zukunftsoptionen. Technikentwicklung in der Wissenschafts- und Risikogesellschaft. Nomos, Baden-Baden

Hesse JJ, Zöpel C (eds) (1987) Zukunft und staatliche Verantwortung. Nomos, Baden-Baden

Hewitt K (ed) (1983) Interpretations of calamity. Allen & Unwin, Boston/London

Hirschhorn L (1980) Scenario writing. A developmental approach. In: Journal of American Planning Association 46 (2): 173–183

Hobfoll SE, de Vries MW (eds) (1995) Extreme stress and communities. Impact and intervention. Behavioural and Social Sciences 80. Kluwer, Dordrecht/ Boston/London

Hoffmann V (1993) Szenarien im strategischen Planungsprozeß der Shell-Gruppe. In: Strategien für Investitionsgütermärkte. Antworten auf neue Herausforderungen. moderne industrie, München

Holling CS (1994) Simplifying the complex. The paradigms of ecological function and structure. In: Futures 26 (6): 598–609

Horgan J (1995) From complexity to perplexity. Can science achieve a unified theory of complex systems? In: Scientific American, June 1995: 74–79

Hubler A, Pines D (1993) Prediction and adaptation in an evolving chaotic environment. http://www.santafe.edu/...racts/93-06-04labs.html

Humphreys P, Svenson O, Vári A (1983) Analysing and aiding decision processes. Advances in Psychology No. 14. North-Holland Publishing Company, Amsterdam/New York/Oxford

IDNDR Secretariat (ed) (1996a) Disaster reduction in urban systems. In: Stop Disasters 2 (28).

IDNDR Secretariat (ed) (1996b) The international decade for natural disaster reduction. Planning for a safer world in the 21st century. Genf

IDNDR Secretariat (1996c) Disaster reduction. A project proposed by the Secretariat for the International Decade of Natural Disaster Reduction – Project description. Genf

Inayatullah S (1994) Life, the universe and emergence. In: Futures 26 (6): 683–695

Inayatullah S (1993) From "Who am I" to "When am I". In: Futures 19 (6): 234–253

Institut für Arbeit und Technik (ed) (1991) Zur Methode der Szenarienwerkstatt. ifat Arbeitspapier No. 8. Institut für Arbeit und Technik eV, Hamburg

Institut für Landes- und Stadtentwicklungsforschung (ed) (1989) Szenarien in der Stadtentwicklung. Zum Stand der Diskussion. waz-Druck, Dortmund

Institut für Landes- und Stadtentwicklungsforschung (ed) (1985) Szenariotechnik in der Stadtplanung. Theorie und Anwendung. Dortmund

Institut für Unternehmens-Kybernetik (ed) (1993) Unternehmenssteuerung bei zunehmender Komplexität? Tüv Rheinland, Köln

International Development Committee (ed) (1998) First special report. Government response to the first report from the committee – Session 1997–1998. Stationary Office, London

International Development Committee (ed) (1997) First report Montserrat. Report together with the proceedings of the Committee, minutes of evidence and appendices. Stationary Office, London

Irish GJA (1991) Life in a colonial crucible. Plymouth, Montserrat

Island Resources Foundation, Montserrat National Trust (eds) (1993) Montserrat. Environmental profile – An assessment of the critical environmental issues facing Montserrat with an action agenda for the future. Bridgetown, Barbados

Johnston RJ, Taylor PJ, Watts MJ (1995) Geographies of global ghange. Remapping the world in the late twentieth century. Blackwell, Oxford/Cambridge

Johnston RJ (1994) The dictionary of human geography. 3 edn. Blackwell, London

Johnston RJ (1993) The challenge for geography. A changing world – A changing discipline. Blackwell, Oxford/Cambridge

Johnston RJ (1991) A question of place. Blackwell, Oxford/Cambridge

Johnston RJ (1991) Geography and geographers. Anglo-american human geography since 1945. Hodders & Stoughton, London

Jungermann H, Rohrmann B, Wiedemann P (eds) (1990) Risiko-Konzepte, Risiko-Konstrukte, Risiko-Kommunikation. Monographien des Forschungszentrums Jülich edn 3. Jülich

Jungk R (1990) Zukunft zwischen Angst und Hoffnung. Ein Plädoyer für die politische Phantasie. Heyne, München

Jungk R, Müllert NR (1989) Zukunftswerkstätten. Mit Phantasie gegen Routine und Resignation. Heyne, München

Kahn H, Pepper T (1980) Will she be right? The future of Australia. University of Queensland Press, Queensland

Kahn H, Wiener AJ (1968) Ihr werdet es erleben. Voraussagen der Wissenschaft bis zum Jahr 2000. Molden, Wien/München/Zürich

Kartez J (1991) Problems and alternatives in the disaster assistance system. In: Federal Emergency Management Agency (ed) Draft study on improving earthquake mitigation. Washington DC

Kates RW (1970) Natural hazards in human ecological perspective. Hypotheses and models. Natural Hazards Research and Applications Information Center, Working Paper No. 14. University of Colorado, Boulder, Colorado

Kates RW (1962) Hazard and choice perception in flood plain management. Research Paper No. 78. University of Chicago, Chicago

Kauffmann S (1995) At home in the universe. The search for laws of self-organization and complexity. Oxford University Press, Oxford/New York/Athen

Kernig CD (1994) Welttrend 2000. Zur Struktur und Eigendynamik moderner Gesellschaftssysteme. Baden-Baden

Kiel DL (1994) Managing chaos and complexity in government. A new paradigm for managing change, innovation and organizational renewal. Jossey-Bass, San Francisco

Klaus D (1991) Vom Sein zum Werden. Räumliche Systeme mit chaotischer Dynamik. In: Geographische Rundschau 43 (2): 110–116

Kleindorfer PR, Kunreuther HC, Schoemaker PJH (1993) Decision sciences. An integrative perspective. Cambridge University Press, Cambridge

Kleining G (1994) Qualitativ-heuristische Sozialforschung. Schriften zur Theorie und Praxis. Fechner, Hamburg

Kluckholm H, Strodtbeck T (1961) Variations in value orientations. Dallas

Klüter H (1986) Raum als Element sozialer Kommunikation. Giessener Geographische Schriften No. 60. Giessen

Knauer P (1986) Ökosystemforschung und ökologische Planung. In: Geographische Rundschau 38 (6): 290–293

Knight N (1991) The problem of uncertainty. Institutional alternatives and planning approaches. University of British Columbia, Vancouver

Krimsky S, Golding D (1992) Social theories of risk. Praeger, New York/London

Kriz J (1992) Chaos und Struktur. Grundkonzepte der Systemtheorie. Quintessenz, München

Kunreuther H (1973) Recovery from natural disasters. Insurance or federal aid? American Enterprise Institute for Public Policy Research, Washington DC

Lane D, Maxfield R (1995) Foresight, complexity and strategy. Manuscript, not published

Langton CC (ed) (1989) Artificial life. Santa Fe Institute studies in the sciences of complexity. Addison-Wesley, Redwood City

Lausche B, Organization of Eastern Caribbean States (eds) (1987) Inventory of national legislation for natural resources. Management and environmental protection. Montserrat, Legal report No. 1. Castries, St. Lucia

Lavell A (ed) (1994) Al Norte del Río Grande. Ciencias Sociales, Desastres – Una Perspectiva Norteamericana. Tercer Mundo Editores, Bogotá

Lazslo E (1972) The systems view of the world. The natural philosophy of the new development in sciences. Braziller, New York

Lembke HH (1992) Umweltpolitik in der Nord-Süd-Dimension. UNCED 1992 und danach. In: Zeitschrift für angewandte Umweltforschung 5 (3): 322–334

Lewin R (1992) Die Komplexitätstheorie. Wissenschaft nach der Chaosforschung. Hoffmann & Campe, Hamburg

Lewis J (1990) The vulnerability of small island states to sea level rise. The need for holistic strategies. In: Disasters 14 (3): 241–249

Lewis J (1984) A multi-hazard history of antigua. In: Disasters 8 (3): 190–197

Lienemann F (1975) Zur Verwendung langfristiger Szenarien als Grundlage für regionalisierte Zielprojektionen. Informationen zur Raumentwicklung. Hannover

Lienemann F, Unholzer G (1975) Die Szenario-Methode als Beitrag zur Fortschreibung des Bundesraumordnungsprogramms. Alternative Perspektiven gesamtgesellschaftlicher Zusammenhänge im Rahmenwerk der Raum- und Siedlungsentwicklung. In: Raumforschung und Raumordnung (5): 240–249

Lo Presti A (1996) Futures research and complexity. In: Futures 28 (10): 891–902

Loda M (1990) Erdbeben, Wiederaufbau und industrielle Entwicklung in Friaul. Münchener Geographische Hefte No. 65. Michael Laßleben, Regensburg

Luhmann N (1991) Soziologie des Risikos. de Gruyter, Berlin/New York

Luhmann N (1988) Ökologische Kommunikation. Kann sich die moderne Gesellschaft auf ökologische Gefährdungen einstellen. Westdeutscher Verlag, Opladen

Lutz R (1981) Einführung in die Zukunftswerkstatt. In: Analysen und Prognosen (77–78): 34–36

Marston SA (1983) Natural hazards research. Towards a political economy perspective. In: Political Geography Quarterly 2 (4): 339–348

Maskrey A (1989) Disaster mitigation. A community based approach. Development Guidelines No. 3. Oxfam, Oxford

Masser I, Svidén O, Wegener M (1992) The geography of Europe's futures. Belhaven Press, London/New York

Maurer B (ed) (1994) Die Zukunft des Raumes. Campus, Frankfurt

Maurer J (1993) Die Annäherung an Möglichkeiten räumlicher Entwicklungen. In: Raumforschung und Raumordnung 51 (2–3): 81–84

May P, Williams W (1986) Disaster policy implementation. Managing programs under shared governance. Plenum Press, New York

McDaniels T, Axelrod LJ, Slovic P (1996) Perceived ecological risks of global change. In: Global Environmental Change 6 (2): 159–171

McElroy J, Albuquerque K (1992) The economic impact of retirement tourism in Montserrat. Some provisional evidence. In: Social and Economic Studies 21 (2): 127–152

McGregor DF, Potter B (1997) Environmental change and sustainability in the Caribbean. In: Ratter BMW, Sahr WD (eds) Land, sea and human effort in the Caribbean. Beiträge zur Geographischen Regionalforschung in Lateinamerika 10. Institut für Geographie, Hamburg, pp 1–16

McGregor DF, Barker D (1995) Environment and development in the Caribbean. Geographical perspectives. University of the West Indies Press, Barbados/Jamaica

Meadows D, Zahn E, Milling P (1973) Die Grenzen des Wachstums. Bericht des Club of Rome zur Lage der Menschheit. rororo-Taschenbuch, Reinbek

Mehrmann E (1994) Moderierte Gruppenarbeit mit Metaplan-Technik. Econ, Düsseldorf

Meise J, Volwahsen A (1980) Stadt- und Regionalplanung. Ein Methodenhandbuch. Friedrich Vieweg, Braunschweig/Wiesbaden

Merriman PA, Browitt CWA (eds) (1993) Natural disasters. Protecting vulnerable communities – Proceedings of the conference held in London 13–15 October 1993. Thomas Telford, London

Merry U (1995) Coping with uncertainty. Insights from the new sciences of chaos, self-organization and complexity. Praeger, Westport

Mertens H (1993) Politische Ökologie und globale Krisenszenarien. In: Peripherie 13 (51–52): 137–154

Meyer–Schönherr M (1992) Szenario-Technik als Instrument der strategischen Planung. Wissen und Praxis, Ludwigsburg/Berlin

Mikulecky DC (1997) Life, complexity and the edge of chaos. Cognitive aspects of communication between cells and other components of living systems. http://griffin.vcu.edu/complex/mikulecky/rev.htm

Ministry of Agriculture, Trade & the Environment (eds) (1994) National physical development plan for Montserrat. Draft version. Plymouth, Montserrat

Mitchell B (1997) Resource and environmental management. Longman, Harlow

Mitchell B (ed) (1995) Resource and environmental management in Canada. Addressing conflict and uncertainty. Oxford University Press, Oxford

Mitchell B (1989) Geography and resource analysis. John Wiley & Sons, New York

Mitchell JK, Devine N, Jagger K (1989) A contextual model of natural hazard. In: The Geographical Review 79 (4): 391–409

Mittelstaedt W (1993) Zukunftsgestaltung und Chaostheorie. Grundlagen einer neuen Zukunftsgestaltung unter Einbeziehung der Chaostheorie. Peter Lang, Frankfurt am Main

Möller F, Clausen L (1993) Bestandsaufnahme im Bereich der Katastrophensoziologie. In: Deutsche Forschungsgemeinschaft (ed) Naturkatastrophen und Katastrophenvorbeugung. Bericht zur IDNDR. VCH Verlagsgesellschaft, Weinheim/Basel/Cambridge

Montserrat Statistical Office (1994) Vital statistics. Plymouth, Montserrat

Montserrat Volcano Observatory (ed) MVO weekly scientific reports from 1995 to 1998. Montserrat

Moore DP, Moore JW (1996) Post-hurricane burnout. An island township's experience. In: Environment and Behavior 28 (1): 134–155

Morfill G, Scheingraber H (1991) Chaos ist überall...und es funktioniert. Eine neue Weltsicht. Ullstein, Frankfurt am Main

Mossler M (1996) Environmental hazard analysis and small island states. Rethinking academic approaches. In: Ratter BMW, Sandner G (eds) (1996) Small islands – large questions. Spezialband der Geographischen Zeitschrift 84 (2): 86–93

Nagel G (1980) Die Rolle strategischer Raumplanung für Entwicklungsländer. In: Raumforschung und Raumordnung 38 (1–2): 1–12

Nicolis G, Prigogine I (1989) Exploring complexity. Freeman, New York

Niznik J (1995) The future of small countries. Cooperation is imperative. In: Futures 51 (8): 891–895

Oakley P (1991) The concept of participation in development. In: Landscape and Urban Planning (20): 115–122

O'Connor M (1994) Complexity and coevolution. Methodology for a positive treatment of indeterminacy. In: Futures 26 (6): 640–645

O'Riordan T, Rayner S (1991) Risk management for global environmental change. In: Global Environmental Change 1 (2): S. 91–108

O'Riordan T (1971) Perspectives on resource management. Pion, London

Oßenbrügge J (1991) Umweltrisiko und Raumentwicklung. Wahrnehmung von Umweltgefahren und ihre Wirkung auf den regionalen Strukturwandel in Norddeutschland. Springer, New York/Berlin/Heidelberg

Palm RI (1990) Natural hazards. An integrative framework for research and planning. John Hopkins University Press, London

Pan American Health Organization (ed) (1992) Natural disaster reduction. A Caribbean IDNDR
meeting. Kingston

Pan American Health Organization (ed) (1989) Gilbert in Jamaica 1988. Disaster Reports No. 5. Washington DC.

Pan American Health Organization (ed) (1984) Report on disasters and emergency preparedness for Jamaica, St. Vincent and Dominica. Disaster Reports No. 2. Washington D.C.

Pan American Health Organization (ed) (1980) The effects of hurricane David, 1979, on the population of Dominica. Disaster Reports No. 1. Washington DC.

Pan American Health Organization, World Health Organization (eds) (1994) A world safe from natural disasters. The journey of Latin America and the Caribbean. Washington DC

Pannell Kerr Forster Associates, Overseas Development Administration (eds) (1994a) Strategic review for the Government of Montserrat. Final report 1 – Executive summary. London

Pannell Kerr Forster Associates, Overseas Development Administration (eds.) (1994b): Strategic review for the Government of Montserrat. Final report 2 – Overview. London

Pannell Kerr Forster Associates, Overseas Development Administration (eds) (1994c) Strategic review for the Government of Montserrat. Final report 4 – Blackburne airport. London.

Parkes D, Thrift N (1980) Times, spaces and places. A chronogeographic perspective. Wiley, Chichester/New York

Peak D, Frame M (1995) Komplexität. Das gezähmte Chaos. Birkhäuser, Basel/Boston/Berlin

Petak WJ, Atkisson AA (1982) Natural hazard risk assessment and public policy. Anticipating the unexpected. Springer, New York/Berlin/Heidelberg

Peters T (1987) Thriving on chaos. The Chaucer Press Ltd, Bungay, Suffolk

Physical Planning Unit, Ministry of Agriculture, Trade & the Environment (eds) (1997) Physical development plan for North Montserrat. Plymouth, Montserrat

Physical Planning Unit, Ministry of Agriculture, Trade & the Environment (1996a) Development strateg for north Montserrat. Volume 1 – Development strategy. Plymouth, Montserrat

Physical Planning Unit, Ministry of Agriculture, Trade & the Environment (eds) (1996b) Development strategy for north Montserrat. Volume 3 – Development programme for the agricultural sector. Plymouth, Montserrat

Physical Planning Unit, Ministry of Agriculture, Trade & the Environment (eds) (1995a) National physical development plan. Reappraisal of development strategy in response to volcanic situation – Draft version. Plymouth, Montserrat

Physical Planning Unit, Ministry of Agriculture, Trade & the Environment (eds) (1995b) National physical development plan. Volume 1 – The survey. Plymouth, Montserrat

Physical Planning Unit, Ministry of Agriculture, Trade & the Environment (eds) (1995c) National physical development plan. Volume 2 – Development proposals. Plymouth, Montserrat

Physical Planning Unit, Ministry of Agriculture, Trade & the Environment (eds) (1995d) National physical development plan. Volume 3 – Project concepts. Plymouth, Montserrat

Physical Planning Unit, Ministry of Agriculture, Trade & the Environment (1995e) National physical development plan. Volume 4 – Summary. Plymouth, Montserrat

Plate E, Kron W (1994) Naturkatastrophen und Katastrophenschutz. Die Internationale Dekade für Katastrophenvorbeugung (IDNDR). In: Geographische Rundschau 46 (7–8): 749–750

Pohlmann HJ (1993) Kommunikationsplanung. Planungstheoretische Perspektive für die Zukunft? In: Raumplanung (61): 93–96

Possekel AK (1997a) Die andere Seite des Paradieses. Naturkatastrophen in der Karibik. In: Geographische Rundschau 49 (11): 656–661

Possekel AK (1997b) Complex resource management on small Caribbean islands. The potential of scenarios in planning approaches. In: Ratter BMW, Sahr WD (eds) Land, sea and human effort in the Caribbean. Beiträge zur Geographischen Regionalforschung in Lateinamerika 10. Institut für Geographie, Hamburg, pp 31–44

Possekel AK (1996) Offshore financial centres in the Caribbean. Potential and pittfalls. In: Caribbean Geography 7 (2): 81–96

Possekel AK, Ratter BMW (1997) Scenarios for Montserrat. A workshop proceedings report. University of the West Indies Center for Environment and Development, Worldbank/Commonwealth Secretariat, Kingston, Jamaica

Priebs A (1995) Von der Planung zur Moderation. Aktuelle Anforderungen für die angewandte Geographie. In: Geographische Rundschau 47 (10): 546–550

Prigogine I (1980) From being to becoming. Time and complexity in the physical sciences. Freeman, San Fransico

Prigogine I, Stengers I (1984) Order out of chaos. Bantam Books, New York

Prigogine I, Nicolis G (1977) Self-organization in non-equilibrium systems. John Wiley & Sons, New York

Quarantelli EL (1989) A review of literature in disaster recovery research. University of Delaware, Newark

Ratter B (1997) Komplexes Ressourcenmanagement in Kanada. Theorieansätze und Umsetzungsproblemtik. Paper presented at the Deutschen Geographentag 1997.

Ratter B (1992) Karibische Netze. Wayasbah, Hamburg

Ratter B, Sandner G (eds) (1996) Small islands – large questions. Spezielband der Geographischen Zeitschrift 84 (2). Franz Steiner, Stuttgart

Ravetz JR (1990) The merger of knowledge with power. Essays in critical science. Mansell Publishing Limited, London/New York

Rayner S (1993) Risikowahrnehmung, Technologieakzeptanz und institutionelle Kultur. Fallstudien für einige neue Definitionen. In: Bayrische Rückversicherung (ed) Risiko ist ein Konstrukt. Wahrnehmungen zur Risikowahrnehmung. Knesebeck, Karlsruhe, pp 213–244

Rayner S (1991) A cultural perspective on the structure and implementation of a global environmental agreements. In: Education Review 15 (1): 75–102

Redclift M (1987) Sustainable development. Exploring the contradictions. Methuen, London/New York

Rees J (1989) Natural resources, economy and society. In: Gregory D, Warford R (eds) Horizons in human geography. Macmillan, Basingstoke, pp 364–394

Rees J (1985) Natural resources. Allocation, economics and policy. Methuen, London/New York

Rhode–Jüchtern T (1996) Den Raum lesen lernen. Perspektivenwechsel als geographisches Konzept. Oldenbourg, München

Rhode–Jüchtern T (1995) Raum als Text. Perspektiven einer konstruktiven Erdkunde. Materialien zur Didaktik der Geographie und Wirtschaftskunde 11. Institut für Geographie der Universität Wien, Wien

Riad JK, Norris FH (1996) The influence of relocation on the environmental, social and psychological stress experienced by disaster victims. In: Environment and Behavior 28 (2): 163–182

Rochford BE, Blocker JT (1991) Coping with natural hazards. Hazards as stressors – The predictors of activism in a flood disaster. In: Environment and Behavior 23 (2): 171–194

Rohlmann M (1993) Integrated natural resource management. A question of property institutions? In: Geo Journal, April 1993, pp 405–412

Rosen R (1978) Fundamentals of measurement. North-Holland Publishing, Amsterdam/New York/Oxford

Rosenhead J (ed) (1989) Rational analysis for a problematic world. Problem structuring methods for complexity, uncertainty and conflict. John Wiley & Sons, Chichester

Rosnay Jd (1979) A new world scientific system. Harper & Row, New York

Rubin CB (1995) Physical reconstruction. Timescale for reconstruction. In: New Zealand Earthquake Commission (ed) Wellington after the quake. The challenge of rebuilding cities. Christchurch, New Zealand, pp 97–110

Rubin CB (1993) Local public capacity to deal with a catastrophic earthquake. In: Natural Hazard Reduction Center (ed) Improving earthquake mitigation. Report to Congress – National earthquake hazards reduction program reauthorization act. Arlington, pp 137–160

Rubin C (1991) Recovery from disaster. In: Drabek T, Hoetmer G (eds) Emergency management. Principles and practises for local government. International City Management Association, Washington DC., pp 23–34

Rubin CB, Yezer AM (1987) The local economic effects of natural disasters. Natural Hazards Research and Applications Information Center, Working Paper No. 61. University of Colorado, Boulder, Colorado

Russo S, Hanrahan M, Valencia I (1992) Strategic long-range planning for environmental and natural resources management. Results from applied experiences in Central America and the Caribbean. In: Environmental Management 36: pp 215–223

Sahr WD (1997) Ville und countryside. Land-Stadt-Verflechtungen im ländlichen St. Lucia. Ein Beitrag zu einer postmodernen Sozialgeographie der Karibik. Wayasbah, Hamburg

Sandner G (1996) Die karibischen Archipele Turks & Caicos Islands und San Blas: Ressourcenkonflikte und Kulturbedrohung. In: Jahrbuch der Marburger Geographischen Gesellschaft, pp 39–59

Sandner G (1993) Über die Schwierigkeiten beim Umgang mit dem Räumlichen im Zusammenhang von Kultur, Identität und Kommunikation. In: Ammon G, Eberhard T (ed) Kultur, Identität und Kommunikation. 2. Versuch. Eberhard Verlag, München, pp 33–51

Sandner G (1984) Zentralamerika und der Ferne Karibische Westen. Konjunkturen, Krisen und Konflikte 1504–1984. Franz Steiner, Stuttgart

Sandner G (1982) Spatial interaction within the Caribbean. In: Fanger U et al (eds) Problems of Caribbean development. Regional interaction, international relations and the constraints of small size. Beiträge zur Soziologie und Sozialkunde Lateinamerikas. W. Fink, München, pp 13–21

Sardar Z (1994) Conquests, chaos and complexity. The other in modern and postmodern science. In: Futures 26 (6): 665–682

Sardar Z, Ravetz J (1994) Complexity – Fad or future? In: Futures 26 (6): 563–567

Sayer A (1983) Notes on geography and the relationship between people and nature. In: The London Group of the Union of social Geographers (ed) Society and nature. London, pp 45–57

Schiff MR (1970) Some theoretical aspects of attitudes and perception. Natural Hazards Research and Applications Information Center, Working Paper No. 15. University of Colorado, Boulder, Colorado

Schlange LE (1995) Linking futures research methodologies. An application of systems thinking and metagame analysis to nuclear energie policy issues. In: Futures 51 (8): 823–838

Schlicksupp H (1977) Kreative Ideenfindung in der Unternehmung. Methoden und Modelle. Berlin

Schmincke HU (1994) Vulkaneruptionen, Vulkangefahren, Vulkankatastrophen. In: Geographische Rundschau 46 (7–8): 440–449

Schmincke HU (1986) Vulkanismus. Wissenschaftliche Buchgesellschaft, Darmstadt

Schmincke HU, Behnke B, Dehn J, Ippach P (1993) Vulkanismus. In: Deutsche Forschungsgemeinschaft (ed) Naturkatastrophen und Katastrophenvorbeugung. Bericht zur IDNDR. VCH Verlagsgesellschaft, Weinheim/Basel/Cambridge, pp 353–408

Schneider ED, Kay JJ (1994) Complexity and thermodynamics. Towards a new ecology. In: Futures 26, (6): 626–647

Schneider G (1980) Naturkatastrophen. Enke, Stuttgart

Schnelle E (1978) Neue Wege der Kommunikation. Spielregeln, Arbeitstechniken und Anwendungsfälle der Metaplan-Methode. Königstein

Schoemaker PJH (1995) Scenario planning. A tool for strategic thinking. In: Sloan Management Winter 1995, pp. 25–40

Schretzenmayr M (1996) Was führt zum Scheitern raumplanerischer Konzepte? In: Raumforschung und Raumordnung 54 (6): 397–409

Schüller A, Schlange L (1994) Komplexität und Managementpraxis. Reale Visionen zum Komplexitätsmanagement. Enke, Stuttgart

Schult J, Wahlers L (1990) Chaos. Das neue Prinzip der Wissenschaft. In: Physis 5: 25–26

Scott GP (1991) Time, rhythms, and chaos in the new dialogue with nature. Iowa State University, Ames, Iowa

Sedlacek PH (1989) Programm und Praxis qualitativer Sozialgeographie. Wahrnehmungsgeographische Studien zur Regionalentwicklung No. 6. Bibliotheks- und Informationssystem der Universität Oldenburg, Oldenburg

Segner M (1976) Szenario-Technik. Methodische Darstellung und kritische Analyse. In: Forschungsreihe Systemtechnik. Bericht 8/1976. Berlin

Shackley S, Wynne B, Waterton C (1996) Imagine complexity. The past, present and future potential of complex thinking. In: Futures 28 (3): 201– 225

Shepard JB, Tomblin JF, Wood E (1988) Volcano-seismic crisis in Montserrat, West Indies 1966–67. In: Bulletin Volcanologique (35): 143–163

Simmons IG (1993) Ressourcen und Umweltmanagement. Eine Einführung in Geo-, Umwelt- und Wirtschaftswissenschaften. Spektrum Akademischer Verlag, Heidelberg

Simon HA (1956) Rational choice and the structure of the environment. In: Psychological Review , (63): 205–216

Simonis UE, Wätzold F (1997) Ökologische Unsicherheit. Über Möglichkeiten und Grenzen von Umweltpolitik. In: Aus Politik und Zeitgeschichte 97 (27): 3–14

Slocombe DS (1990) Complexity, change and uncertainty in environmental management. From the Great Lakes to the Kluane/Wrangells. Dissertation, University of Waterloo, Waterloo

Slovic P, Fischhoff B, Lichtenstein S (1987) Behavioral decision theory perspectives on protective behavior. In: Weinstein, ND (ed) Taking care. Understanding and encouraging selfprotective behavior. Cambridge University Press, Cambridge

Slovic P, Fischoff B, Lichtenstein S (1982) Facts versus fears. Understanding perceived risks. In: Kahnemann D, Slovic P, Tversky A (ed) Judgement under uncertainty. Heuristics and biases. Cambridge University Press, Cambridge, pp 463–489

Smith K (1994) Environmental hazards. Assessing risk and reducing disaster. Routledge, London/New York

Spirgi EH (1979) Disaster management. Comprehensive guidelines for disaster relief. Hans Huber, Bern/Stuttgart/Wien

Stein DL (ed) (1989) Lectures in the science of complexity. Addison Wesley, Reading

Steuer M (1979) Wahrnehmung und Bewertung von Naturrisiken am Beispiel zweier ausgewählter Gemeindefraktionen in Friaul. Münchener Geographische Hefte No. 43. Michael Laßleben, Regensburg

Stewart I, Cohen J (1994) Why are there simple rules in a complicated universe? In: Futures 26, (6), pp 648–664

Stiens G (1996) Prognostik in der Geographie. Das Geographische Seminar. Westermann, Braunschweig

Stiens G (1992) Prognostische Geographie. Bisherige Ausprägungen und künftige Anforderungen im Überblick. Berlin.

Stiens G (1989) Geographische Prognostik aus der Sicht bestehender und möglicher Anwendungsfelder: Neue Anforderungen an eine Disziplin mit Tradition. In: Geographica Helvetica (4): 187–195

Stiens G (1982) Langfristszenarien zur Raumentwicklung. Raumbezogene Zukunftsforschung im Rahmen des MFPRS. Informationen zur Raumentwicklung (8): 573–585

Stiens G (1977) Zur Verwendung von Szenarien in der Raumplanung. In: Raumforschung und Raumordnung 35 (112): 69–73

Stierand R (1993) Neuorientierung in der Planungstheorie? In: Raumplanung (61): 141–147

Stoddart DR (1987) To claim the high ground. Geography for the end of the century. In: Transaction 12 (3): 327–336

Stout D (1995) Technology foresight. A view from the front. In: Business Strategy Review 6 (4): 1–16

Sträter D (1988) Szenarien als Instrument der Vorausschau in der räumlichen Planung. Hannover

Street P (1997) Scenario workshops. A participatory approach to sustainable urban living. In: Futures 29 (2): 139–158

Susmann P, O'Keefe P, Wisner B (1983) Global disasters. A radical interpretation. In: Hewitt K (ed) Interpretations of calamity. Allen & Unwin, Boston/London, pp 274–276

The United Nations University, World Institute for Development and Economics Research (eds) (1995) Small islands, big issues. Crucial issues in the sustainable development of small developing states. Forssa Printing House, Helsinki

Thompson M, Ellis R, Wildavsky A (1994) Cultural theory. Westview Press, San Francisco/Oxford:

Tilling RI, Lipman PW (1993) Lessons in reducing volcano risk. In: Nature 36: 277–280

Toulmin L (1987) Disaster dreparedness and regional training on nine Caribbean islands. A long-term evaluation. In: Disasters 11 (3): 221–234

Trischler J (1991) NRW 2000. Wirtschaft, Beschäftigung, Qualifikation und neue Techniken – Szenarien und Gestaltungsmöglichkeiten gesellschaftlicher Entwicklung. Westdeutscher Verlag, Opladen

Tversky A, Kahnemann D (1974) Judgement under uncertainty. Heuristics and biases. In: Science (185): 1124–1131

United Nations Department of Humanitarian Affairs (ed) (1996) Disasters and development. DHA Issues in Focus Series No. 6. Genf

United Nations Disaster Relief Organisation (ed) (1984) Disaster prevention and mitigation. Volume 11. Preparedness aspects. United Nations, New York

United Nations Economic Commission for Latin America and the Caribbean (ed) (1993) A draft plan of action for sustainable development of small Caribbean island developing states. Doc. No. CCSUNECLAC/ACM/RTM1/21

Unwin, T (1992) The place of geography. Longman, Harlow, Essex

Uphoff N (1986) Approaches to participation in agriculture and rural development. In: Bamberger M (ed) Readings in community participation. Worldbank, New York

van Asselt MB, Rotmans J (1996) Uncertainty in perspective. In: Global Environmental Change 6 (2): 121–157

van Bertalanffy L (1972) General systems theory. Braziller, New York

Varley A (1994) Disasters, development and environment. John Wiley & Sons, Chichester/New York/Brisbane

Vereinte Nationen (ed) (1989) Resolution 44/236 der Generalversammlung der Vereinten Nationen zur Internationalen Dekade für die Reduzierung von Naturkatastrophen vom 22.12.1989, Bonn

Vermeiren J (1991) Natural disasters. Linking economics and environment with a vengeance. In: Girvan N, Simmons D (eds) Caribbean ecology and economics. Caribbean Conservation Association, Barbados, pp 127–142

Vester F (ed) (1985) Zukunftswerkstätte Kraftfeld. Menschwärts. Böhlau, Wien/Köln/Graz

Vester F (1983) Unsere Welt. Ein vernetztes System. dtv-Verlag, München

von Bieberstein VR () Die Szenariomethode als Mittel der wissenschaftlichen Politikberatung. In: Bundesminister für Raumordnung, Bauwesen und Städtebau (ed) Alternativen für die von Verstädterung Europas. Bonn, pp A37–A50

von Bieberstein VR, Bormann W (1975) Die Szenariomethode als Verfahren zur zukunftsorientierten Untersuchung komplexer sozialer Probleme. In: Analysen und Prognosen (38): 21–24

von Furstenberg GM (1990) Acting under uncertainty. Multidisciplinary conceptions. Boston/Dordrecht/London: Kluwer.

von Ilsemann W (1980) Die geteilte Zukunft. Szenarioplanung bei Shell. In: Manager Magazin (5): pp 115–123

von Reibnitz U (1992) Szenario-Technik. Instrumente für die unternehmerische und persönliche Erfolgsplanung. Gabler, Wiesbaden

Wadge G, Isaacs MC (1989) Volcanic hazards from Soufriere Hills Volcano Montserrat, West Indies. University of Reading, Reading

Waldrop MM (1992) Complexity. The emerging science at the edge of order and chaos. Simon & Schuster, New York

Warfield JN (1976) Societal systems. Planning, policy and complexity. Wiley, New York/London/Sydney

Warrick RA (1981) Four communities under ash. After Mount St. Helens. Program on Technology, Environment and Man No. 34. Institute of Behavioral Science, Boulder

Wegener A (1994) Die Übergangszone. Ein vergessener Raumtyp in der Stadtentwicklung – Szenarien zur Zukunft des Hamburger Stadtteils Wandsbek. Diploma Thesis, University of Hamburg

Wentz M (ed) (1992) Planungskulturen. Campus, Frankfurt/New York

Werlen B (1987) Gesellschaft, Handlung und Raum. Grundlagen handlungstheoretischer Sozialgeographie. Erkundliches Wissen 89. Franz Steiner, Stuttgart

Wheeler Marion (1988) Montserrat, West Indies. A chronological history. Montserrat

White CM (1991) Risk and economic retardation – The long-run impact of natural disasters on the greater Caribbean Basin. Economics and Commerce Discussion Paper No. 8. Bundoora, Australia

White GF, Cohen J, Jeggle T (1998) The International Decade for Natural Disaster Reduction. A look back and ahead. Research and Applications Workshop. Boulder, Colorado, pp 98–15

White GF, Haas JE (1975) Assessment of research on natural hazards. Massachusetts Institute of Technology, Cambridge

White GF (1961) The choice of use in resource management. In: Natural Resources Journal (1): 23–40

White GF (1945) Human adjustment to floods. A geographical approach to the flood problem in the United States. Research Paper No. 29. University of Chicago, Chicago

White P (1996) Volcanic crisis. Development in the north. In: Montserrat Reporter 12 (28): 10

Wiener N (1948) Cybernetics, or control and communication in the animal and the machine. Wiley, New York

Wildavsky AB (1993) Vergleichende Untersuchung zur Risikowahrnehmung. Ein Anfang. In: Bayrische Rückversicherung (ed) Risiko ist ein Konstrukt. Wahrnehmungen zur Risikowahrnehmung. Knesebeck, Karlsruhe, pp 191–212

Wildavsky AB (1988) Searching for safety. Studies in social philosophy & policy No. 10. Social Philosophy & Policy Center, Boston

Williams T (1994) The end of strategy. In: Futures 26 (4): 365–381

Wiman BLB (1991) Implications of environmental complexity for science and policy. In: Global Environmental Change 1 (3): 235–247

Wiman IMB (1990) Expecting the unexpected. Some ancient roots to current perceptions of nature. In: Ambio 19 (2): 62–69

Wischnewski HJ (1994) IDNDR. Eine interdisziplinäre Aufgabe für Wissenschaft und Politik. In: Geographische Rundschau 46 (7–8): 392–393

Wisner B (1993) Disaster vulnerability. Scale, power and daily life. In: Geo Journal 30 (2): 127–140

World Commission on Environment and Development (ed) (1987) Our common future. Oxford University Press, Oxford/New York

Wynne B (1992) Uncertainty and environmental learning. Reconceiving science and policy in the preventive paradigm. In: Global Environmental Change 2 (2): 111–127

Yohe G (1996) Exercises in hedging against extreme consequences of global change and the expected value of information. In: Gloval Environmental Change 6 (2): 87–101

Zimmermann EW (1951) World resources and industries. Harper & Brothers, New York

Zimmermann M (1997) Lokale Agenda 21. Ein kommunaler Aktionsplan für die zukunftsbeständige Entwicklung der Kommune im 21. Jahrhundert. In: Aus Politik und Zeitgeschichte 97 (27): 25–38

Appendix

Montserrat 1996 (7 Seiten) **Interview Nr.** ______

1. Female ____ Male ____ 2. Age _______

3. Where do you live ?

4. Where were you born ?

5. How many members does your family have (are you living with) ?

6. Which school did you attend ? no school ____ primary school ____ high school ____

7. Which church do you belong to ?

Anglican ____ Methodist ____ Baptist ____ Catholic ____

7th-Day-Adventist ____ no church ____ other:

M1. Where did you live before the eruption ?

area: street:

M2. Where do you live right now ?

moved in with friends ____ moved in with relatives ____

went to an official shelter ____ built my own shelter ____

M3. How are the living conditions right now ?

in shelters:	good ____	moderate ____	poor ____
with friends:	good ____	moderate ____	poor ____
with relatives:	good ____	moderate ____	poor ____
in same location as before:	good ____	moderate ____	poor ____
other:	good ____	moderate ____	poor ____

General questions with regard to the time before the eruption

8. How do you earn your living ?

tourism ____ civil servant ____ trade ____ personal services ____

financial services ____ fishing ____ farming ____ transport ____ construction ____

retired ____ school/university ____ others:

9. Do you know what your parents and grandparents have been doing for living ? don´t know ____

farming ____ fishing ____ trading ____ civil servant ____ handicraft ____

housewife ___ saltraking ____ special answer:

10. What do you consider to be the „Natural Resources" of the island ? don´t know ____

water ____ ocean ____ beach ____ people ____ animals ____

landscape ____ vegetation ____ waterfall ____ volcano ____ farming ____ fishing ____

handicraft ____ special dishes ____ special answer:

14. What do you think are the benefits and problems of tourism for the development of the island? don´t know ____

good to have ____ a problem ____

benefits:

problems:

15. What do you think are the benefits and problems of offshore banking for the development of the island? don´t know ____

good to have ____ a problem ____

benefits:

problems:

16. What do you think are the benefits and problems of agriculture for the development of the island? don´t know ____

good to have ____ a problem ____

benefits:

problems:

17. What do you think are the benefits and problems of fishing for the development of the island? don´t know ____

good to have ____ a problem ____

benefits:

problems:

18. What do you think about the economic contribution of expatriats ? don´t know ____

good to have ____ a problem ____ some are good but not too many ____

benefits:

problems:

19. Do you think political decisions have an impact on your daily life ? don´t know ____

no impact ____ positive impact ____ negative impact ____ others:

20. Do you think that the political parties respond to the needs of the local people ? don´t know ____

respond ____ respond not ____ others:

2

21. Do you think that politicians are capable of solving problems ? don´t know ____

capable ____ not capable ____ others:

22. Do you think it is an advantage being a British Dependent Territory or would you prefer to be independent ? don´t know ____

prefer independency ____ prefer to be dependent ____ others:

23. Do you think that the people here have a strong national identity ? don´t know ____

strong national identity ____ no national identity ____ others:

M4. Do you think there is a competition between the islands in the Eastern Caribbean ? don´t know ____

competition/jealousy exists ____ no competition/jealousy ____ others:

25. What do you think of the condition of the natural environment here on the island ? don´t know ____

good condition ____ normal condition ____ bad condition ____

problem areas:

26. Are there any environmental problems which have an influence on your daily life ? don´t know ____

no problems ____ garbage ____ human beings ____ animals ____

ocean problems ____ sand extraction/beach problems ____ others:

27. How important is the protection of the environment to you ? don´t know ____

important ____ not important ____

28. What can you personally do to protect the environment ? don´t know ____

no litter ____ education of children ____ advise other people ____ other:

31. What should be the overall priorities for development on the island in the future ? don´t know ____

economy ____ politics ____ government ____ education ____ tourism ____

agriculture ____ environment ____ other:

32. What is more important: economic development or environmental protection ? don´t know ____

economic development ____ environmental protection ____ both ____

33. Do you think that resources can be used and protected at the same time ? don´t know ____

yes ____ no ____ other:

34. Which possibilities do you see to use and protect resources at the same time ? don´t know ____

protection in general ____ regulations ____ institutions ____ certain areas ____

people power ____ agriculture ____ others:

35. Do you know the „National Trust" ? don´t know ____

yes ____ no ____

General questions with regard to the present situation

M5. What are the causes for the actual volcanic activity ? don´t know ____

destiny ____ extreme event from nature ____ act of god ____ planning defficiencies ____

naturally unsound behaviour ____ others:

M6. What, in your opinion, is the most likely development of the recent volcanic crisis ? don´t know ____

continuation of the recent form of activity ____ a major eruption ____

volcanic activity calms down without a major eruption ____ others:

M7. What kind of destruction happened to your property due to the volcanic activity ?
(Please name the following categories to the interviewee:)

no destruction ____ destruction on furniture ____ destruction on house ____

destruction on lifestock ____ destruction on crops ____ lost my job ____

injuries ____ others:

M8. Considering the time before the eruption in July 1995, where you aware of the risk?

yes ____ no ____ other:

M9. Where do you get the information about the actual volcanic activity ?

via TV ____ via newspaper ____ via radio ____ via neighbours, friends ____

directly via MVO ____ via Internet ____ other:

M10. Apart from the actual volcanic activity, have you experienced further natural hazards here in Montserrat ? don´t know ____

yes ____ no ____

How many disasters have you experienced (time frame) ?

Which was the most disastrous event ?

M11. Please name possible future hazardous events on Montserrat ? don´t know ____

volcanic eruption ____ hurricane ____ landslide ____ earthquake ____

other:

M12. How do you judge the possibilities of these events (event named by interviewee) ? don´t know ____

very high ____ high ____ moderate ____ low ____ very low ____

4

M13. Which hazardous event has the strongest impact on your life and property ? don´t know ____

volcanic eruption ____ hurricane ____ landslide ____ earthquake ____

other:

M14. How can you protect yourself against possible risks of a hazard ? don´t know ____

(Please name the following categories to the interviewee:)

building guidelines ____ insurance ____ to leave the island ____

to move in areas which are less vulnerable to hazards on the island ____

M15. Have you already used or will you use any of these measures ?

I used ____ I will use ____ measure:

I didn´t use ____ I won´t use ____

Recovery Phase

M16. How do you judge the overal recovery effort of the GOM ? don´t know ____

(Please name the following categories to the interviewee:)

reconstruction of medical facilities:	very good ____	good ____	moderate ____	poor ____	very poor ____
reconstruction of schools:	very good ____	good ____	moderate ____	poor ____	very poor ____
providing shelter:	very good ____	good ____	moderate ____	poor ____	very poor ____
providing food:	very good ____	good ____	moderate ____	poor ____	very poor ____
providing water and electricity:	very good ____	good ____	moderate ____	poor ____	very poor ____
providing transportation:	very good ____	good ____	moderate ____	poor ____	very poor ____
honesty in handling foreign aid:	very good ____	good ____	moderate ____	poor ____	very poor ____
volcano monitoring and information:	very good ____	good ____	moderate ____	poor ____	very poor ____

M17. How do you judge the overal recovery effort of the GOM,
UK, the red cross and other institutions ? don´t know ____

institution:	very good ____	good ____	moderate ____	poor ____	very poor ____
institution:	very good ____	good ____	moderate ____	poor ____	very poor ____

M18. Do you get any financial help ? yes ____ no ____

M19. What kind of help do you demand ? don´t know ____

M20. Do Montserratians help each other through the crisis or
is there a growing competition ? don´t know ____

they help ____ competition ____ other:

M21. Have you thought about moving to other places ? don´t know ____

yes ____ no ____

M22. If yes, where would you like to go ? don´t know ____

in Montserrat: to other islands in the Caribbean:

to the US ____ to the UK ____ other:

M23. Have any of your friends and relatives left the island ? yes ____ no ____

M24. If yes, how many and where to ? number: ________ don´t know ____

in Montserrat ____ to other islands in the Caribbean ____ to the US ____

to the UK ____ other:

Reconstruction and Prepardness

M25. Considering your neighbours, friends and relatives, what feeling is stronger ? don´t know ____
(Please name the following categories to the interviewee:)

reconstruction ____ resignation ____ other:

M26. How would you judge the long-term perspectives for reconstruction following the volcanic crisis ? don´t know ____
(Please name the following categories to the interviewee:)

very good perspectives ____ good perspectives ____ moderate ____

problematic ____ no perspectives ____ others:

M27. Do you think one could perceive the disaster as a possibility for structural changes or do you prefer to restore what had been ? don´t know ____

change ____ restore ____

M28. How important are the following measures for structural changes in the future ?
(Please name the following categories to the interviewee:)

settlement structure change:	very necessary ____	necessary ____	not necessary ____	don´t know ____
economic development:	very necessary ____	necessary ____	not necessary ____	don´t know ____
building guidelines:	very necessary ____	necessary ____	not necessary ____	don´t know ____
seismic zoning:	very necessary ____	necessary ____	not necessary ____	don´t know ____
insurance schemes:	very necessary ____	necessary ____	not necessary ____	don´t know ____
planning process:	very necessary ____	necessary ____	not necessary ____	don´t know ____
public participation in politics:	very necessary ____	necessary ____	not necessary ____	don´t know ____
information policies:	very necessary ____	necessary ____	not necessary ____	don´t know ____
foreign aid:	very necessary ____	necessary ____	not necessary ____	don´t know ____
cooperation with other islands in the Caribbean:	very necessary ____	necessary ____	not necessary ____	don´t know ____
to help the most vulnerable:	very necessary ____	necessary ____	not necessary ____	don´t know ____

6

M29. What do you think how long will it take until the reconstruction phase is finally over ?
don´t know ____ one year ____ up to five years ____ up to ten years ____
up to twenty years ____ more than twenty years ____

M30. What are your personal whishes for the future ? don´t know ____

36. Where do you get your food stuff? don´t know ____
Supermarket ____ little nearby shop ____ market ____ Self growing ____

37. Where do you get your fresh water from and how much do you spend per month/average ?
public H2O ____ pipe H2O ____ other: $ per month:

38. How much money do you spend on electricity per month/average ?
don´t know ____ $ per month:

39. How is the garbage disposal of your household handled ? don´t know ____
truck collects ____ dump place ____ other:

40. How is the sewage of your household handled ? don´t know ____
septic tank ____ sink or sespit ____ throw it away ____ pipe system ____ others:

41. Do you have a car? yes ____ no ____

43. Have you heard of recycling ? yes ____ no ____

44. Have you heard about alternative technology like solar or wind energy ? don´t know ____
solar energy ____ wind energy ____ thermic energy ____ other:

45. Where else have you been in the Caribbean area ?

Printing (computer to plate): Mercedes-Druck, Berlin
Binding: Buchbinderei Lüderitz & Bauer, Berlin